AF516169

New models in geography

Volume I

New models in geography

The political-economy perspective

Volume I

edited by

Richard Peet & Nigel Thrift

Clark University & University of Bristol

RAWAT PUBLICATIONS

Jaipur • New Delhi • Bangalore • Guwahati • Kolkata

ISBN 978-81-316-0863-0

Indian reprint, 2017

Published by
Prem Rawat for **Rawat Publications**
Satyam Apts., Sector 3, Jawahar Nagar, Jaipur - 302 004 (India)
Phone: 0141 265 1748 / 7006 Fax: 0141 265 1748
E-mail : info@rawatbooks.com
Website: rawatbooks.com

New Delhi Office
4858/24, Ansari Road, Daryaganj, New Delhi 110 002
Phone: 011-23263290

Also at *Bangalore, Guwahati* and *Kolkata*

Printed and bound in India.

Contents

Foreword

It seems such a long time ago, another age – yet it is a mere twenty-odd years since the original *Models in Geography* was published. It is an even shorter time since the first tentative steps were taken towards an alternative formulation of what might constitute a geographical perspective within the social sciences. It all seemed very daring at the time, and it began with critique and with an eager reading of basic texts. But what came to be called the political-economy perspective has progressed with remarkable speed and energy to generate its own framework of conceptualization and analysis, its own questions and debates.

The papers in these two volumes are witness to the richness and range of the work which has developed over this relatively short period within the political-economy approach. Moreover, from being a debate within an institutionally-defined 'discipline of geography', to introducing into that discipline ideas and discussions from the wider fields of philosophy and social science and the humanities more generally, it has now flowered into a consistent part of enquiries that span the entire realm of social studies. Not only has 'geography' increasingly become an integral part of the study of society more widely, but a geographical perspective is contributing to, as well as learning from, that wider debate.

The political-economy approach has been of central importance in this move. Indeed, debate within political-economic approaches to geographical studies has reflected, in its different phases, that reintegration of geography within social sciences. The form of this integration is still an issue today, but it is striking how many of the chapters in these volumes, while often talking about quite distinct empirical areas of concern, document in broad outline a similar trajectory on this issue.

The path has not always been smooth. There have been difficult and sometime confusing debates, which have involved the reformulation of questions as well as of answers. Many of the longer-running (and in the main continuing) debates are reflected in these papers; again it is striking how different authors in distinct fields frequently agree on which discussions have been of central importance. Perhaps most fundamental to the reformulation of geography's place within the social sciences has been the thoughtful and productive debate (productive in the sense that it really has moved on and has made progress from stage to stage) concerning the relation between the social and the spatial, and whether it is in any case an impossible dichotomy which should be dissolved. (Maybe we ought to be conducting a similar debate about the relation between the social and the equally difficult concept of the natural?) That debate is documented here from a number of angles. It is also clear that, even if we have understood a few things better, there are still important issues unresolved. There is still debate over exactly what it means to say that space makes a difference. Is it that particular time-space contexts trigger the realization of causal powers embedded in the social or, alternatively, do not trigger them? That is one position persuasively argued here. Or is it that some social phenomena cannot be adequately conceptualized without some degree of spatial content? And how does this tie in to the notion locality effects? Whatever 'the answers' are, it is an important debate which connects directly to the philosophical foundations of the subject and which links geography inextricably to other areas of social science.

The relation between theoretical and empirical work has also been a consistent preoccupation, both in terms of the priority that should be given to one or the other and in terms of the relation between them. A number of the papers here document the debate and wrestle with the problem. The importance of the contribution of realism to the discussion is evident, even if its usefulness, and even its form, is not always agreed. Many of the papers call for the development of 'middle-level' theories or concepts. One fascinating thing here is how different authors, from the evidence presented in these volumes, seem to mean quite distinct things by these terms. There is clearly debate here which could perhaps be addressed more directly.

Much of the early political-economy writing within geography grew up within studies of uneven development and of industrial geography. Perhaps for that reason, but also reflecting the contemporary character of political economy more widely, it had a heavily economic, at times economistic, bent and one which often accompanied a greater attention to structures than to agency. Once again there is agreement in a very wide set of the papers in this collection that this is changing. The newer research, which focuses on cultural forms and on representation and interpretation, is breathing a different kind of life into the debates. It also, very importantly, promises to help us establish closer links between areas of geographical studies which might have become too separated. But here, too, the debate is not finished; indeed in geography it has only just begun.

There are also debates between the contributors to these volumes which the careful reader will detect, but which are not addressed directly. There are contrasting understandings of the meaning of basic terms, such as 'theory'; I suspect that there are also variations in what people would include under the rubric 'political economy approach'! And I am sure that the meaning, if not the use, of the term 'model' would make a lively topic of debate! But the persistence of debate should not be seen as deeply problematical, nor necessarily as a weakness. It certainly proves the political-economy perspective is alive; over two decades it has moved from being a few isolated voices to being one of the major influences on the current development, and richness, of human geography. That implies responsibilities too, as well as a pat on the back for our collective achievement. It means pursuing these debates in a constructive, friendly and un-pompous (if there is such a word) manner; it means writing in a way which is accessible to participants outside the immediate discussion. The contributions to these volumes, in my view, achieve that aim.

Another thing many of them achieve very well is to set the development of these discussions in their (also developing) historical contexts, both societal and theoretical. Sometimes we have clearly been guilty of bending the stick too far, but often that has been the importance of stressing particular arguments at particular moments. The emphasis in the early years on social causes at the expense of the spatial dimension is a case in point. Arguments are not developed in vacuums. Today both social and theoretical contexts are presenting new challenges. Similarly, a particular focus for research may be appropriate, even urgently needed, in some contexts, without any implication that it should be a priority in some absolute, eternal, sense. The current emphasis on locality research in the UK is, in my opinion, a case in point. Amidst a flood of writing on national structural change (in which, for example, 'the end of the working class' figured prominently) it was among other things important to point out that, and to analyze how, the picture varied dramatically between different parts of the country; how some of the social dynamics in which people were caught up were often quite different from what one might divine from a national picture. This focus also gelled with questions of theory and methodology which had been raised at the same time. But none of it means to say that a focus on locality research will always and everywhere be important.

Which raises another point: the political import and impact of our work. How much difference has it made that the political-economy perspective has blossomed academically? I think the editors of these volumes are right to say in their introduction that here the record is mixed. There have been other shortcomings, too, inevitably. Some are mentioned by contributing authors, but two in particular struck me. There is a UK-US focus to much of the work, which is reflected not only in the object of study, but also in tendencies to universalize from their particularities, and also to be less than aware of academic work going on in other parts of the world. And there is still a million miles to go before the full impact of the feminist critique has been taken on board.

Nonetheless, what these volumes incontrovertibly establish is the enormous progress which has been made since those early days. They are something of a monument to years which have been energetic, full of debate, and often fun. What is more, the possibilities now opened up by geography's more fruitful relations with the rest of social science mean that the future looks set to be equally productive.

Doreen Massey

Introduction

The publication of *Models in geography* (Chorley & Haggett 1967) presaged a sea change in the practice of Anglo-American geography. Since that date, the practice of geography has changed again. A set of new models – based upon a political-economy perspective – now peppers the geographic landscape. This book provides a summary of the nature of these models, their spirit, and purpose.

The new models often took their original inspiration from Marx and Marxism.[1] That original inspiration has by now been overlain with many other layers of influence, so that this book reports on what is an increasingly diverse body of work, but one which still holds to the critical vision of society which was at the heart of Marx's project. Of course, these new political-economy models do not form the only approach to geography, but they have certainly been influential in the subject over the past few years. Their influence can be measured in three ways.

First, there is the quality of the work that the models have generated; on this count, they can surely be judged a success. As the following pages record, the approach has generated a flood of substantive theoretical and empirical work in geography, ranging all the way from class to culture, from gentrification to geopolitics, from restructuring to the urban–rural shift.

A second gauge of the influence of an approach is its ability to move outside narrow disciplinary boundaries and influence other disciplines. On this count again, the new political-economy models can surely be judged a success. Geographers who subscribe to these models now feature regularly in books and journals the length and breadth of social science, where not too many years ago it would have been very difficult to find any work by geographers at all. Geographers have also had notable success in participating in certain debates in the social sciences as a whole, for example, on subjects such as realism, structuration theory, deindustrialization and industrial restructuring. Geographers using political-economy models are also disseminating their work to a wider audience.[2] As a result geography is now surely held in greater respect in the social sciences.

Finally, there is judgement to be made about the practical import of the political-economy models in terms of active intervention. Here, the record is mixed. But the responsibility is greater because the political economy approach does, after all, encapsulate an avowedly critical approach to society (Johnston 1986). Yet, in a world where millions of people are dying in famines or war, where more millions live in acute poverty and fear, and where there is an ecological crisis of grave proportions, it is surely important to hold on to that emancipatory vision. Here, at the cutting edge of capitalism, much new thinking and ideological face work remains to be done.

Organization of the book

The book is split into two volumes, each consisting of four parts. Both volumes have a common introduction and first introductory part. Subsequently, volume 1 consists of the second, third and fourth parts of the book while volume 2 consists of the fifth, sixth and seventh parts. Each part of the book except the introductory part is prefaced with an introduction written by one of the editors.

The first introductory part provides essential background to the book. It sketches the history of political-economic models in geography and their chief characteristics. In addition, the changing fortunes of the original models in geography are documented. The second part of the book is devoted to the natural environment. It is true to say that the natural environment has received less attention than its due from geographers interested in the new models, although there are signs that these omissions are now being righted. The third part is concerned with models of the geography of production. These have been at the hub of the new models in geography and they are therefore given considerable attention. The fourth part considers models of the state and politics in all their manifestations. The fifth part explores the struggle to provide political-economic models of the city in a time of considerable social and economic change. The sixth part is concerned with models of civil society, ranging from gender through race to landscape and locality. The seventh and final part of the book moves to the links currently being forged between political-economic models and social theory. Appropriately, the book ends on an open note.

Most of the chapters have two emphases. The first consists of a review of the work of the past 20 years. A good chapter tells the reader what main ideas have developed, in which order, and where they fit in terms of the changing social structures. But we are also concerned with a second emphasis: where the political-economy approach is going. Most of the review-type chapters conclude with prospects for future research and several are almost exclusively concerned with expanding the frontiers of political-economic theory. The book, then is intended as both retrospect and prospect.

As in any edited collection, there are omissions which we have not, because of pressures of time, circumstance and (most especially) space, been able to rectify. To a degree we have tried to minimize these omissions by pointing to them in the introductions to each part of the book. Nevertheless, there are still omissions of which we are particularly conscious, especially in four areas of work. The first of these is the Third World. Some of this work is documented in these volumes, but not enough. A second area omitted concerns the socialist countries. It is striking how little work in political-economic geography has been directed towards the socialist countries – a case, perhaps, of capitalism becoming an obsession. The third omission concerns historical geography. One of the most important elements of political-economic models is their sensitivity to the importance of history, so that most work of this kind includes a strong sense of change and process. Hence we have not included a specific section on historical geography. Suffice it to say that a book which went beyond the contemporary era was likely to become monumental in size. The final omission is of physical geography. Clearly, unlike the original *Models in geography*, this is not a book that includes the work of physical geographers. This is chiefly because, whether for good or ill, in the years since the publication of the original *Models in geography* human and physical geography have drifted further apart (Johnston 1983). Human geography now lies firmly in the camps of the social sciences and the humanities. There are encouraging signs of a renewal of the *entente cordiale* between human and physical geography (Peake & Jackson 1988), but as yet they hardly constitute sufficient grounds for an integrated volume.

In what follows we have tried to retain the initial sense of criticism and excitement about new approaches to new and old topics which pervaded the early 'radical geography' while also displaying the more sophisticated work of recent years, which no longer needs to criticize the conventional to establish a position. Here is what we have done, with hints at how we felt; there is where we are going. The struggle continues.

Richard Peet
1989 Nigel Thrift

Notes

1 Marx was mentioned in the original *Models in geography* by Hamilton, Harries & Pahl, if only in passing. That epitaphs should never be written can be seen in the examples of the return to popularity in the late 1980s of Talcott Parsons and Althusser.

2 As one instance only, the new Open University course D314 *Restructuring Britain* is made up of units, many of which are explicitly concerned with the political-economy approach.

References

Chorley, R. J. & P. Haggett, (eds) 1967. *Models in geography*. London: Methuen.

Johnston, R. J. 1983. Resource analysis, resource management, and the integration of human and physical geography. *Progress in Physical Geography* **7**, 127–46.

Johnston, R. J. 1986. *On human geography*. London: Edward Arnold.

Peake, L. & P. Jackson 1988. The restless analyst: an interview with David Harvey. *Journal of Geography in Higher Education* **12**, 5–20.

Acknowledgments

Richard Peet would like to thank ETC (United Kingdom) for a grant enabling him to visit Britain in 1988 for consultation with Nigel Thrift. He also acknowledges the comradely support of several generations of students at Clark University and the tolerance of his faculty peers for views which must frequently appear extreme and doctrinaire. Finally, for her warm companionship during the editing of this book, thanks to Kathy Olsen. Nigel Thrift thanks Lynda, Victoria, and Jessica for their continuing forebearance.

Both editors wish to acknowledge Roger Jones at Unwin Hyman for his encouragement and patience during the course of an enterprise which sometimes seemed to have no end.

Contributors

Contributors to this volume are shown in bold type.

Sophia Bowlby, Lecturer, Department of Geography, University of Reading, Reading, RG6 2AB, UK.

Lata Chaterjee, Associate Professor, Department of Geography, Boston University, Boston, Massachusetts 02215, USA.

Gordon Clarke, Professor, School of Urban and Public Affairs, Carnegie-Mellon University, Pittsburgh, Pennsylvania 15213, USA.

Martin Clarke, Lecturer, School of Geography, University of Leeds, Leeds, LS2 9JT, UK.

Paul Cloke, Reader, Department of Geography, Saint Davids University College, Lampeter, Dyfed, SA48 7ED, UK.

Philip Cooke, Reader, Department of Town Planning, University College of Wales, Cardiff, CF1 3EU, UK.

Stuart Corbridge, Lecturer, Department of Geography, University of Cambridge, Cambridge, CB2 3EN, UK.

Stephen Daniels, Lecturer, Department of Geography, University of Nottingham, Nottingham, NG7 2RD, UK.

Simon Duncan, Lecturer, Department of Geography, London School of Economics, London, WC2A 2AE, UK.

Jacque Emel, Assistant Professor, Graduate School of Geography, Clark University, Worcester, Massachusetts 01610, USA.

Ruth Fincher, Department of Geography, University of Melbourne, Parkville, Victoria 3052, Australia.

Jo Foord, Principal Policy and Research Officer, Borough of Southwark, London, UK.

Derek Gregory, Professor, Department of Geography, University of British Columbia, Vancouver V6T 1W5, Canada.

Peter Jackson, Lecturer, Department of Geography, University College London, London, WC1H 0AP, UK.

R. J. Johnston, Professor, Department of Geography, University of Sheffield, Sheffield, S10 2TN, UK.

Helga Leitner, Associate Professor, Department of Geography, University of Minnesota, Minnesota 55455, USA.

Jane Lewis, Economic Development Unit, Borough of Ealing, London, UK.

John Lovering, Research Fellow, School for Advanced Urban Studies, University of Bristol, Bristol, BS8 4EA, UK.

Linda McDowell, Senior Lecturer, Faculty of Social Sciences, Open University, Milton Keynes, MK7 6AA, UK.

Suzanne Mackenzie, Assistant Professor, Department of Geography, Carleton University, Ottawa K1S 5B6, Canada.

Doreen Massey, Professor, Faculty of Social Sciences, Open University, Milton Keynes, MK7 6AA, UK.

Timothy O'Riordan, Professor, School of Environmental Sciences, University of East Anglia, Norwich, NR4 7TJ, UK.

Richard Peet, Professor, Graduate School of Geography, Clark University, Worcester, Massachusetts 01610, USA.

Geraldine Pratt, Assistant Professor, Department of Geography, University of British Columbia, Vancouver, British Columbia V6T 1W5, Canada.

Erica Schoenberger, Assistant Professor, Department of Geography and Environmental Engineering, Johns Hopkins University, Baltimore, Maryland 21218, USA.

Eric Sheppard, Professor, Department of Geography, University of Minnesota, Minneapolis, Minnesota 55955, USA.

David Slater, CEDLA, 1016 EK Amsterdam, The Netherlands.

Neil Smith, Associate Professor, Rutgers University, New Brunswick, New Jersey 08855, USA.

Edward Soja, Professor, Graduate School of Architecture and Urban Design, University of California, Los Angeles, California 90024, USA.

Nigel Thrift, Reader, Department of Geography, University of Bristol, Bristol, BS8 1SS, UK.

John Urry, Professor, Department of Sociology, University of Lancaster, Lancaster, LA1 4YL, UK.

Alan Wilson, Professor, Department of Geography, University of Leeds, Leeds LS2 9JT, UK.

Part I
NEW MODELS

1 *Political economy and human geography*

Richard Peet & Nigel Thrift

Introduction

Since the publication of the original *Models in geography* (Chorley & Haggett 1967) some 20 years ago, human geography has changed dramatically. It has matured theoretically, it is more directly oriented to social problems, and it has achieved an awareness of politics without sacrificing its advance as a 'science'. This transformation can be traced to the emergence, and the widespread acceptance, of a new set of models which have a common root in the notion that society is best understood as a political economy.

We use the term 'political economy' to encompass a whole range of perspectives which sometimes differ from one another and yet share common concerns and similar viewpoints. The term does not imply geography as a type of economics. Rather economy is understood in its broad sense as social economy, or way of life, founded in production. In turn, social production is viewed not as a neutral act by neutral agents but as a political act carried out by members of classes and other social groupings. Clearly, this definition is influenced by Marxism, the leading class-orientated school of critical thought. But the political-economy approach in geography is not, and never was, confined to Marxism. Marxism was largely unknown to early radical geographers. Humanists and existentialists, who had serious differences with Marxism, have definitely been members of the political-economy school. At present, there are several critical reactions to Marxism, particularly in its stucturalist form, which nevertheless remain broadly within the political-economy stream of geographic thought. So, while political economy refers to a broad spectrum of ideas, these notions have focus and order: political-economic geographers practise their discipline as part of a general, critical theory emphasizing the social production of existence.

A number of themes related to the development and present contents of this school of thought are examined in this introductory chapter. We begin by tracing, in barest outline, the history of radical or critical geography. We then consider the development of the structural Marxist conception of society in the 1970s and early 1980s which provided the chief guiding theoretical influence over this development. We follow by noting some of the critical reactions to this conception in the discipline in the mid-1980s which have strongly influenced the current direction of the political-economy approach. Finally, we conclude with a statement of the present position of political-economic geography in the late 1980s. It is important to note that the chapter makes no claim to be all inclusive, noting every byway that the political-economy approach has taken. Rather, we will examine a few of the more important *theoretical* debates that have taken place in and around the political-economy approach to human geography since it first became of consequence.

The development of a political-economy approach

The critical anti-thesis to the thesis of conventional geography developed unevenly in time and space, so unevenly, indeed, that its various phases have frequently emerged independently rather than in linked sequence. Each phase had its distinct character, its own unique reaction to the events of its time. Each phase was also a particular reaction to themes in conventional explanation of geography at the time. Here we examine three of these phases in the recent development of conventional, geographic, thought and their critical counterparts: environmental determinism and its anarchist and Marxist critics; areal differentiation and its (limited) opposition; and, in more detail, conventional quantitative–theoretical geography and the radical geography movement.

Environmental determinism and its critics

It has been argued that modern geography first emerged as a justification for the renewed Euro-American imperial expansion of the late 19th century (Hudson 1977, Harvey & Smith 1984, Peet 1985b, Stoddart 1986). The need to explain Euro-American dominance compounded with the biological discoveries of Darwin, and Spencer's ideology of social Darwinism, to produce an explanation of social conquest cast in terms of the varying natural qualities and abilities of different racial groups. In the new modern geography this took the particular form of environmental determinism: differences in humans' physical and mental abilities, and in the level of their cultural and economic potential and achievement, were attributed to regionally differing natural environments. Euro-American hegemony was the natural, even god-given, consequence of the superior physical environments of Western Europe and North America.

Social Darwinism, and its geographic component environmental determinism, were opposed by the anarchist Russian geographer Kropotkin (1902). Kropotkin agreed that interaction with nature created human qualities, but differed on what these might be. As opposed to the social Darwinists' theory of inherent competitiveness and aggression as behaviours suggesting capitalism and imperialism as the natural modes of human life, he argued for cooperativeness and sociability as the natural bases for an anarchist form of communism. Only in the 1920s did Wittfogel (1985), a Marxist with geographical interests and training, criticize the environmental thesis from a position opposed to the direct natural causation of inherent human characteristics. For Wittfogel, human labour, organized in different social forms, moulded nature into the different material (economic) bases of regional societies. These in turn were the productive bases of different human personalities and cultures; that is, humans made themselves, rather than were made by nature. Yet Wittfogel remained within the environmental tradition by concluding that nature differentially directed the development of regional labour processes. Specifically, he argued that the climatically determined need for irrigation in the East (India, China) yielded a line of social development greatly different from that followed by rainfall-fed agriculture in the West (Wittfogel 1957). Hence, entirely different kinds of civilization developed in East and West.

Kropotkin and Wittfogel both achieved political and intellectual notoriety outside geography, but they were peripheral to the main lines of development of the discipline. Conventional geography tended to stand firm in support of the current social order. This was certainly one of the reasons for its widespread adoption in schools and universities.

Areal differentiation and its opponents

The 30 years between the late 1920s and the late 1950s must be characterized as the period of conventional geography's retreat from its position as a science of the origins of human nature, in the light of internal and external critiques of environmental determinism. Possibilism, a leading school of thought of the time, was so vague a formulation of environmental causation as to preclude systematic, theoretical, or even causal generalizations. In the United States, geography turned into areal differentiation (Hartshorne 1939, 1959): the description of the unique features of the regions of the Earth's surface. Critical reactions to this extremely conservative position, which began to surface in the 1940s and 1950s, were muted by the rampant anti-communism of the Cold War. Some regional geographies carried isolated, critical statements. The Lattimores' (1944) regional history of China, for example, says of late 19th-century United States foreign-policy makers that they 'did not propose a cessation of imperialist demands on China; they merely registered a claim of "me too"'. (A few years later Lattimore (1950, p. vii, Harvey 1983, Newman 1983) found himself labelled 'the top Russian espionage agent in this country' by US Senator McCarthy.) Hartshorne's conception of geography as a unique integrating science which, however, precluded generalization in the form of universal laws, also began to be opposed on theoretical grounds. Schaefer (1953) mildly proposed instead that geography explains particular phenomena as instances of general laws. In reply Hartshorne, philosopher-general of geography at the time, had merely to label Schaefer's criticisms 'false representation' to dismiss them. Hartshorne commented on a brief (and critical) mention of Marx in Schaefer's article:

> Whether the analysis of Karl Marx is sound, few readers of the *Annals* would be competent to judge. They should be competent to judge the appropriateness of including the analogy [between Marx and the geographer Hettner] in a geographic journal (Hartshorne 1955, p. 233).

After such broadsides, criticism was limited to less directly political arenas in the purely quantitative 'revolution' (Burton 1963) of the late 1950s and early 1960s.

Quantitive theoretical geography and the radical geography movement

We must leap into the late 1960s to find a widespread critical *and* political geography continuously responding to social crises and conventional geography's analysis of them. **Radical geography** originated as a critical reaction to two crises of capitalism at that time: the armed struggle in the Third World periphery, specifically United States involvement in the Vietnamese War, and the eruption of urban social movements in many cities, specifically the civil rights movement in the United States and the ghetto unrest of the middle and late 1960s in the United States, Great Britain, and elsewhere. Conventional geography's response to these momentous events lacked conviction, in more ways than one.

However, in the late 1960s some geographers already active in broader sociopolitical movements began to turn their attention inwards, towards their own discipline. The Detroit Geographical Expedition, led by William Bunge (Horvath 1971), used its conventional geographical skills on behalf of the black residents of the city's ghettos. At Clark University, in Worcester, Massachusetts, the radical journal *Antipode* began publication in 1969, carrying articles on socially relevant geographic topics (Peet 1977a).

But, it soon became apparent that conventional geographic theories and methodologies were inappropriate for a more relevant geography. The search for an alternative theoretical approach is exemplified by the intellectual biography of radical geography's leading theorist. David Harvey (1969) had previously written a conventional treatise on geographical methodology, but in the early 1970s began exploring ideas in social and moral philosophy – topics neglected in his earlier work. The journey took him through a series of liberal formulations, based on social justice as a matter of eternal morality, to Marxism with its analysis of the injustices built into specific societies; and from an interest in material reality, merely as the place to test academic propositions, to the transformation of capitalist society through revolutionary theory (Harvey 1973, pp. 9–19, 286–314). Harvey's journey was made by many other young radical geographers in the 1970s. For a few years in the early part of the decade radical geography explored, still from a liberal-geographical perspective, the many social injusticies of advanced capitalism (Peet 1977a). But increasingly, as the 1970s wore on, and environmental crises and economic recession were added to political problems of the 1960s critical liberal formulations were found lacking and radical geographers increasingly turned to the analysis of Marx.

The mid-1970s saw a flowering of radical culture in geography celebrated by the publication of *Radical geography* (Peet 1977b). Here radical geographers critically examined almost every geographic aspect of life in modern capitalism: the geography of women, the ghetto, the mentally ill, housing, rural areas, school busing, planning, migrant labour, and so on. The period was notable for a series of increasingly sophisticated critiques of conventional geography by Anderson (1973), Slater (1973, 1975, 1977), and Massey (1973). A series of exegetical writings (e.g. Harvey 1975) explored areas of Marx's writing most applicable to geographical issues. The growing interest in Marxism was broadened to include a comprehension of social anarchism (Breibart 1975, Galois 1976). The geographical expeditionary movement, which had spread to the Canadian cities of Toronto and Vancouver and over the Atlantic to London, was joined in 1974 by the Union of Socialist Geographers, which organized leftist faculty and students in the discipline. In the late 1970s *Antipode* published issues on the environment and anarchism which, in retrospect, were the last bursts of colour in the fall of its 1960s-style radicalism (Peet 1985a).

The radical geography movement changed again in the 1980s. In general, it became more sober and less combative for at least four reasons. First, the mainstream of Marxist thought was subjected to a number of more or less powerful critiques. Second, the disciplining effect of the 1979–83 economic recession and a greater knowledge of existing socialist countries made revolutionary politics a less certain quantity. Third, the laid-back academic style of the 1970s was replaced by the narrower professionalism of the 1980s. Finally, some of the Young Turks who had battled against the human geography establishment now found themselves part of it.

Yet, such a momentum had been built up in the 1970s that Marxist and related scholarship continued to flourish in geography. For example, major works were published by Harvey (1982, 1987a & b) and Massey (1984). In some areas of research, such as industrial geography, views influenced by Marxism had become engrained (e.g. Massey 1984, Massey & Meegan 1986, Peet 1987, Scott & Storper 1986, Storper & Walker 1988), and even in the last bastion of the traditionalist approach, cultural geography, Marxism and other interpretations of political economy were accepted as at least one valid viewpoint (e.g. Cosgrove 1985, Cosgrove & Jackson 1987). New journals such as *Society and Space*, founded in 1983, were still springing up, and important collections, such as *Social relations and spatial structures* (Gregory & Urry 1985) have continued to appear.

Thus, the political-economy approach to human geography now stretches through more than two decades. It has survived counterattack, critique, and economic and professional hard times, and has matured into a leading and, for many, *the* leading school of contemporary geographic thought.

The history of the approach can be roughly split into phases. The first phase, the 1970s and early 1980s, covers a period when structural Marxism was particularly influential. The second period, beginning in the late 1970s but peaking in the mid-1980s, sees a greater diversity of concerns, especially the relative potency of social structure and human agency, realism, and the study of localities. Finally, the latest period, the late 1980s, finds such issues as postmodernism and its critiques coming to the fore.

The 1970s and early 1980s: structural Marxism

The most dramatic event in the intellectual Odyssey of the political-economy approach was the turn to Marxism in a discipline in which, as Hartshorne's earlier remarks suggest, the very mention of Marx had certainly been unusual, and sometimes even anathema. Not only did geographers now read Marx, they were influenced by a particularly powerful version of Marxism, the structural ideas of Louis Althusser and his followers. To appreciate Althusser's version of Marxism, however, we must first briefly outline some of the basic theses of Marx and Engels themselves.

Marx on social and natural relations

Marxism is simultaneously politics and science. The political purpose of Marxism is social transformation on behalf of the oppressed people of the world. Communism proposes that power be placed in the hands of the workers and peasant masses in the belief that economic and political democratization will produce a higher order of society and a new kind of human being (Peet 1978–9). This proposal does not stem from utopian optimism alone. It results from a whole way of knowing the world, the science of existence called **dialectical materialism**.

Dialectics is a way of theoretically capturing interaction and change, history as the struggle between opposites, with a conception of long-term dynamics in the form of non-teleological historical laws. **Materialism** proposes that matter precedes mind, consciousness results from experience, and experience occurs primarily in the material reproduction of life. Combining the two, dialectical materialism analyzes societies in terms of modes of production, the struggle within them of the forces and relations of production, and the succession of modes of production through time towards the eventual achievement of a society characterized by high levels of development, socialized ownership of the means of production, economic democracy, and freedom of consciousness within a system of social responsibility. In the following paragraphs we emphasize the geographical aspect to the Marxian idea that social production is fundamental to human existence. By 'geographical' we mean an emphasis on the social transformation of nature followed by non-geographer Marxists (Schmidt 1971, Timpanaro 1975) as well as Marxist geographers (Burgess 1978, Smith & O'Keefe 1980, Smith 1984). We shall follow Marx directly rather than interpreting his interpreters.

Marx's view of the human relation to nature was fundamentally different from that of the classical economists. Smith and Ricardo began their analysis of production and exchange with the individual already formed by nature. Marx begins with production by individuals who form their personalities as they transform nature through the labour process. For Marx, the fact that all people are involved in broadly similar natural and social relationships makes possible a discussion of human nature in general side by side with a set of abstract (transhistorical) analytical categories (Horvath & Gibson 1984). Thus Marx always regards nature as the 'inorganic body' of the human individual, the source of the means of continued existence and locational context in which life unfolds. Matter is always exchanged between the inorganic and the organic bodies, described by Marx (1976 p. 209) as the 'universal condition for the metabolic interaction between man and nature, the everlasting nature-imposed condition of human existence . . . common to all forms of society in which human beings live'. During this necessary interaction, humans develop themselves as particular kinds of social beings. The distinguishing feature of this history is an increasingly conscious direction of labour and natural relations by human subjects.

However, Marx spends little time at the transhistorical level of analysis – elaborating the production of life in general – preferring a more concrete, historical understanding. Implicit in the above relation with nature is a second relation essential to life: the social relation among people, especially co-operation in the labour process: 'All production is appropriation of nature on the part of an individual within and through a specific form of society' (Marx 1973 p. 87). Here the essential analytical category is the **property relation**. At the dawn of human history, nature was communally owned; over time, parts of it became the private property of certain individuals; communism envisages a return to the social ownership of nature at a high level of economic development.

In this interpretation, **mode of production**, the central category of Marxian analysis, appears 'both as a relation between the individuals, and as their specific active relation to inorganic nature' (Marx 1973, p. 495; see also Godelier 1978). These relations form the economic structure of society, the foundation on which arises 'a legal and political superstructure, and to which correspond definite forms of social consciousness' (Marx 1970, p. 20). The type and level of social and natural relations correspond with a 'specific stage in the development of the productive forces of working subjects' (Marx 1973, p. 495).

Development of the productive forces fundamentally changes the economic structure and, through it, the entire society. However, this productive forces–social-relations framework should be understood as a very general conception for long-term historical analysis. The productive forces make up a structure of limitations and probabilities within which class struggles, resulting from opposition to the prevailing relations of production, actively bring about social change. Marx provides at least two accounts of the historical succession of modes of production: a 'broad outline' in which 'the Asiatic, ancient, feudal and modern bourgeois modes of production may be designated as epochs marking progress in the economic development of society' (Marx 1970, p. 21) and a more complete version (Marx 1973, p. 471–514) in which universal primitive communism decomposes into classical antiquity (based on slavery) and then feudalism and capitalism in the West and an Asiatic mode in the East. In this second version, we see the potential for an historical and geographical theory. History may be interpreted as the development and interaction of regional social formations characterized by different modes of production, each mode being further characterized by dominant social relations, including the social relation to nature.

Structural Marxism

The particular version of Marxism that was dominant in the West in the 1960s and for much of the 1970s grew in the fertile intellectual and political soil of France in the postwar years. The orthodox Marxism of the French Communist Party took the Stalinist position that all human natural history could be replicated in the scientific laws of dialectical materialism. A critical reaction to this notion in the late 1950s led many West European intellectuals towards a new synthetic form of Marxism, drawing on diverse systems of non- and neo-Marxist thought. One source was the existential and phenomenological ideas developed by Merleau-Ponty and Sartre in postwar France, particularly their critique of Stalin's insistence on the unity of the natural and human worlds. This unity, they claimed denies the specificity of the human being – her social and creative potential, his subjectivity in the historical process – and thus destroys Marxism as a theory of revolutionary self-emancipation. In opposition to Stalin's iron laws of history, Merleau-Ponty and Sartre proposed a subject-centred history with lived experience as the source of consciousness.

Stalinism, however, was not the only theoretical tradition that saw subjectivity as constitut*ed* rather than constitut*ive*. The various functionalist and structuralist streams of thought emerging from 19th century biology and sociology saw the human being made by her social milieu. In the late 1950s and early 1960s, intellectual attention (particularly in France) shifted from existential phenomenology towards structuralist ideas developed in linguistics, anthropology, and psychology. In the structural linguistics of Saussure, a coercive sign system bestows meaning on the speech of the subject. In Levi-Strauss's anthropology, the meaning of history is imparted not by historical actors but by the totality of rule systems within which actors are located. And in Lacan's psychology, the phases by which Freud's human individual achieves identity are reinterpreted as stages in the subjection of the personality to the authority of culture (Benton 1984, Callinicos 1976, 1985, Elliott 1987).

The French philosopher Althusser (1969), responding critically to Sartre and more positively to structuralism, reworked Marx's theoretical schema and analytical categories. For Althusser, as for Stalin, Marxism was indeed science. But in contradiction to Stalin's *direct* economic and technical determination, determination by the economy was, for Althusser, a thesis of the *indirect* causal relations between elements of society – relations, however, which he theorized in abstraction from actual history. In Althusser's formulation, 'non-economic' elements, such as consciousness and politics, were relatively autonomous in an *overdetermined* social structure (i.e. one in which there are diverse elements interacting one with another). For Althusser, society was a complex 'structure in dominance', yet human beings were bearers, rather than makers, of social relations.

The details of this structuralist position were elaborated by Althusser's collaborator Etienne Balibar (Althusser & Balibar 1970). Balibar argued that mode of production, the central category of Marxism, had two distinct roles: as a principle for identifying periods of history and as a means of conceptualizing the relationship between the economic, political, and ideological 'levels' of societies. In the second, synchronic role, mode of production assigned each social element its place in a hierarchy of dominance and subordination. Economic class relations (between owners and workers) always determine the structure of society in the last instance. But determination takes an indirect form; the economic level assigns to the non-economic levels their place in a hierarchy of dominance and the kinds of connection or articulation between them. Historical materialism so conceived became a theory of connections or articulations between, and the dynamic of, the main social elements. As such, structural Marxism claimed a status as a true, theoretical science (Benton 1984 p. 115).

Structural Marxism in geography

Marxist theorists influenced by Althusser subsequently applied this version of science to a range of problems, many of particular interest to geography, such as the structures of precapitalist societies (Meillasoux 1981, Terray 1972, Hindess & Hirst 1975) the historical transition and articulation of modes of production (Rey in Wolpe 1980), the state (Poulantzas 1975, 1978), and critical analysis of culture, ideology and consciousness. This work, with its potential to yield a theory of regional social structures, thought of as particular interconnected modes of production, went only a limited way before being replaced by more diverse 'post-structuralisms' as the leading frontier of Marxist and neo-Marxist geography. Structuralist geographers became bogged down defining details of space and nature, rather than applying the broad conceptions of mode of production and social formation to geohistorical development. We cannot, therefore, report on a sophisticated structuralist geography with a rich history of conceptualization and application. We shall, however, follow one line of theoretical development that has received sustained attention: the connection between society and space, or social structure and spatial structure.

Society and space

The most direct importation of structuralist ideas into geography was undoubtedly Manuel Castells's *The urban question* (1977; originally published in French in 1972). Castells saw the city as the projection of society on space: people in relation one with another give space 'a form, a function, a social signification' (Castells 1977, p. 115). The theory of space, he insists, is an integral part of a general social theory. For this theory, Castells turns to Althusser's conception of modes of production and their constituent elements, for instance:

> To analyse space as an expression of the social structure amounts, therefore, to studying its shaping by elements of the economic system, the political system and the ideological system, and by their combinations and the social practices that derive from them (Castells 1977, p. 126).

Under capitalism, the economic system is dominant and is the basic organizer of space. (By economic Castells means economic activities directly producing goods located at certain places, activities that reproduce society as a whole such as housing and public services, and exchange activities such as communications and commerce.) The political system organizes space through its two essential functions of domination/regulation and integration/repression. The ideological system marks space with a network of signs, with ideological content. Over and above this the social organization of space is determined by each of the three instances: by structural combinations of the three, by the persistence of spatial forms created in the past, and by the specific actions of individual members of social and spatial groups. As an Althusserian, Castells believed that the analysis of space first required abstract theorization of the mode of production and then concrete analysis of the specific ways structural laws are realized in spatial practice (Castells 1977, Ch. 8).

But when it came to empirical research, Castells's own formulation of the urban question did not turn on production directly (i.e. the city as a locus of production and class struggle between workers and owners in the factory). Rather he turned to consumption or the 'reproduction of labor power', and the increasing intervention of the state in such areas of *collective consumption* as housing and social services. Through state intervention, Castells argued, collective consumption is made the political arena for the struggles of the urban social movements its deficiencies produce. The reasons for Castells's diversion from production to consumption were not made clear, for he shared with Althusser an imprecise mode of expression and an argumentative style. Even so, his ideas were extremely influential in work on urban services, politics, and social movements in the 1970s, particularly in France, but also in Britain and the United States (Castells 1977, pp. 465–71).

Writing in France in the 1960s and 1970s, Castells was immediately exposed to Althusserian thought; indeed, we can call *The urban question* a direct, if idiosyncratic, application of structural Marxism to urban space. But in the Anglo-American world, structural Marxism has always been more eclectic, especially in geography. David Harvey's *Social justice and the city*, written at about the same time as Castells's book, used a concept termed 'operational structuralism', drawn from Piaget (1970), Ollman (1971), and Marx directly, rather than Althusser. It emphasized the relations between the constituent elements of governing structural change. Elements such as social classes, frequently in contradiction, force changes in society. For Harvey (1973, p. 290), structure is defined as a 'system of internal relations which is being structured through the operation of its own transformation rules'. Contradictions occur between structures as well as within them. For example, Harvey says that the political and ideological structures have their own contradictions and separate revolutions, as well as being in contradiction with the economic base of society. But some structures are regarded as more basic than others. Thus Harvey follows the Marxist view that the reproduction of material existence forms the starting point for tracing the relations within society. So for Harvey (1973, p. 302): 'Any attempt to create an interdisciplinary theory with respect to a phenomenon such as urbanism has perforce to resort to the operational structuralist method which Marx practices and which Ollman and Piaget describe.'

Harvey's (1973, p. 205) approach to the city was quite general: 'some sort of relationship exists between the form and functioning of urbanism . . . and the dominant mode of production.' Cities are economic and social forms capable of extracting significant quantities of the social surplus created by people. For Harvey, the central connections lay between the mode of economic integration in history,(whether reciprocity, redistribution, or market exchange), the subsequent creation of social surplus, and the form of urbanism. The transformation from reciprocity to redistribution precipitated a separated urban structure with, however, limited powers of inner transformation. Born of a contradiction between the forces and relations of production, the city functioned as a political, ideological, and military force to sustain a particular set of relations of production – especially property rights. The movement from this early political city to a commercial city based on market exchange was interpreted, following Lefebvre (1972), as an inner transformation of urbanism itself. The industrial city resulted from a reorganization of the industrial forces of production. But urbanism is not simply created by the forces and relations of production; it both expresses and fashions social relations and the organization of production.

Even so, urbanism is channelled and constrained by forces emanating from the economic base and, to be understood, it has ultimately to be related to the reproduction of material existence. Thus, industrial society dominates urbanism, by creating urban space through the deployment of fixed capital investments, by disposing of products through the process of urbanization, and by appropriating and circulating surplus value through the device of the city. Cities, for Harvey, are founded on the exploitation of the many by the few. Therefore:

> It remains for revolutionary theory to chart the path from an urbanism based in exploitation to an urbanism appropriate for the human species. And it

remains for revolutionary practice to accomplish such a transformation (Harvey 1973, p. 314).

Such was the optimistic tenor of the times! Such is the conclusion inherent in structural Marxism: changing any part of society, such as the city, involves changing the relations of production that guide human development.

Other works written in the 1970s saw a more direct 'one-way' relation between mode of production and the social organization of space. For example, Buch-Hansen & Nielsen (1977) coined the term 'territorial structure' to refer to the totality of production and consumption localities, their external conditions, and the infrastructures linking the whole together. Again, the most crucial determinant of territorial structure was the mode of production, both the forces of production that directly form the material contents of space and the relations of production that condition development of the productive forces in space. The social and political superstructure, which has some independence from the productive base, also makes and transforms territorial structure. In turn, however, territorial structure also conditions the further development of production.

Harvey's major work, *The limits to capital* (1982), declared its intention of steering a middle course between spatial organization as a mere reflection of capitalism and a spatial fetishism which treats the geometric properties of space as fundamental. But in reality, Harvey emphasized the first tendency, extending the analytical categories of Marx's *Capital* (use and exchange value, competition, etc.) to the explication of spatial organization through an extended theory of crisis. Although he no longer employed the explicitly structuralist language of parts of *Social justice and the city*, Harvey nevertheless integrated social and spatial structure in a 'landscape that has been indelibly and irreversibly carved out according to the dictates of capitalism'. This position was carried to its extreme by Harvey's student, Smith (1984), who termed the connections between society and environment 'the production of space' and 'the production of nature'. For Smith, the transformation of environment by capitalism implies an end to the conceptual division between the natural and the social:

> In its constant drive to accumulate larger and larger quantities of social wealth under its control, capital transforms the shape of the entire world. No god-given stone is left unturned, no living thing is unaffected. To this extent the problems of nature, of space and of uneven development are tied together by capital itself. Uneven development is the concrete process and pattern of the production of nature under capitalism . . . . There can be no apology for the anthropomorphism of this perspective: with the development of capitalism, human society has put itself at the centre of nature, and we shall be able to deal with the problems this has created only if we first recognize the reality (Smith 1984, p. xiv).

With this recognition of uneven development as a structural imperative of capitalism, as a necessary outcome of the unfolding of capital's inherent laws, the structural approach to space reached its ultimate conclusion.

It should be clear from the above discussion that while there were obvious parallels between Castells and Harvey, Marxist geography in the English-speaking world was influenced by Althusser only indirectly; a much broader and more fluid conception of 'structure' and of 'structuralism' was employed. By the early 1980s, this conception was broadening ever further. One index of this change is the work of Doreen Massey. Massey's key work, *Spatial divisions of labour* (1984) expresses the transition she had made from using geographical space as a passive surface, expressing the mode of production, to a conception of space as an active force. As she put it 'geography matters' too.

> The fact that processes take place over space, the facts of distance, of closeness, of geographical variation between areas, of the individual character and meaning of specific places and regions – all these are essential to the operation of social processes themselves. Just as there are no purely spatial processes, neither are there any non-spatial processes. Nothing much happens, bar angels dancing, on the head of a pin . . . . Geography in both its senses, of distance/nearness/betweenness and of the physical variation of the earth's surface (the two being closely related) is not a constraint on a pre-existing non-geographical social and economic world. It is constitutive of that world (Massey 1984, p. 52).

As a Marxist, Massey emphasized variation in the social relations of production over space. Thus as social classes are constituted in places, so class character varies geographically. But Massey went on to argue that the social structure of the economy necessarily develops in a variety of local forms which she termed 'spatial structures of production'. The archetype she developed was the hierarchy of functions of the multi-locational company, different stages in production (organization, research, assembly, parts-making) being assigned in different combinations to various regions, although other ways of conceptualizing spatial structures are possible. Massey maintained that spatial structures not only emerge from the dictates of corporate initiative but are also established and changed through political and economic battles on the part of social groups (i.e. through class struggle). In turn, spatial structures, through differential employment possibilities, create, maintain, or alter class and gender inequalities over space. Her main point was that 'spatiality is an integral and active condition' of the production process (Massey 1984, p. 68).

The mid-1980s: the structure–agency debate, realism and locality

Before the 1980s began, Althusser's influence had already sparked off a furious debate throughout the social sciences about the relative contributions of economic structure and human agency to the making of history. The 'structure–agency' debate, as it has come to be known, is documented in a whole series of responses and counter-responses to E. P. Thompson's original critique of Althusser, *The poverty of theory* (1978), which ranged widely across all social science, taking in life, the universe, and everything along the way (see, for example, Anderson 1980, 1983).

Human geography's version of this debate was prefigured in the work of

Gregory (1978), but did not fully take off until the exchanges that followed Duncan & Ley's critique of structural Marxism in 1982 (see Duncan & Ley 1982, Chouinard & Fincher 1983). Like the debate in the social sciences as a whole, human geography's version of the structure–agency debate was wide ranging, but, in particular, it intertwined three themes; the relative importance of structure and agency, and how they might be reconciled in a single approach; the efficacy of a realist methodology; and the importance of localities. However, in essence, the structure–agency debate in human geography was multifaceted because of the several rather different impulses that fuelled it. What were these impulses? Five of them seem particularly relevant.

Human geography

First of all, the debate came about because of the peculiar circumstances of human geography. Almost alone amongst other human sciences, in the 1950s human geography still had a poorly developed base in social theory. One only has to compare such work as C. Wright Mills's (1959) *The sociological imagination* with the extant human geography books of the period to grasp the differences in range and depth. Thus, as Urry points out in this volume, when Marxism began to have an influence on the subject in the 1960s and into the 1970s it was successful in part because there was so little in the way of social theory in human geography with which it had to contend. The discourses of Marxism and social theory were almost synonymous.

Therefore the structure–agency debate was in many ways a parade of traditions of social theory well known in the social sciences but which hitherto had received scant attention in human geography, traditions such as phenomenology, symbolic interactionism, even hermeneutics (Gregory 1978). Marxism provided both the space for these different traditions to be introduced into the subject and, at the same time, a suitable theoretical orthodoxy, in its structuralist form, against which to battle (see Billinge 1978, Ley & Samuels 1978, Duncan & Ley 1982).

Of course, the Marxist tradition in human geography did not remain unchanged in the face of the assaults that were mounted upon it. The different traditions encouraged a more eclectic approach to political economy. There were parallels here with what had been happening in the Marxist tradition anyway, especially in its European incarnation. There was Gramsci, Sartre, and the Frankfurt School to discuss, and later Habermas's and Giddens's reconstructions of historical materialism. Each of these authors had drawn on traditions outside Marxism to strengthen their analysis (see Held 1980, Habermas 1979, Giddens 1981).

Marxism and change in society

A second impulse came from the changing course of history itself. In the 20th century, Marxism has strained to account for far-reaching changes in the nature of society. The list is almost infinite. To start with, there are the continuing twists and turns of capital itself, including new regimes of capital accumulation based first upon mass consumption and latterly upon segmentation of the mass market and the flexible accumulation that goes with it, the growth of service industries, and the spread of a new international division of labour. Far-reaching social changes have also occurred and especially the rise of a service class of managers and professionals and the greater participation of women in the labour force. That is not all. Then there has been the rise of a large and comprehensive state apparatus with extensive disciplinary, welfare, and socialization functions. There has been the growth of socialist societies (Forbes & Thrift 1987, Thrift & Forbes 1986) with social dynamics which are often only marginally based upon capital accumulation and owe much more to the growth of bureaucracies. All round the world there are developing countries that have generated important social forces opposed to capitalism; the growth of a radical Islam is a case in point. The list of changes goes on and on.

The Marxist tradition has not found it easy to accommodate all of these changes. In order to survive it has been necessary to revise extant theory (Harvey 1982) and to broaden it to include the 'non-economic' factors of state and civil society.

The political impulse

A third impulse was political. By the end of the 1970s, the forces of the new right were asserting themselves in many countries. Certain of the demands of the new right, especially a radical individualism, clearly appealed to large sections of the population. The reaction of many on the left was to study such developments in detail with a view to formulating effective counter-strategies. But this was not all. New forms of politics started to come into existence in the 1970s and into the 1980s which were not based on the old axes of support such as class, but cut across them. The ecological movement, or what O'Riordan (1981), rather more generally calls 'environmentalism', has taken off worldwide; it directly challenges a number of the 19th-century ideas of industrial progress which have cast their shadow over the 20th century with often disastrous consequences for the environment. Similarly, the feminist movement has laid down a challenge to old ways of thinking , laying bare the subordination of women of all classes in patriarchal structures. None of these developments could be ignored for they were constitutive of new ways of thinking which were both political and personal, that is they involved not just planning out programmes to change social structures but also a deep commitment to changing 'oneself'.

The realist approach and attitudes to theory

A fourth impulse was theoretical. By the end of the 1970s some human geographers were beginning to catch the scepticism about the power of theory that had already infected the other social sciences, especially through the work of more extreme writers such as Foucault and Rorty. This is not to say that theory was to be consigned to the rubbish bin. Rather it was that there were limits to the applications of theory and its ability to illuminate geographic practice (Thrift 1979). Depending upon where these limits were placed, it was possible to argue for a thoroughgoing relativism or a thoroughgoing rationalism. But most commentators seemed happiest with the compromise formulations of realism.

In its present incarnation, realism has been associated primarily with the names of Bhaskar (1975, 1979, 1986) and Harré (1987). Its main routes of entry into human geography were through the work of Keat & Urry (1981) and Sayer (1984). Realism is a philosophy of science based on the use of abstraction as a means of identifying the causal powers of particular social structures, powers which are released only under specific conditions. In many ways, realism consists of a state-of-the art philosophy of science allowing for structural explanation but incorporating the scepticism about the powers of theory characteristic of late 20th century philosophy (Baynes *et al.* 1987). Certainly, realism has become the major approach to science in human geography, with special attractions for those pursuing the political-economy approach. But, as Sayer makes clear, it is not automatically radical:

> The changes are clearest in radical and Marxist research in geography . . . . However, this association of realism is not a necessary one: some radical work has been done using a nomothetic deductive method . . . and acceptance of realist philosphy does not entail acceptance of a radical theory of society – the latter must be justified by other means (Sayer 1985, p. 161).

Perhaps realism's greatest impact has been in promoting the thoughtful conduct of empirical research. The realist approach makes for a level-headed appraisal of what is possible. Thus Sayer points to what a viable 'regional geography' might ordinarily consist of:

> The best we can normally manage is an incomplete picture consisting of a combination of descriptive generalisations at an aggregate level (e.g. on changes in population and standards of living), some abstract theory concerning the nature of basic structures and mechanisms (e.g. concerning modes of production) and a handful of case studies involving intensive research showing how in a few, probably not very representative, cases these structures and mechanisms combine to produce concrete events. (Sayer 1985, p. 172).

This empirical connection was important. By the mid-1970s it had become crucial for those sponsoring a political-economy approach to demonstrate that they could do good empirical or concrete research, both in order to find out more about what was happening 'on the ground' and to demonstrate their skills in this area to colleagues sceptical about the work of abstract theorists. This meant that more careful attention had to be paid to how abstract theory could be applied in particular contingent situations. This, in turn, lead to more careful formulations of theory by realists aimed at eliciting the causal powers of particular social relations in a whole range of contingent situations. Thus, more attention was paid to such matters as the distinction between 'internal' and 'external' relations and the relative merits of 'intensive' and 'extensive' research (Sayer 1984). However, the problems of the relations between theory and empirical work have hardly been solved. For example, it is all very well to talk of levels of abstraction forming a neat hierarchy all the way from abstract theoretical proposals to empirical complexities, but it is difficult to find a mechanism which unambiguously allows the researcher to detect which theoretical objects occupy which levels (see Urry 1985).

Such problems begin to explain why rather more room was also given to the **hermeneutic** dimension (to the theory of the interpretation and clarification of meaning by those promoting political-economy perspectives. The hermeneutic tradition finally lapped upon the shores of human geography. Hermeneutics had been introduced into geography in the 1970s in a number of guises, including phenomenology and existentialism (see Pickles 1986). Its most important function was to underline the necessity for taking the act of interpretation seriously, whether as an awareness of theory as representative of a set of interpretive acts or as a set of procedures for explaining the interpretations which people give to their world (Jackson & Smith 1984). The results of its adoption by those involved in political-economy perspectives have been twofold. First, the contextual dimension of theory – its rootedness in particular times and places – is taken more seriously. Theories are themselves historical and geographical entities. Second, there is much greater awareness of the validity of a qualitative geography, made up of methods for getting at peoples' interpretations of their worlds in as rational and ordered a way as possible (Eyles & Smith 1988). For example, ethnography has gained a new respectability (Geertz 1973).

The importance of space as a constituent of the social

The last impulse was in many ways the most significant. It was a growing realization that space was rather more important in the scheme of societies than was envisaged at the farthest swing of the structural Marxist pendulum. Space is not just a reflection of the social but *a constitutive element of what the social is*; Massey (1984, p. 4) summarized these concerns well in *Spatial divisions of labour* where she pointed out that: 'It is not just that the spatial is socially constructed: the social is spatially constructed too.' She went on:

> The full meaning of the term 'spatial' includes a whole range of aspects of the social world. It includes distance, and differences in the measurement, connotations and appreciation of distance. It includes movement. It includes geographical differentiation, the notion of place and specificity and of differences between places. And it includes the symbolism and meaning which in different societies, and in different parts of given societies, attach to all of these things (Massey 1984, p. 5).

The renewed emphasis on the importance of space connects back to the realist project sketched above. In this project, space clearly makes a difference to whether the causal powers of particular social relations are activated, and the forms which these social relations can take. In other words, important social relations are necessary. For example, for the wage–labour relation to exist it is necessary for both capitalists and labourers to exist. But the existence and expression of such social relations in particular places relies upon the web of contingencies that is woven by the spatial fabric of society. The picture is immeasurably complicated by the fact that social relations, in their diverse locally contingent forms, continually constitute that spatial fabric. In Pred's (1985) terms, the social becomes the spatial, the spatial becomes the social'.

These five different impulses came together in the so-called 'structure–agency' debate. There were two main problems with this debate as it took place in geography. First of all, as is now hopefully apparent, its participants had quite different impulses motivating their participation in it. The opportunities for confusion were, therefore, legion. Second, especially in its initial stages, it was easier to point to what was unsatisfactory about extant theory and research than to cite examples of theory and research that met the standards being prescribed; critique piled upon critique in a wasteful duplication of effort. Five years later these two problems have become much less prominent; the different impulses have been negotiated and have even merged to produce new lines of thought, while a body of theory and research has been built up which can act as a template for further endeavours.

Responses

What, then, were the main foci of the structure–agency debate? Three foci were particularly important. First, there was a general concern with the individual. Parts of Marx (see Geras 1983) and Marxist writers such as Sartre show sensitivity to the question of the individual but, in practice, many commentators concede that the balance of Marxist theory has tipped away from the individual towards social structure. Of course, simply stating the need for more concern for the individual would, by itself, be an act of empty rhetoric. There was a need actively to expand this concern.

The chief efforts came from a group of human geographers who were interested in Hägerstrand's time-geography (Thrift & Pred 1981, Pred 1981). This interest rapidly transmuted into a concern with Giddens's structuration theory, the development of which has been partly influenced by human geography (see Giddens 1979, 1981, 1984, 1985, Thrift 1983). Structuration theory had a major influence on the political-economy approach in human geography for three main reasons. First of all, in Giddens's earlier works, it offered geographers a way out of the problem of structure and agency, precisely by concentrating on the importance of geography. Giddens's (1979, 1981, 1984) conceptions of locale and time–space distanciation (the stretching of societies over space) were meant to show how social structures were 'instantiated' in a particular geography, so that at any time social structure did not have to exist everywhere in order to have influence. Second, structuration theory emphasized the importance of hermeneutics at all scale levels, from the areas of day-to-day communicative interaction between individuals to the structures of communication (signification), power (domination), and sanction (legitimation) underpinning society as a whole. Third, especially in Giddens's later work, structuration theory offered a coherent and sympathetic critique of historical materialism, based in part on its lack of spatiality and in part on its lack of attention to matters of signification and legitimation.

Structuration theory has been attacked by some geographers for its schematic form, and for its glossing over of some major problems (Gregson 1986, 1987). But it remains one of the few examples of an advance made in social theory with explicit connections to human geography and with, in contrast to comparable schemes such as that of Habermas, an appreciation of geography as socially constitutive as well as socially constituted (Pred 1987).

A second general focus of research on structure and agency was on the reproduction of social structure. Quite clearly, with the impetus provided by structuration theory, the debate on structure and agency could not stay at the level of the individual and individual agency. It had to move towards analysis of social structure, and especially of how institutions come into being which are aimed at enforcing a particular order and a particular vision of social reality. There is no doubt that the reproduction of capitalism involves a good deal of crude coercion aimed at keeping workers disciplined. Nor is it to deny that many of the institutions of state and civil society have, as one of their functions, transmitting the multiple disciplines of capitalism: to produce, to reproduce, to consume, and so on. Rather is it to suggest that the ways in which capitalism is reproduced within these institutions are less direct than was once thought, leaving a number of social relations relatively untouched, and providing all kinds of sites from which it is possible to generate opposition and change. More than this, the processes by which the reproduction of capitalism is assured are not just negative ones of constraint but also processes in which people become positively involved. They are based on consent as well as coercion (Gramsci 1971).

Many strategies have been constructed by researchers in the social sciences to deal with these indeterminate elements of capitalist reproduction. In human geography three main strategies have been followed. The first of these is theoretical and still quite abstract. It is to use a theoretical system influenced by realism which explicitly invests social objects other than the capitalist economic system with causal powers, and especially the state. For example, Urry (1981) invested three 'spheres' – the state, certain entities within civil society (e.g. the family), and capital – with causal powers. The outcomes in any society – of the multiple determinations flowing into and out of these three spheres – will be complex, with different societies producing different resonances in capitalist social relations and different degrees to which capital is able to penetrate the state and civil society. Lovering (1987), in similar vein, provides an analysis of the way in which the state can direct the course of capital through the defence industry. Foord & Gregson (1986) provide patriarchy (the structures by which men oppress women) with a causal existence which is independent of capital although intertwined with it in various ways.

A second strategy is less abstract. It involves detailed study of how capitalist social relations overlap in societies as value systems and as symbols. This kind of work has focused in general upon the mechanics of cultural production and has been of two main types. First, there is a considerable amount of work involved with the varying modes of reproduction of the meanings attached to landscape (see Cosgrove 1985, Cosgrove & Daniels 1988). Second, there is all manner of work on communications media and the way they are used to promote capitalist and/or establishment values, with especial reference to use and manipulation of ideas of places (Burgess & Gold 1985). More recently, this work has burgeoned into consideration of how commodities are sold through the conscious use of symbolic systems which both draw on and reproduce particular lifestyles. Part of this process consists of the setting-up of places within which consumption and lifestyle can come together, reinforcing one another (Thrift & Williams 1987).

A third strategy is to study subcultures and ideologies which conflict with the dominant ones. These cultures are distinguished by their resistance to all or, more likely, a part of the capitalist system and the state. There are, of course, some oppositional cultures so strongly stigmatized that a muted but continuous opposition is their only choice if the integrity of the group is to survive. Such continually harassed groups as travellers are a case in point (Sibley 1981). However, most cultures choose a mixture of conflict and compromise. The classic cases of resistance can be found amongst 19th century working-class communities where the battle-lines between labour and capital were tightly drawn, amongst the 'Manchesters, Mulhouses, and Lowells' (Harvey 1985, p. 9). These communities have continued on in to the 20th century with the lines of battle often being drawn even more starkly, when a distinctive ethnic or religious composition strengthened community ties, rather than weakened them. However, these studies are not the only possibility. In the 20th century there has been an expansion in the number of urban movements, many of which are of middle-class, not working-class, composition (Castells 1983). The range of these movements is now very great indeed, and geographers have been studying all of them. There is the ecological movement, black, feminist and gay movements, the forces of nationalism, and so on. Each and every one of these oppositional cultures has a distinctive geography which is a vital part of their ability to survive and contest dominant orders.

These three research strategies have come together in certain literatures, and especially in that investigating the gentrification of urban neighbourhoods (Smith & Williams 1986, Rose 1988). But literature such as this also points towards one vexed question that cuts across all three research strategies – the question of class. Class has been, and continues to be, a focal point of the structure–agency debate since Thompson's interventions on the nature of class galvanized all manner of writers' pens into actions (Thompson 1963, 1978). What seems certain is that the Marxian depiction of class was too 'thin', concentrating too much on class struggle at work, important though this undoubtedly is. The social and cultural dimensions of class were neglected, even though they provide important forces dictating the intensity and direction of struggle at work, as well as being domains of class conflict in their own right. This bias has now been corrected and the full range of the permutations of conflict between capital and labour is now being revealed by a coalition of social historians, sociologists, and geographers. This is not to say all the problems have been solved; far from it. Many questions remain only partially conceptualized (Thrift & Williams 1987). In particular, space can now be seen to be a crucial determinant of class formation, but its exact role in particular situations requires much further work of both theoretical and empirical elaboration. The organization of space clearly alters the ability of classes to coalesce and pursue a class politics, rather than remain as separate islands of community (Harvey 1985).

The mention of space leads on to the third major focus of research on structure and agency: the place of space in the relations between human geography and social structure. Two particular areas of research have been developed here. First of all, there has been an interest in how structures are tied together in space by transport and communications innovations, from the invention of writing through the burst of new media of communication in the 19th and early 20th centuries (the train, the telegraph, the telephone, and so on) to the new instruments of mass communication and processing that dominate our worldview now: radio, television, video, even the computer (Gregory 1987). As a result, social structure has moved from reliance on face-to-face communication to reliance on indirect communication, from 'social' to 'system' integration (Giddens 1984). The notion of time space distanciation captures the uneven spatial dimensions of this integration. These changes have been crucial to the constitution of society in all kinds of ways. Economically, they have allowed multinational corporations and international finance to exist. Socially, they have allowed the state to spread its influence into all the corners of everyday life. Culturally, they have produced 'imagined communities', including nationalist and religious movements (Anderson 1983, Gellner 1985).

The second area of research, and one which has become very important indeed, has focused on the idea of locality. The idea of locality research sprang out of Massey's work in *Spatial divisions of labour* (1984) and a subsequent British research programme, sponsored by the Economic and Social Research Council, called the Changing Urban and Regional System Initiative, in which many of the chief proponents (and critics) of locality research were involved. At its simplest, locality research was an enquiry into the effects of international industrial restructuring on local areas, and especially into why different local areas produced different responses (Massey 1984). But the research soon ranged outside this initial area of enquiry, taking in issues of gender, class, and politics, as well as the consideration of flexible production and the rise of an economy based on the service industries (Murgatroyd *et al.* 1986, Cooke 1989). A complex theoretical debate soon began to rage about the degree to which localities could be defined and considered as independent actors with their own 'proactive' capabilities (Savage *et al.* 1987, Urry 1987). This debate has produced an enormous amount of heat, but it remains to be seen whether it will produce any light. However, the substantive pieces of locality research coming from Britain and the United States are clearly important.

In conclusion, what did the structure–agency debate achieve, in all its different guises? Three main things, perhaps. First, it focused attention on that old Marxian dictum, 'people make history but not in circumstances of their own choosing'. This was always a notoriously opaque statement, saying both everything and nothing. Now, however, it is possible to say more about almost every aspect of this statement. More is known about what people are, the social institutions they make, and the geographies within which they must make them. But second, none of this denies the power of the political-economy approach. What it does is extend and enrich the approach in all kinds of ways. Most particularly, against the background of the continuing effort to understand the shifting contours of capitalism, it makes a contribution to the social, cultural, and political knowledge necessary to withstand capitalism's depredations and understand capitalism's successes. Third and finally, the structure–agency debate underlined the fact that capitalism is not just a phenomenon of economic geography. It is also at one and the same time a social, cultural, and political geography which is equally made and disputed in each of these other realms.

The late 1980s: postmodernism and purity

By the late 1980s, a new issue had arisen within the political-economy approach. It can be summarized under the heading of **postmodernism** although this is a term which is currently used to exess (Punter 1988).

Postmodernism is a confusing term because it represents a combination of different ideas. It is, perhaps, most often seen as concerned with issues of method. As method, it is critical of the idea of totality that is typical of structural Marxism. Instead, it takes its cue from so-called post-structuralist theory, especially the work of those such as Derrida, Lacan, Kristeva, and Foucault (and, ironically, Althusser), which, although it is very different in a number of ways, shares common assumptions about the matters of language, meaning, and subjectivity (Dews 1987, Weedon 1987). In particular, this body of work assumes the following: that meaning is produced in language, not reflected by it; that meaning is not fixed but is constantly on the move (and so the focus of fierce political struggle); and that subjectivity does not imply a conscious, unified, and rational human subject but instead a kaleidoscope of different discursive practices. In turn, the kind of method needed to get at these conceptions will need to be very supple, able to capture a multiplicity of different meanings without reducing them to the simplicity of a single structure. Derrida's deconstruction, Foucault's genealogy, Lyotard's paralogism, the postmodern ethnography of anthropoligists such as Clifford (1988), the discourse analysis of various social psychologists – all these are attempts to produce a method that can capture history as a set of overlapping and interlocking fields of communication and judgement (discursive fields).

Postmodernism has also been used to describe the culture of a new phase of capitalism. Such commentators as Dear (1986), Jameson (1984), Davis (1985), and Harvey (1987a) have built on a variety of sources from the 'situationist' analysis of the consumer spectacle, through the power of financial capital, to the rise of 'flexible' methods of accumulation, to produce an analysis of a new phase of capitalist culture based upon a constant, self-conscious play with meaning and leading to the increased usage in everyday life of historical eclecticism, pastiche, and spectacle. There may be a dispute between those commentators about the point in time at which modern culture gave way to postmodern culture, even about the defining characteristics of postmodern culture, but all share a desire to link its advent to recent changes in the capitalist mode of production, in one way or another.

In contrast to these postmodern excursions, the end of the 1980s also saw the signs of a possible resurgence of a 'traditional' Marxist approach. Some commentators clearly felt that things had gone too far in the direction of eclecticism and that the Marxist core of the political-economy approach was under threat. It was time for the experimenters to return to the Marxist fold. Thus, locality research was subject to a sustained critique for its lack of grounding in grand theory and its apparently empiricist bent (Harvey 1987, Smith 1987a). Similarly, postmodern methodological approaches, although not the epochal developments, were lambasted by Harvey and others, who argued for a return to a more solid Marxism (Harvey 1987).

Other commentators have constructed theoretical halfway houses between the radical uncertainties of postmodernism and the radical certainty of the fully fledged structural approach. For example, some writers have commended the works of the French Regulationist School, whose members include Aglietta, Boyer, and Lipietz, which has developed an approach that holds on to notions which look suspiciously like base and superstructure, suitably altered for less rigid times (de Vroey 1984). Another approach has been to argue for a 'post-enlightment Marxism' in which it is possible to place the analysis of 'civil society on an equal footing with political economy in the theorisation of capital and the explanation of history and geography, while not insisting on subjecting them all to a dialectical totalisation' (Storper 1987, pp. 425–6). This approach would be close to the reconstitutions of historical materialism of such writers as Giddens and Habermas. Thus, in the 1980s, the political-economy approach in geography has continued in a state of flux. it continues to show signs of a healthy self-criticism. Hopefully, it continues to develop and grow.

Conclusions

The political-economy perspective in geography is barely 20 years old. Yet already we are able to chart its several periods of development. Radical ideas grew slowly and late because the discipline was conservative and because geographers had little experience in understanding and debating social theory. At first, therefore, radical geography was merely doing socially relevant work, with Marxism learned the painful way, through reading and interpreting the original, classical works. In the 1970s, structuralism in geography existed more as the reconstructive notions of eager critics than as a distinct and sophisticated school of thought. However, some time around the late 1970s and early 1980s the pace of change increased, the interaction between geography and social theory intensified, and, not coincidentally, fragmentation appeared in what still remained a relatively coherent perspective. Ideas of structuration theory, realism, and locality, towards which many geographers turned, were usually imported from points of origin outside the discipline, but political-economic geographers quickly began giving them new twists, applying them differently, then adding new dimensions. The quality of theorietical discourse improved as space and environment became burning issues of the day. In many ways the conflicts between different positions which typified the 1980s were a necesary part of improving the intellectual product in an era when people were beginning to listen.

Finally, at the end of the 1980s, we found some geographers pushing on through the postmodernist frontier, while others considered it more fruitful to improve on what had already been discovered. Of course, it is not the case that everyone joined each wave of interest, being carried along with the wave until it broke under criticism, then jumping to the next upswell of concern. Each new interest has left a residue of knowledge in all and made committed adherents out of some. This should be the case. For we are not talking here of knowledge as adornment but as interventionary ability. In the end, this is the original contribution of the political-economy approach. Knowledge for its own sake is unconvincing. Knowledge to make the world a better place becomes the only acceptable purpose.

References

Althusser, L. 1969. *For Marx* (Translated by Ben Brewster). London: Penguin.
Althusser, L. & E. Balibar. 1970. *Reading Capital*. London: New Left Books.
Ambrose, P. 1976. British land use planning: a radical critique. *Antipode* **8**, 2–14.
Anderson, B. 1973. *Imagined communities*. London: Verso.
Anderson, J. 1973. Ideology in geography: an introduction. *Antipode* **5**, 1–6.

Anderson, J. 1978. Ideology and environment (special issue). *Antipode* **10**, 2.
Anderson, P. 1980. *Arguments within English Marxism*. London: Verso.
Anderson, P. 1983. *In the tracks of historical materialism*. London: Verso.
Baynes, K., J. Bohman, & T. McCarthy (eds) 1987. *After philosophy. End or transformation?* Cambridge, Mass.: MIT Press.
Benton, T. 1984. *The rise and fall of structural Marxism: Althusser and his influence*. New York: St Martin's Press.
Berry, B. J. & A. Pred 1965. *Central place studies: a bibliography of theory and applications* (revised ed). Philadelphia: Regional Science Institute.
Bhaskar, R. 1975. *A realist theory of science*, 2nd edn. Brighton: Harvester.
Bhaskar, R. 1979. *The possibility of naturalism*. Brighton: Harvester.
Bhaskar, R. 1986. *Scientific realism and human emancipation*. London: Verso.
Billinge, M. 1977. In search of negativism: phenomenology and historical geography. *Journal of Historical Geography*. **3**, 55–67.
Blaut, J.M. 1974. The ghetto as an internal neo-colony. *Antipode* **6**, 37–41.
Boddy, M. 1976. Political economy of housing: mortgage-financed owner-occupation in Britain. *Antipode* **8**, 15–24.
Breitbart, M. 1975. Impressions of an anarchist landscape. *Antipode* **7**, 44–9.
Breitbart, M. (ed.) 1978–9. Anarchism and environment (special issues). *Antipode* **10**, 3 & **11**, 1.
Buch-Hansen, M. & B. Nielsen. 1977. Marxist geography and the concept of territorial structure. *Antipode* **9**, 1–12.
Burgess, J. & J. Gold. 1985. *Geography, the media and popular culture*. London: Croom Helm.
Burgess, R. 1978. The concept of nature in geography and marxism. *Antipode* **10**, 1–11.
Burton, I. 1963. The quantitative revolution and theoretical geography. *Canadian Geographer* **7**, 151–62.
Callinicos, A. 1976. *Althusser's Marxism*. London: Pluto Press.
Callinicos, A. 1985. *Marxism and philosophy*. Oxford: Oxford University Press.
Carney, J. 1976. Capital accumulation and uneven development in Europe: notes on migrant labour. *Antipode* **8**, 30–8.
Castells, M. 1977. *The urban question: a Marxist approach* (translated by Alan Sheridan). Cambridge Mass.: MIT Press.
Castells. M. 1983. *The city and the grassroots*. London: Edward Arnold.
Chorley, R. J. & P. Haggett (eds) 1967. *Models in geography*. London: Methuen.
Chouinard, V. & R. Fincher 1983. A critique of 'Structural marxism and human geography'. *Annals of the Association of American Geographers*. **73**, 137–46.
Clifford, J.C. 1988. *The predicament of culture. Twentieth century ethography, literature and art*. Cambridge, Mass.: Harvard University Press.
Cooke, P. 1986. The changing urban and regional system in the UK. *Regional Studies* **20**, 243–51.
Cooke, P. 1987. Clinical inference and geographical theory. *Antipode* **19**, 69–78.
Cooke, P. (ed.) 1989. *Localities*. London: Unwin Hyman.
Corbridge, S. 1986. *Capitalist world development: a critique of radical development geography*. London: Macmillan.
Cosgrove, D. 1985. *Social formation and symbolic landscape*. Totowa, NJ: Barnes & Noble.
Cosgrove, D & S. Daniels (eds) 1988. *The inconography of landscape*. Cambridge: Cambridge University Press.
Cosgrove, D. & P. Jackson 1987. New directions in cultural geography. *Area* **19**, 95–101.
Crompton, R, & M. Mann (eds) 1986. *Gender and stratification*. Cambridge: Polity Press.
Davis, M. 1985. Urban renaissance and the spirit of post-modernism. *New Left Review* **151**, 106–13.

Dear, M.J. 1986. Postmodernism and planning. *Environment and Planning D, Society and Space.* **4**, 367–84.
Dear, M. J. & J. V. Wolch 1987. *Landscapes of despair.* Cambridge: Polity Press.
de Vroey, M. 1984. A regulation approach of contemporary crisis. *Capital and Class* **23**, 45–66.
Dews, P. 1987. *Logics of disintegration: poststructuralist thought and the claims of critical theory.* London: Verso.
Duncan, J & D. Ley 1982. Structural marxism and human geography: a critical assessment: *Annals of the Association of American Geographers.* **72**, 30–59.
Elliott, G. 1987. *Althusser. The detour of theory.* London: Verso.
Eyles, J & D. Smith (eds) 1988. Quantitative methods in human geography. Cambridge: Polity Press.
Foord, J. & N. Gregson 1986. Patriarchy: towards a reconceptualisation. *Antipode* **18**, 186–211.
Forbes, D. K. & N. J. Thrift (eds) 1987. *The socialist Third World.* Oxford: Basil Blackwell.
Galois, B. 1976. Ideology and the idea of nature: the case of Peter Kropotkin. *Antipode* **8**, 1–16.
Geertz, C. 1973. *The interpretation of cultures.* New York: Basic Books.
Gellner, E. 1985. *Nations and nationalism.* Oxford: Basil Blackwell.
Geras, N. 1983. *Marx and human nature.* London: Verso.
Giddens, A. 1979. *Central problems in social theory.* London: Macmillan.
Giddens, A. 1981. *A contemporary critique of historical materialism.* London: Macmillan.
Giddens, A. 1984. *The constitution of society.* Cambridge: Polity Press.
Giddens, A. 1985. *The nation state and violence.* Cambridge: Polity Press.
Godelier, M. 1978. Infrastructures, societies, and history. *Current Anthropology* **19**, **4**.
Gramsci, A. 1971. *The prison notebooks.* London: Lawrence & Wishart.
Gregory, D. 1978. *Ideology, science and human geography.* London: Hutchinson.
Gregory, D. 1987. The friction of distance? Information circulation and the mails in early nineteeth-century England. *Journal of Historical Geography* **13**, 130–54.
Gregory, D. & J. Urry 1985. *Social relations and spatial structures.* New York: St Martin's Press.
Gregson, N. 1986. On duality and dualism: the case of structuration and time-geography. *Process in Human Geography.* **10**, 184–205.
Gregson, N. 1987. Structuration theory: some thoughts on the possibilities for empirical research: *Environment and Planning D, Society and Space.* **5**, 73–92.
Habermas, J. 1979. *Communication and the evolution of society*, London: Heinemann.
Harré, R. 1987. *Varieties of realism.* Oxford: Basil Blackwell.
Hartshorne, R. 1939. *The nature of geography.* Lancaster, PA.: Association of American Geographers.
Hartshorne, R. 1955. 'Exceptionalism in geography' re-examined. *Annals of the Association of American Geographers* **43**, September, 205–44.
Hartshorne, R. 1959. *Perspective on the nature of geography.* Chicago: Rand McNally.
Harvey, D. 1969. *Explanation in geography.* London: Edward Arnold.
Harvey, D. 1973. *Social justice and the city.* Baltimore: Johns Hopkins University Press.
Harvey, D. 1975. The geography of capitalist accumulation: a reconstruction of the Marxian theory. *Antipode* **7**, 9–21.
Harvey, D 1982. *The limits to capital.* Chicago: University of Chicago Press.
Harvey D. 1983. Owen Lattimore – a memoire. *Antipode* **15**, 3–11.
Harvey, D. 1985. *The urbanisation of consciousness.* Oxford: Basil Blackwell.
Harvey, D. 1987a. Flexible accumulation through urbanisation: reflections on post-modernism in the American city. *Antipode.* **19**, 260–86.

Harvey, D. 1987b. Three myths in search of a reality in urban studies. *Environment and Planning D, Society and Space* **5**, 367–76.
Harvey, D. & N. Smith 1984. Geography: from capitals to capital. In *The left academy: Marxist scholarship on American Campuses*, vol. 2, B. Ollman, & E. Vernoff. (eds), 99–121. New York: Praeger.
Hayford, A. 1974. The geography of women: an historical introduction. *Antipode* **6**, 1–19.
Held, D. 1980. *Introduction to critical theory*. London: Hutchinson.
Hindess, B. & P. Q. Hirst 1975. *Pre-capitalist modes of production*. London: Routledge & Kegan Paul.
Horvath, R. 1971. The 'Detroit geographical expedition and institute' experience. *Antipode* **3**, 73–85.
Horvath, R. J. & K. Gibson 1984. Marx's method of abstraction. *Antipode* **16**, 23–36.
Hudson, B. 1977. The new geography and the new imperialism: 1870–1918. *Antipode* **9**, 12–19.
Jackson, P. & S. J. Smith 1984. *Exploring social geography*. London: Allen & Unwin.
Jameson, F. 1984. Postmodernism, or the cultural logic of late capitalism. *New Left Review* **146**, 53–92.
Johnston, R. J. 1986. *On human geography*. London: Edward Arnold.
Keat, R. & J. Urry 1981. *Social theory as science*. London: Routledge & Kegan Paul.
Kropotkin, P. 1902. *Mutual aid: a factor of evolution*. London: Heinemann.
Lancaster Regionalism Group 1986. *Localities, class and gender*. London: Pion.
Lattimore, O. 1950. *Ordeal by slander*. Boston: Little, Brown.
Lattimore, O. & E. Lattimore 1944. *The Making of modern China: a short history*. New York: Franklin Watts.
Lefebvre, H. 1972. *La Pensée marxiste et la ville*. Paris: Gallimard.
Ley, D. & M. Samuels (eds) 1978. *Humanistic geography*. London: Croom Helm.
Lovering, J. 1987. Militarism, capitalism, and the nation–state: towards a realist synthesis. *Environment and Planning D, Society and Space* **5**, 283–302.
MacKenzie, S., J. Foord & M. Breitbart (eds) 1984. Women and the environment, (special issue). *Antipode* **16**, 3.
Marx, K. 1970. *A contribution to the critique of political economy*. Moscow: Progress Publishers.
Marx, K. 1973. *Grundrisse*. Harmondsworth: Penguin
Marx, K. 1976. *Capital*, Vol. 1. Harmondsworth: Penguin.
Marx, K. & F. Engels 1975. *On Religion*. Moscow: Progress Publishers.
Massey, D. 1973. Towards a critique of industrial location theory. *Antipode* 5, 33–9
Massey, D. 1976. Class, racism and busing in Boston. *Antipode* **8**, 37–49.
Massey, D. 1984a. *Spatial divisions of labour: social structures and the geography of production*. London: Methuen.
Massey, D. 1984b. Geography matters. In *Geography Matters! A Reader*, D. Massey & J. Allen (eds), 1–11. Cambridge: Cambridge University Press.
Massey, D. & R. Meegan 1982. *The anatomy of job loss*. London: Methuen.
Massey, D. & R. Meegan (eds) 1986. *Politics and method: contrasting studies in industrial geography*. New York: Methuen.
Meillasoux, C. 1981. *Maidens, meal and money: Capitalism and the domestic community*. Cambridge: Cambridge University Press.
Newman, R. P. 1983. Lattimore and his enemies. *Antipode* **15**, 12–26.
Ollman, B. 1971. *Alienation: Marx's conception of man in capitalist society*. New York: Cambridge University Press.
Peet, R. 1977a: The development of radical geography in the United States. *Progress in Human Geography* **1**, 64–87.
Peet, R. 1977b. *Radical geography: Alternative Veiwpoints on contemporary social issues*. Chicago: Maaroufa Press.

Peet, R. 1978–9. The geography of human liberation. *Antipode* **10**, nos. 3 and 11, 1, 126–34.
Peet, R. 1985a. The destruction of regional cultures. In *A world in crisis: geographical perspectives*, R. J. Johnston and P. J. Taylor (eds), 150–72. Oxford: Basil Blackwell.
Peet, R. 1985b. Radical geography in the United States: a personal history. *Antipode* **17** 1–8.
Peet, R. 1985c. The social origins of environmental determinism. *Annals of the Association of American Geographers* **75**, 309–33.
Peet, R. (ed.) 1987. *International capitalism and industrial restructuring: a critical analysis*. Boston: Allen & Unwin.
Piaget, J. 1970. *Structuralism*. New York: Basic Books.
Pickles. J. 1986. *Phenomenology, science and geography: spatiality and the human sciences*. Cambridge: Cambridge University Press.
Poulantzas, N. 1975. *Classes in contemporary capitalism*. London: New Left Books.
Poulantzas, N. 1978. *State, power, socialism*. London: New Left Books.
Pred, A. 1981. Social reproduction and the time-geography of everyday life. *Geografiska Annaler*, Series B, **63**, 5–22.
Pred, A. 1985. The social becomes the spatial; the spatial becomes the social. Enclosure, social change and the becoming of places in the Swedish province of Skane. In *Social relations and spatial structures*, D. Gregory and J. Urry (eds), 337–365. London: Macmillan.
Pred, A. 1987. *Place, practice, structure*. Cambridge: Polity Press.
Punter, J. 1988. Post-modernism. *Planning Practice and Research*. **4**, 22–8.
Rees, J. (ed). 1986. *Technology, regions and policy*. Totowa, NJ: Rowman & Littlefield.
Rey, P. 1973. *Les Alliances des classes*. Paris: Maspero.
Rose, D: 1988. Homeownership, subsistence and historical change: the mining district of West Cornwall in the late nineteenth century. In *Class and space*, N. Thrift & P. Williams (eds), 108–53. London; Routledge & Kegan Paul.
Savage, M., J. Barlow, S. Duncan & P. Sanders 1987. 'Locality Research': the Sussex programme on economic restructuring, social change and the locality. *Quarterly Journal of Social Affairs* **1**, 27–51.
Sayer, A. 1984. *Method and social science*. London: Hutchinson.
Sayer, A. 1985. Realism, in geography. In *The future of geography*. R. J. Johnston (ed.), 159–73. London: Methuen.
Schaefer, F.K. 1953. Exceptionalism in geography: a methodological examination. *Annals of the Association of American Geographers* **43**, 226–49.
Schimidt, A. 1971. *The concept of nature in Marx*. London: New Left Books.
Scott, A. J. & M. Storper (eds) 1986. *Production, work, territory: the geographical anatomy of industrial capitalism*. Boston: Allen & Unwin.
Sibley, D. 1981. *Outsiders in an urban society*. Oxford: Basil Blackwell.
Slater, D. 1973. Geography and underdevelopment – Part I. *Antipode* **5**, 21–32.
Slater, D. 1975. The poverty of modern geographical inquiry. *Pacific Viewpoint* **16**, 159–76
Slater, D. 1977. Geography and underdevelopment – Part II. *Antipode* **9**, 1–31
Smith, N. 1984. *Uneven development: nature, capital and the production of space*. Oxford: Basil Blackwell.
Smith, N. 1987. Dangers of the empirical turn: the CURS initiative. *Antipode* **19**, 59–68.
Smith, N. & P. O'Keefe 1980. Geography, Marx and the concept of nature. *Antipode* **12**, 30–9.
Smith, N. & P. Williams (eds) 1986. *Gentrification of the city*. London: Allen & Unwin.
Sopher, D.E. 1967. *Geography of religions*. Englewood Cliffs, NJ: Prentice Hall.
Stea, D. & B. Wisner (eds) 1984. The Fourth World (special issue). *Antipode* **16**, 2.
Stoddart, D. 1986. *On geography*. Oxford: Basil Blackwell.

Stone, M. 1975. The housing crisis: mortgage lending and class struggle. *Antipode* **7**, 22–37.
Storper, M. 1987. The post-Englightenment challenge to Marxist urban studies. *Environment and Planning D, Society and Space* **5**, 418–27.
Storper, M. & R. Walker 1988. *The capitalist imperative*. Oxford: Basil Blackwell.
Tathem, G. 1933. Environmentalism and possibilism. In *Geography in the twentieth century*, G. Taylor (ed.), 128–62. New York.
Terray, E. 1972. *Marxism and 'primitive societies.'* New York: Monthly Review Press.
Thompson, E. P. 1963. *The making of the English working class*. London: Weidenfeld & Nicolson.
Thompson, E. P. 1978. *The poverty of theory and other essays*. London: Merlin Press.
Thrift, N. J. 1979. On the limits to knowledge in social theory: towards a theory of practice. Camberra: Australian National University, Department of Human Geography, Seminar Paper.
Thrift, N. J. 1983. On the determination of social action in space and time. *Environment and Planning D, Society and Space*. **1**, 23–57.
Thrift, N. J. & D. K. Forbes 1986. *The price of war: urbanisation in Vietnam 1954–1965*, London: Allen & Unwin.
Thrift, N. J. & A. Pred. 1981. Time-geography: a new beginning. *Progress in Human Geography*. **5**, 277–86.
Thrift, N. J. & P. Williams (eds) 1987. *Class and space*. London: Routledge & Kegan Paul.
Timpanaro, S. 1975. *On materialism*. London: New Left Books.
Urry, J. 1981. *The anatomy of capitalist societies. The economy, civil society and the state*. London: Macmillan.
Urry, J. 1985. Social relations, space and time. In *Social relations and spatial structures*, D. Gregory & J. Urry (eds), 20–48. London: Macmillan.
Urry, J. 1987. Society, space and locality. *Environment and Planning D, Society and Space*. **5**, 435–44.
Vogeler, I. (ed.) 1975. Rural America (special issue). *Antipode* **7**, 3.
Walker, R. (ed.) 1979. Human–environment relations (special issue). *Antipode* **11**, 2.
Walker, R. A. & D. A. Greenberg 1982a. Post-industrial and political reform in the city: a critique. *Antipode* **14**, 17–32.
Walker, R. A. & D. A. Greenberg 1982b. A guide for the Ley reader of marxist criticism. *Antipode* **14**, 38–43.
Weedon, C. 1987. Feminist practice and poststructuralist theory. Oxford: Basil Blackwell.
Wittfogel, K. 1957. *Oriental despotism: a comparative study of total power*. New Haven,: Yale University Press.
Wittfogel, K. 1985. Geopolitics, geographical materialism and marxism. *Antipode* **17**, 21–72. Geopolitik, geograhischer materialisumus und marxismus. Translation from G. O. Ulmen 1929. *Unter dem banner des marxismus* 3, 1, 4 and 5.
Wolpe, H. (ed.) 1980. *The articulation of modes of production*. London: Routledge & Kegan Paul.
Wolpert, J. & E. Wolpert 1974. From asylum to ghetto. *Antipode* **6**, 63–76.
Wright Mills, C. 1959. *The sociological imagination*. New York: Oxford University Press.

2 *Mathematical models in human geography: 20 years on*

Martin Clarke & Alan Wilson

Whatever happened to mathematical models?

In a book which is related to the publication of Haggett's and Chorley's *Models in geography* 20 years ago, but which is mostly not about quantitative models, it is useful to ask the question posed in the title of this section. The quantitative revolution in geography is usually formally dated from Ian Burton's 1963 paper; radical geography can perhaps be similarly dated from the 1973 publication of David Harvey's *Social justice and the city*. Fashions can change rapidly! One consequence of such change is that only a relatively small core of modellers have continued to work with the appropriate levels of technical expertise on the major research problems. Relatively few have attempted to engage in anything but knockabout debates on the relationship between modelling and radical geography. In this chapter we argue that it is important to understand what has happened to mathematical modelling, and that it does have a substantial contribution to make in the long term. Indeed, it can be argued that modelling (which also provides much of the conceptual basis of information systems) and what might be called *critical* (rather than radical) human geography form the two main strands of the subject for the forseeable future. Because of the differences in expertise between the two populations of practitioners, it is likely that much of this development will be separate. However, there is no intrinsic need for this subdisciplinary apartheid, and one of the arguments of this chapter is for greater mutual understanding as a basis for possibile future collaboration. It is useful to note in this context that some social theorists are arguing that analytical modelling and mathematics should have a rôle in contemporary studies which goes beyond the old arguments about positivism. What all sides have in common is a recognition that, even if the basis is heuristic rather than scientific, they have a part to play in handling complexity. For example, Turner (1987) argues that 'analytical models provide an important supplement to abstract propositions because they map the complex causal connections – direct and indirect effects, feedback loops, reciprocal effects'; and to quote from the same volume 'mathematical models have an essential place in our efforts to untangle the complexities of social realities' (Wilson in Giddens & Turner 1987).

However, it must first be appreciated that all the elements of quantitative geography must not and ought not to be lumped under one heading. It is not our purpose here to discuss essentially inductive, that is statistical, methodology; we restrict ourselves to *mathematical* modelling on the basis that this has a more direct contribution to make to the evolution of geographical theory in the long term. This distinction, between the statistical and the mathematical, has usually not been well understood. Another area of weakness has been that there appeared to be little explicit connection between what mathematical modelling had to offer to geographical theory and what might be called the classical contributions of such authors as von Thünen, Weber, Burgess and Hoyt, Christaller and Lösch – and these authors have provided the basis of much geographical textbook writing both before and after the advent of radical geography. (Perhaps their works constitute a neoclassical geography?) This was partly because the classical modellers had ventured into areas where the 1960s modellers had not the expertise to tread; and, more simply, because the effort of understanding what each perspective contributed to the other was not made. It could be argued that the contributions of modelling in the 1960s and early 1970s, exciting though they were at the time and useful though they remain in many ways, did not in fact address the central problems of geographical theory. However, this position has now changed and a brief articulation of the new contribution is a major purpose of our discussion.

A further complication then arose in the quantitative geography–radical geography debates: the arguments were conducted in rather simplistic terms (based on the outmodedness of positivism, and the perceived corollary that the positivist label could be used to dismiss anything to do with mathematical modelling) without the issues raised above being fully understood. In other words, debates were presented as arguments between incompatible paradigms with neither of the new paradigms being very closely related to the classical or neoclassical ones. No wonder that geography as a discipline seemed to be in a fragmented state.

It is now useful to try to improve upon this: to understand the development of mathematical modelling; its historical connections to classical theory; the levels of expertise that have now been achieved; and the possibilities for using it in the future in the light of the radical critique. We aim to show that modelling has a contribution in relation to geographical theory in general, but also that expertise is available for a wide range of applications which throw light on a variety of problems. We begin, therefore, by briefly outlining the state of the art and the history of modelling; next we look at the relationship of modelling to geographical theory; then we outline some illustrative problems in applied human geography; and, finally, we discuss the role of modelling, as we see it, in the future of geography.

What have modellers achieved?

There is a rich variety of approaches to modelling in ways which are relevant to geography – perhaps best distinguished in the first instance by a variety of disciplinary backgrounds. For example, this ranges from the 'new urban economics' school (cf. Richardson 1976) to ecological approaches (Dendrinos & Mullaly 1985). Here, we illustrate the argument with models based on spatial interaction concepts – initially rooted in entropy maximizing methods (Wilson 1970), but with a recently extended range of application through the use of methods of dynamical systems theory (Wilson 1981). This restriction both corresponds to our expertise but also, none the less, serves to illustrate most of the general points which have to be made: the kinds of development described within this subparadigm have also been achieved, or have to be achieved, in any of the alternative approaches. In other words, the gist of the argument would be preserved if it was rewritten as though from the viewpoint of another modelling perspective.

In the 1960s, a broad ranging family of spatial interaction models was defined and applied (cf. Wilson (1974) for a broad review). It was also recognized at an early stage that many such models, particularly the so-called singly-constrained varieties, also functioned in an important respect as *location* models. In the case of retailing flows, for example, the models could be used to calculate total revenue attracted to the shopping centre, as a sum of flows, and this of course is an important locational variable. However, such modelling exercises could only be carried out if a number of important geographical variables were taken as given: in particular, the spatial distribution of physical structures of, as in this example, shopping centres. In other words, no attempt was being made to model the main geographical structural variables – and this is where modellers had failed to tackle one of the problems of the classical theorists – in this case, Christaller and central place theory.

The situation was rectified in the late 1970s and this has led to dramatic advances in modelling technique as well as in the understanding of the contribution of modelling to geographical theory in a wider sense. The argument was first set out in Harris & Wilson (1978) in relation to the singly-constrained spatial interaction model and in particular to retailing, but it was realized from the outset that it had a much wider application. The modelling advance involved the addition of an hypothesis to spatial interaction models which specified whether particular centres at particular locations would grow or decline. It was then possible to model, in a dynamic context, not only the spatial flows within a geographical system but also the evolution over time of the underlying physical and economic structures. This method can be applied to any geographical location system which depends on spatial interaction as an underlying basis. These include agriculture (relating crop production to markets), industrial location (related to flows from input sources or to markets), residential location and housing (in relation to the journey to work and services), retailing, and a whole variety of services. In principle, subsystem models can be combined into whole system models and then comprehensive models such as those of Christaller (from an earlier generation) or Lowry (from a later generation) can be rewritten. This programme of rewriting has now been carried through and can be used to illustrate the application to the main areas of geographical theory:

(a) agriculture (Wilson & Birkin 1987);
(b) industrial location (Birkin & Wilson 1986a, b);
(c) residential location (Clarke & Wilson 1983);
(d) retailing (Harris & Wilson 1978, Wilson & Clarke 1979, Clarke & Wilson 1986, Wilson 1988a);
(e) health services as an example of a different kind of service (Clarke & Wilson 1984, 1985);
(f) comprehensive modelling (Birkin *et al.* 1984).

There are two kinds of achievement from these advances: first, there is a contribution of general insights from modelling to the development of geographical theory, and we take these up next; the second arises when the model developments can be fully operationalized and data are available for testing, and the models can then be applied in planning contexts. We take this up afterwards.

Modelling and geographical theory

We take the argument forward in two steps: first, the relationship of modelling in its current form to classical theory; and, second, the relationship of modelling to radical geography.

It is possible in each of the fields of classical theory which have been mentioned to take the classical problem and to reproduce it in the new modelling framework. But then, it is possible to use the model to progress beyond the restrictions of the traditional approach and to tackle more complicated problems. Indeed, what emerges is that the classical theorists were limited by technique. It did not help that they were fixated on a continuous space representation: in the modelling era, discrete zones were more natural for computer data bases and turned out to have intrinsic advantages in mathematical terms. We consider what can now be achieved in each of the major fields, summarizing in broad terms the arguments presented in more detail in the references listed above, at the end of the preceding section.

In the agricultural case, it is possible to reproduce von Thünen's rings for the example of a single market centre and uniform plain. However, with the new model there is no problem in having as many market centres as is appropriate and building in variable fertility on the plain (and also coping with the distorting effects of transport networks and so on). It is difficult to make more than theoretical progress in this field (though with hypothetical numerical examples) because real-world data are not systematically available.

A similar argument applies to the industrial location case: Weber's Triangle can be reproduced (with the different situations of the single firm in relation to the vertices), but the model can be extended to handle the competitive relationships of a set of firms. However, new complications also have to be built into the model. As soon as many firms are included, it is recognized that they should be classified into a number of centres, and the model also has to represent the input–output relationships between industrial sectors as well as the spatial relationships between all firms which are consistent with them. This, needless to say, is a very complex task. However, it can be accomplished, but again only using numerical examples rther than real data. In the residential location and housing case, the residential location part of the model has been available since the 1960s (though only developed in empirical applications relatively slowly because of the complexity of the problem). The new insights now allow housing to be added and modelled. What is achieved in this case is a rich generalization of Burgess's rings and Hoyt's sectoral patterns.

The retail case, building on the work of authors like Reilly, is interesting because spatial interaction models were used together with the equivalent of discrete zone systems. However, the models used were essentially unconstrained (and must have produced rather silly predictions for flows) and so the main use was to demarcate market areas between shopping centres – essentially a continuous space use of a discrete zone model. What is clear from modern spatial interaction models (and from all relevant empirical data) is that market areas do overlap substantially and it is better to focus on the flows directly and to model these rather than to worry about boundaries or market-area demarcations.

Interestingly enough, there is little or no classical work on public services. So the applications of models in fields such as health services analysis represent a new gain. This reflects the historical importance of the different sectors.

The subsystems can be combined and comprehensive models developed which can then be considered to replace central place theory (Wilson 1978, Birkin *et al.* 1984). That the state of the art is now highly developed can be seen from a number of reviews which have been compiled in the last few years. Examples are Weidlich & Haag (1983), Wilson & Bennett (1985), Bertuglia *et al.* (1987), Bertuglia *et al.* (1989), Nijkamp (1986), and Dendrinos & Mallaly (1985).

The new dynamic modelling methods also offer a different kind of insight for geographical theory, and this is the sense in which the argument applies to any modelling style. It turns on the existence of nonlinearities (from externalities, scale economies, or whatever) in these models. Analysis then shows that while in some sense the models represent general laws, there are in any application a large number of possible equilibrium states and modes of development. A particular one chosen, say in a particular city or region, will depend on the particular behaviour of local agents (or historical accidents). In other words, the modelling insights integrate the two sides of the uniqueness–generality debate (which still manifests itself in various forms in a variety of paradigms). Analysis also has a bearing on the agency–structure problem. It enables real world complexities to be understood and illustrates that it is impossible to forecast the future in a deterministic way. However, it does provide detailed accounts of the past, and is therefore of great importance in the context of historical geography; and it provides insights, but not precise forecasts, for the future in terms of the modes of possible development in different circumstances.

A further property of nonlinear models is that their structures are subject to instantaneous (or, in practice, rapid) change at certain critical parameter values. This is a phenomenon now very widely recognized in many situations in many other disciplines. An interesting research task in human geography in the future will be to identify empirically, and to model, rapid structural change of this type. These observations are relevant to the application of modelling ideas in the theory developed under the application of radical geography as well as to the alternatives.

How can we relate contemporary mathematical modelling to the radical critique? The first point to make is that it is important to distinguish alternative hypotheses or theories from issues of technique for representing those theories in models. Once this is achieved, then any disagreement can be shifted to where it ought to be: between theories rather than in terms of the validity of certain kinds of technique. It does not follow, as has sometimes been naïvely argued, that any piece of work involving mathematics is positivist. In practice, many models are based on the assumptions of neoclassical economics and are therefore subject to the criticisms which can be brought to bear on that perspective. However, most of the basis of the models used to illustrate the argument in this chapter are not so dependent: the entropy maximizing base does not depend on such economic assumptions – it is more reasonably seen as a combination of accounting and statistical averaging notions. Indeed, it can be argued that any good piece of geographical analysis should be underpinned by the appropriate accounts. In the case of the study of economic structure in a region, for example, this has been done both by neoclassical and by Marxist researchers. The first perspective leads to the input–output model, the second, through the work of Sraffa, leads to an alternative. But they both have, in principle, the same underlying set of accounts. The future of modelling could well be seen in this way: what contribution can it make to operationalize hypotheses?

Some might argue that there are deeper structuralist questions involved, that modelling inevitably engages with surface phenomena and as such fails to offer adequate in-depth explanation. This was to an extent true of the modelling techniques of the 1960s (though much of the information generated was useful in a variety of practical situations). This is much less true (at least in terms of potential) of the modelling methods of today. On the whole, whatever kind of theory can be clearly articulated can also be modelled. However, when we attempt to carry through this argument in relation to some examples of radical geography then the complexity of some of the issues raised becomes apparent, and this raises a new generation of modelling problems (cf. Wilson 1988b, on the potential for configurational analysis in this kind of situation). None the less, progress is being made and the works of Webber (1987), Webber & Tonkin (1987), and Sheppard (1987) all provide important examples of how modelling skills can be deployed in critical or political-economic approaches.

A contemporary view of applied mathematical modelling

We hope that we have illustrated the actual and potential contribution of model-based methods to various aspects of location theory. We now move on to examine how mathematical modelling can be used in an applied problem-solving context. Before doing so, it is worth making three general points. First, a distinction can be made between the use of models as frameworks for understanding and the use of models in some prescriptive way. It is the latter role that is most often attributed to modellers, perhaps because of the relationship in the 1960s between modelling and planning and the attempts to make planning model-based. What has emerged is that models have an important contribution to make in the understanding of how systems operate and in particular of their dynamics. Without this understanding, of course, prescription becomes a dangerous and difficult task. The second point, made earlier but worth repeating, is that through the work on dynamic modelling it has become clear that the conventional use of modelling in urban and regional planning – conditional forecasting – has to be replaced by a more qualitative approach where models are used to identify the possible range of developments rather than to specify fairly precisely the exact form of change. A final preliminary comment related to this is that while urban or town planning faces a real difficulty from the fact that there is only a small element of control, in other public sector systems, notably health and education, a much greater degree of control is possible. For example, in health care the size and location of new hospitals, the level of service provision, the setting of priorities, and so on are all within the power of health authorities and managers; this provides a much more promising opportunity for the use of model-based methods in planning (Clarke & Wilson 1986).

Given these comments, what differentiates applied modelling in the late 1980s from that in the 1960s? Is it possible to be confident about the contribution modellers can make and if so why? A list of points emerges in attempting to answer these questions:

More experience There is now a considerable amount of experience in applying models in practical contexts. This relates both to the technical aspects of model application, such as calibration and validation, as well as to the more strategic issues, such as model design and policy representation. This has resulted in a reduction of the naïve, simplistic applications that, often rightly, drew most criticism and an increase in more sophisticated, but more realistic studies. There has also been a recognition that model-based analysis is but part of a wider process of management and planning rather than the central feature of planning. It may still be the *critical* phase, however.

Better methods Although we focused on just one methodological approach, that of spatial interaction (see pp. 31–5), it was shown that developments in that method, for example, through the introduction of dynamics, had significantly improved the range of applicability. The same has been true in other areas such as optimization and also new methods, such as Q-analysis and microsimulation, which have been developed. The modellers' kit-bag of methods has, therefore, been improved and extended. This results in the availability of more appropriate methods for particular applications.

Better information In the late 1960s and early 1970s almost every paper written on applied modelling concluded that the full potential of a particular approach would be achieved only when better information and faster and bigger computers became available. Computing power is dealt with below. Information systems have improved, not necessarily in terms of the quality of data collected, but in the ways in which such data can be accessed and manipulated. For example, the 1981 Census is freely available on-line to academics through SASPAC (Small Area Statistics Package). There is a worrying trend, however, which Goddard & Openshaw (1987) term 'the commodification of information', whereby information becomes a valuable and traded commodity, collected and supplied by private organizations. With the present government's adherence to marketforce principles, this could reduce the quality and amount of data traditionally located within the public realm.

Better computers Increasing computer power in itself does little to improve the application of model-based methods. It does, however, remove certain types of barriers and creates opportunities. Perhaps surprisingly, it is not the increased power of computers that has heralded a change in mathematical modelling but the advent of the microcomputer, most notably the IBM PC and its clones. This modest computer has two distinct advantages over its mainframe brothers. First, it allows a model system developed on a PC in, say, Leeds, to be transferred with ease to any other compatible PC elsewhere in the world, this was simply not practical with programs developed on mainframes. Second, PCs have superb colour graphical facilities for displaying information and results that can vastly improve the quality of presentation – an important aspect of popularizing and selling modelling to both the initiated and the unconvinced, and a point to which we return later. The availablity of a new generation of PCs based on the Intel 386 chip means that computing will never present constraints for modellers, only opportunities.

Better packaging and presentation of outputs There was a time when modellers were instantly recognizable on a university campus. They trudged back and forth from the geography department to the computer centre, returning with vasts swathes of computer output, most of which was immediately dispatched to the bin. While the odd modeller might still engage in this practice, he or she is an endangered species. What the end user of a modelling system typically requires, whether this is a public sector planner or a marketing director of a private firm, is a succinct, informative, and well presented analysis of what is likely to happen if. In response to this a number of developments have occurred. One which we have already mentioned is computer graphics, where, to paraphrase an old saying, a colour map is worth a thousand lines of computer output. A second development has been in interactive computing, where the changes to the system are input at the terminal, the model run, and the results presented at the screen, to be selected, say, from a menu; if another run of the system is required this can be performed immediately. In a system called HIPS (Health Information and Planning Systems) which has been developed as a planning tool for health authorities (Clarke & Wilson 1985), a number of variants of the strategic plan could be examined in an afternoon using the interactive system. Finally, to allow the outputs of models to be interpreted meaningfully we have seen the development of Performance Indicators (Clarke & Wilson 1984), which can be seen as the outcome of transformations on either data or model outputs that relate stock or activity variables to consistent denominators.

More interest While the academic community has expended much energy in discussing the intricacies and merits of model-based methods, a number of people engaged in market research and management consultancy recognized the potential contribution of geographical models to problem solving in the public and private sector. Spurred on by the commercial success of small-area profiling systems such as ACORN (A Classification Of Residential Neighbourhoods) and an evident demand for locational analysis, these types of companies have been undertaking what we would recognize as applied human geography for a number of years. In another paper (Clarke & Wilson 1987) we have developed an explanation for why this has happened and the potential for the future. Suffice it to say that many organizations in both the public and private sector take location analysis – from locating a new supermarket, marketing a product, to allocating public funds – very seriously and wish to employ useful and appropriate methodologies.

The above six points do not, in themselves, either defend the appropriateness of mathematical models or suggest a transformation in the role of models in human geography. What it is hoped that they do illustrate is the maturity of the discipline and a concern with both usefulness and understanding combined with an interest from outside geography in their application. It is now possible to articulate a much longer list of model application areas which extends considerably beyond the realm of urban planning. We mentioned health and education earlier, but interesting applications exist in retailing, financial services, utilities (e.g. the water industry), leisure, and so on. Where progress is still most difficult is in economic and industrial analysis, although progress is being made. Evidence suggests that model-based analysis within human geography can retain a vital role within the discipline.

Conclusion

In this chapter we have attempted to describe the position of mathematical models within the discipline of human geography 20 years after the publication of *Models in geography*. We hope that we have illustrated that modelling does not exist in a technical vacuum (as many would like to think), but has strong links with traditional geographical location theory on the one hand and important contemporary applications on the other. In terms of application, model-based analysis in the 1960s and 1970s was strongly associated with urban planning. This association has weakened and the new relationships outlined above have emerged. There is every indication that these relationships are much stronger than those of the past and are, therefore, likely to be more enduring.

Given our bullishness about the prospects for model-based geography how can we rekindle this enthusiasm amongst colleagues and, perhaps most importantly, our students? The most promising way forward appears to rest in the development of an applicable human geography based on case studies and examples (Clarke & Wilson 1987) in which modelling plays a central role but one which is also based upon firm theoretical foundations.

References

Bertuglia, C. S., G. Leonardi, S. Occelli, G. A. Rabino, R. Tadei, & A. G. Wilson (eds) 1987. *Urban systems: contemporary approaches to modelling*. London: Croom Helm.

Bertuglia, C. S., G. Leonardi, & A. G. Wilson (eds) 1989. *Urban dynamics: towards an integrated approach*. London: Routledge.

Birkin, M., M. Clarke, & A. G. Wilson 1984. Interacting fields: comprehensive models for the dynamical analysis of urban spatial structure. Paper presented to the 80th Annual AAG meeting, Washington DC, April 1984. Also Working Paper 385, University of Leeds: School of Geography.

Birkin, M. & A. G. Wilson 1986a. Industrial location models I: a review and an integrating framework. *Environment and Planning A* **18**, 175–205.

Birkin, M. & A. G. Wilson, 1986b. Industrial location theory II: Weber, Palander, Hotelling and extensions in a new framework. *Environment and Planning A* **18**, 293–306.

Burton, I. 1963. The quantitative revolution and theoretical geography. *Canadian Geographer* **7**, 151–62.

Clarke, G. P. & A. G. Wilson 1985. Performance indicators within a model-based approach to urban planning. Working Paper 446. University of Leeds: School of Geography.

Clarke, M. & A. G. Wilson 1983. Exploring the dynamics of urban housing structure: a 56 parameter residential location and housing model. Working Paper 363. University of Leeds: School of Geography.

Clarke, M. & A. G. Wilson 1984. Models for health care planning: the case of the Piedmonte region. Working Paper 38. Turin, Italy: IRES.

Clarke, M. & A. G. Wilson 1985. Developments in planning models for health care policy analysis in the UK. In *Progress in medical geography* **10**, 427–51.

Clarke, M. & A. G. Wilson 1986. The dynamics of urban spatial structure: the progress of a research programme. *Transactions of the Institute of British Geographers* **10**, 427–51.

Clarke, M. & A. G. Wilson 1987. Towards an applicable human geography: some developments and observations. *Environment and Planning A* **19**, 1525–42.

Dendrinos, D. S. & H. Mullaly 1985. *Urban evolution: studies in the mathematical ecology of cities*. Oxford: Oxford University Press.

Giddens, A. & J. H. Turner (eds) 1987. *Social theory today*. Oxford: Polity Press.

Goddard, J. & S. Openshaw. Some implications for the commodification of information and the emerging information economy for applied geographical analysis in the United Kingdom. *Environment and Planning A* **19**, 1423–40.

Harris, B. & A. G. Wilson 1978. Equilibrium values and dynamics of attractiveness terms in production constrained spatial-interaction models. *Environment and Planning A* **10**, 371–88.

Harvey, D. 1973. *Social justice and the city*. London: Edward Arnold.

Nijkamp, P. (ed.) 1986. *Handbook of regional and urban economics* Vol. 1: *Regional economics*. Amsterdam: North Holland.

Richardson, H. 1976. *The new urban economics and alternatives*. London: Pion.

Sheppard, E. 1987. A Marxian model of the geography of production and transportation in urban and regional systems. In *Urban dynamics: towards an integrated approach*, C. S. Bertuglia, G. Leonardi & A. G. Wilson 189–250. London: Routledge.

Turner, J. H. 1987. Analytical theorizing. In *Social theory today*, A. Giddens & J. H. Turner (eds), 156–94. Oxford: Polity Press.

Webber, M. J. 1987. Quantitative measurement of some Marxist categories. *Environment and Planning A* **19**, 1303–21.

Webber, M. J. & Tonkin 1987. Technical changes and the rate of profit in the Canadian food industry. *Environment and Planning A* **19**, 1579–96

Weidlich, W. & G. Haag 1983. *Concepts and models of quantitative sociology: the dynamics of interacting populations*. Berlin: Springer.

Wilson, A. G. 1970. *Entropy in urban and regional geography*. London: Pion.

Wilson, A. G. 1974. *Urban and regional models in geography and planning*. Chichester: Wiley.

Wilson, A. G. 1978. Spatial interaction and settlement structure: towards an explicit central place theory. In *Spatial interaction theory and planning models*, L. Lundqvist, F. Snickars & J. W. Weibull (eds), 137–56. Amsterdam: North Holland.

Wilson, A. G. 1981. *Catastrophe theory and bifurcation: applications to urban and regional systems*. London: Croom Helm.

Wilson, A. G. 1988a. Store and shopping centre location and size: a review of British research and practice. In *Store choice, store location and market analysis*, N. Wrigley (ed.), 160–86. London: Routledge.

Wilson, A. G. 1988b. Configurational analysis and urban and regional theory. *Sistemi Urbani*.

Wilson, A. G. & R. J. Bennett 1985. *Mathematical models in human geography and planning*. Chichester: Wiley.

Wilson, A. G. & M. Birkin 1987. Dynamic models of agricultural location in a spatial interaction framework. *Geographical Analysis* **19**, 31–56.

Wilson, A. G. & M. Clarke 1979. Some illustrations of catastrophe theory applied to urban retailing structures. In *Developments in urban and regional analysis*, M. Breheny (ed.). London: Pion.

Wilson, T. P. 1987. Sociology and the mathematical method. In *Social theory today*, A. Giddens & J. H. Turner (eds), 383–404. Oxford: Polity Press.

Part II
NEW MODELS OF THE CITY

Introduction

Nigel Thrift

Modern towns and cities often seems to have little in common but their diversity. There are the mega-cities of the world, vast assemblages of people and concrete with their pockets of extreme wealth and extreme misery. There are still a few single-industry, working-class towns in poignant contrast, and there are opulent stockbroker-belt towns and villages, rural representatives of a new international financial system. There are desert retirement communities. There are heritage towns and cities pandering to the tourist dollar. There are cities which are centres of government and the military. The list goes on and on. How can we meld all these towns and cities into a single, coherent vision?

Urban processes

Towns and cities can perhaps best be viewed as the loci of a number of processes interacting with one another in all manner of ways (Logan & Molotch 1987). First of all, and most fundamentally, the city can be viewed as a focus of production. As Leitner and Sheppard, and Pratt point out in their chapters, the mode of commodity production is an essential first step in understanding the nature of towns and cities. It provides the *raison d'être* for many of the activities that take place in a city. But production is a difficult business for geographers; each moment of production has its own set of geographies which interact with other geographies in a complex synthesis. There is the point of commodity exchange. That will include a complex geography of transport and distribution. It will also include the buying in of labour power which must be done within a labourmarket geography which is continually in flux (Clark 1981). Then there is the point of production. That will include, most especially, the labour process, which can be a complex geography in its own right. It will consist of detailed functions like design, planning, production, and marketing which can be spatially differentiated. Production of parts of a commodity can also take place at many places in the world before the product is finally made up. In sum, cities are spatial divisions of labour, embedded in even larger spatial divisions of labour. They are, and they are part of, a mosaic of uneven development and redevelopment, the result of a constant dialogue between the location of production and the location of labourforces.

Currently, there appear to be monumental shifts occurring in the structure of commodity production, in the urban systems of many countries. For example, new flexible labour processes and technologies seem to be leading to clusters of light industry in certain cities of the world, with interesting social consequences

(Scott 1988, Storper & Scott 1989). At the same time, the growth of high-level producer services has brought a reconcentration of office activity into the downtown areas of some larger cities. As Leitner and Sheppard point out, such developments are conditioned by the supply of property by commercial institutions, mediated by the state. Whether these and other new developments add up to a new regime of flexible or disorganized accumulation is still – for all the talk of a *fin de siècle* shift in the nature of capitalism – a moot point. The effects of these new developments on cities and city systems are even harder to foresee, simply because modern towns and cities are not just loci of production.

Thus we come to other ways in which the city is a locus of processes. Second, the city is a locus of money, finance and credit, of the manipulation of money by banking and commercial capital (Thrift 1987). The city has too rarely been seen in this way by geographers, yet it might be argued that the time has never been more ripe for such a vision to take hold. For example, it might be argued that much of the growth of producer services in cities has had very little to do with production *per se*, and rather more to do with the growth of banking and finance capital. Harvey (1985a, p. 190) argues that financial markets can 'acquire a certain autonomy *vis à vis* production' and currently that autonomy seems very great indeed. Many cities have become 'centres of mercantilist endeavour' (Harvey 1985a, p. 217) in which banking and finance have an identifiably separate spatial division of labour with important social consequences (Friedman & Wolff 1982) such as social polarization.

A third and related way in which the city can be viewed is as a locus of consumption. Hence the state can be seen as an important progenitor of the city through its rôle as a provider of public goods and services, of collective consumption. There is a very large literature now on the geography of collective consumption (see Pinch 1985), following the original work of Castells (1977), on the state as provider of schools, housing, health care facilities, and the like, and as guarantor of the reproduction of labour power. However this work is now beginning to show its age. It is work borne of the late 1960s and early 1970s when state provision and corporatist solutions were at their height in many Western societies. A comparable body of work on private sector consumption, of schools, housing, health care as well as entertainment, leisure, and the like has still to be constructed. Certain objects of consumption like housing are very well covered (Ball 1985, Ball *et al.* 1988), but other elements like schooling and health care are not, whilst the explosion of luxury consumption in many countries has hardly been touched upon. Yet, as Harvey (1985a) pointed out, the mode of consumption and the spatial division of consumption surely deserve more attention:

> Production is typically separated from consumption under capitalism by market exchange. This has enormous implications for urbanisation and urban structure. Work spaces and times separate out from consumption spaces and times in ways unknown in artisan or peasant culture. The moment of consumption, like that of production, stands to be further fragmented. Vacation, leisure, and entertainment places separate from spaces of daily reproduction, and even the latter fragment into the lunch counter near the office, the kitchen, the neighbourhood drugstore or bar. The spatial division of consumption is as important to the urban process as the spatial division of

labour – the qualities of New York, Paris and Rome as well as the internal organisation of these and other cities could not be understood without consideration of such phenomena. This is however, a theme that remains underexplored in Marxian theory, in part because of the tendency to focus exclusively on production because it is the hegemonic moment in the circulation of capital (Harvey 1985a, p. 189).

The mode of production and the mode of consumption are more and more closely linked together by the increasing tendency for towns and cities to be designed to sell commodities. From the invention of the department store (Benjamin 1973) to the consumer spectacles of postmodern cities, this tendency seems to be becoming stronger, more intrusive, and more internalized.

Fourth, the city is a locus of the reproduction of subjects. Harvey (1982) counts the synthesis of the varied and varying work on the reproduction of labour power as perhaps the most urgent task facing Marxian theory, and it is difficult to demur:

> It is . . . a task that must be undertaken in the clear knowledge that the reproduction of labour power throughout the lived life of the working classes is a quite different dimension to the analysis of the capitalist mode of production. It is not a mere addendum to what we already know, it constitutes a fundamentally different point of departure to that upon which the theory of *Capital* is based. The starting point is not the commodity, but a simple event – the birth of a working class child. The subsequent processes of socialisation and instruction, of learning and being disciplined may transform that human being into someone who also has a certain capacity to labour and who is willing to sell that capacity as a commodity. Such processes deserve the closest possible study (Harvey 1982, p. 447).

The socialization of human beings, from cradle to grave, is the result of a complex interdigitation of influences which cannot easily be reduced to a theory. But what is clear is that any such theory will have to take into account the competing or integrating influences of the economy, state, and civil society (Urry 1981). Thus it will involve disciplinary processes and processes of resistance at work and in the labourmarket. It will also involve the state's rôle in schooling, law, and other means of control – all those practices that are part of Foucault's biopolitics, his demography of power. It will include, as well, the various dimensions of civil society and the way that they structure subjects in households and elsewhere, most especially gender, ethnicity, and religion.

The chief importance of such a tripolar format is to point to the way that diverse social institutions can *interact* to form active and creative subjects by producing systems of social positioning which are simultaneously objective conditions of existence and categories of perception, what Bourdieu (1977, 1984, 1987) calls 'habitus'. Thus Willis (1978) in his famous work in a Midlands town, Jenkins (1982, 1983) in his work in Northern Ireland, and Macleod (1987), have all shown how working-class school children are piloted into manual working-class jobs by categories of perception which quite realistically reflect their life chances, but which also assure that their acts of resistance to the prevailing order will only reinforce these categories.

Leading out of and back into the process of reproduction of subjects are all manner of important topics. In her chapter Mackenzie points to the importance of gender as a distinctive subfield of work on the political economy of urban processes. In some of its manifestations this work links to work in geography on the regulation of the subject by the state, much influenced by the interest in the work of Weber and Foucault on processes of rationalization. In such a scheme, the city appears as an enormous disciplinary technology for the production of human beings and their distribution into various categories – worker or capitalist, man or woman, for example – especially through the stereotyping of groups as deviant (the technology of normalization). The city is a carceral machine 'for distributing individuals, ordering them along a graded scale in any of a number of institutional settings' (Dreyfus & Rabinow 1982, p. 142).

Last, the city can be seen as a locus of meaning, as a means of representation, as a source of understanding, as an anchor for culture (in its widest sense) (Williams 1973). It is striking the degree to which the structure of the city is used in literature as a metaphor for social life (Timms & Kelley 1985). In the opening to *Bleak House* Dickens (1853, p. 1) provides a famous example:

> London. Michaelmas Term lately over, and the Lord Chancellor sitting in London's Inn Hall. Implacable November weather. As much mud in the streets as if the waters had been newly retired from the face of the earth, and it would not be wonderful to meet a Megalosaurus, forty feet long or so, waddling like an elephantine lizard up Holborn Hill. Smoke lowering down from chimney pots, making a soft black drizzle, with flakes of soot in it as big as full-grown snowflakes – gone into mourning, one might imagine, for the death of the sun. Boys, undistinguishable in mire. Horses, scarcely better – splashed to their very blinkers. Foot passengers, jostling one another's umbrellas, in a general infection of ill-temper, and losing their foothold at street corners, where tens of thousands of other foot passengers have been slipping and sliding since the day broke (if this day broke), adding new deposits to the crust upon crust of mud, sticking at those points tenaciously to the pavement, and accumulating at compound interest.

Here Dickens conjures up London as a gloomy purgatory of mud and soot and money, creating a lasting interpretation which would colour its inhabitants' perception in time to come. Yet Dickens provided another interpretation as well, of London transformed by locomotive power into a quintessentially modern metropolis, exciting, threatening, even unstoppable (Schwarzbach 1978, Welsh 1971). Here mud and soot have turned into money:

> As to the neighbourhood which has hesitated to acknowledge the railroad in its straggling days, that had grown wise and penitent, as any Christian might in such a case, and now boasted of its powerful and prosperous relation. There were railway patterns in drapers' shops, and railway journals in the windows of its newsmen. There were railway hotels, office-houses, lodging houses, boarding houses; railway plans, maps, views, wrappers, bottles, sandwich-boxes and timetables; railway hackney coach and cabstands; railway omnibuses, railway streets and buildings, railway hangers on and

possible. Reproduction specifically includes relations with nature, relations of production, and gender relations, and has the logical conclusion that all such relations must be transformed in the restructuring of society. The term is broad enough to encompass many ideas presently existing as fragments of the imagination. In terms of politics, it may serve to unify diverse, loosely connected reactions to a world in crisis. Reproduction is an idea which strains the systematizing ability of critical theory. Such a strain on the imagination is imperative, however. Thought must expand to precede reality if we are to prevent calamity and achieve harmony between people and Earth.

References

Berry, B. 1961. *Central place studies: a bibliography of theory and applications*. Philadelphia, PA: Regional Science Research Institute.

Burgess, R. 1978. The concept of nature in geography and Marxism. *Antipode* **10**, 1–11.

Carson, R. 1962. *Silent spring*. Boston: Houghton Mifflin.

Giddens, A. 1979. *Central problems in social theory*. London: Macmillan.

Glacken, C. 1973. *Traces on the Rhodean shore: nature and culture in western thought from ancient times to the end of the eighteenth century*. Berkeley, CA: University of California Press.

Leiss, W. 1974. *The domination of nature*. Boston: Beacon Press.

Peet, R. 1981. Spatial dialectics and Marxist geography. *Progress in Human Geography* **5**, 105–10.

Peet, R. 1985. Review of N. Smith, *Uneven Development*. *Environment and Planning A* **17**, 1560–2:

Pepper, D. M. 1984. *The roots of modern environmentalism*. London: Croom Helm.

Quaini, M. 1982. *Geography and Marxism*. Oxford: Basil Blackwell.

Redclift, M. 1987. The production of nature and the reproduction of the species. *Antipode* **19** 222–30.

Sayer, A. 1979. Epistemology and conception of people and nature in geography. *Geoforum* **10**, 19–43.

Schmidt, A. 1971. *The concept of nature in Marx*. London: New Left Books.

Smith, N. 1979. Geography, science and post positivist modes of explanation. *Progress in Human Geography* 3, 356–83.

Smith, N. 1981 Degeneracy in theory and practice: spatial interactionism and radical eclecticism. *Progress in Human Geography* **5**, 111–18.

Smith, N. 1984. *Uneven development: nature, capital and the production of space*. Oxford: Basil Blackwell.

Smith, N. & P. O'Keefe 1980. Geography, Marx and the concept of nature. *Antipode* **12**, 30–9.

Thomas, W. L. (ed.) 1956. *Man's role in changing the face of the earth*. Chicago: University of Chicago Press.

3 *Resource management and natural hazards*

Jacque Emel & Richard Peet

The unity of geography never comes closer to realization than in resource geography. At the junction between society analyzed by human geography and nature studied by physical geography, resource geography has the potential to impart coherence even to a reluctant academic discourse. For those political-economic geographers who still believe in disciplinary coherence, the question of resources also assumes a central philosophical position. Beyond this question lies the increasing significance of environmental and resource issues in the contemporary world. It would be difficult to find a set of issues which symbolizes more vividly the torment of a way of life gone astray, which captures more exactly the transformative urge propelling political-economic work. This is a research path with the potential to unify diverse critical perspectives and apply them to issues of intense interest and mass political engagement. In this chapter we review the extent to which, and the ways in which, this potential has been realized. We first review theories of resource management, moving from critiques of neoclassical economic approaches in the conventional literature, to institutional analyses, to the new political ecology literature. Turning to natural hazards, we follow a similar trajectory, beginning with a quick outline of the conventional view, summarizing various lines of criticism of this view, and pointing to an emerging alternative conception of the socio-nature origins of hazardous events.

These are both diffuse literatures, strewn over several academic disciplines and published in a broad array of journals and books. We focus on the contributions of geographers to this literature, particularly where their arguments depart from more dominant conceptualizations. In part we are imposing order where little existed. But, isn't this the purpose of such a synthetic review?

Resource management theories

Resource management research concentrates on the allocation and development of resources; the biophysical, technological, economic, social, political, and legal variables which account for patterns of allocation and development; the impacts of these patterns; and the decisions, controls, or policies that direct allocation and development (Mitchell 1979). The field is interdisciplinary and highly fragmented, in part a reflection of its strong empirical and policy-oriented emphasis. The term resource management is troublesome because of its technocratic and positivistic overtones, and because it is often narrowly interpreted to mean conscious, rational decision making.[1] Typically, socialized knowledge, ideology, and contradictions in economic and political rationality have been neglected by the dominant resource management paradigms rooted in neoclassical economics and pluralistic political science. In addition, management implies practice but not necessarily theory or explanation. Although many argue that theory is a precondition of practice, this has not been a pre-eminent theme in resource geography (Wescoat 1987).

Beginning with White's (1945) floodplain analysis, resource geography has been an applied area of geography. In the tradition engendered by the Chicago resource/environmental School, in which White is the leading figure, resources, natural hazards, technological hazards, and wastes are examined with a view toward problem solving. The research question is 'what can we do now to remedy this problem?' Confronting those taking neoclassical economic or engineering approaches to problem solving, resource geographers were able to broaden the discourse to include non-economic considerations (White 1961, 1969). This contribution, most apparent in floodplain development and river basin planning (Wescoat 1987), is no small feat given the hold that neoclassical economics and the engineering sciences continue to exercise on resource management.

Much recent work in resource and environmental geography continues the tradition of broadening the discourse on human use of the environment. Geographers have contributed substantially to discussions of sustainable development (Redclift 1987), assessment of the social impacts of technology (Kasperson *et al.* 1980, Kasperson 1983, Kates *et al.* 1985), and analyses of specific resource management problems (for example, Walker & Storper 1978, Mitchell & King 1984, Soussan & O'Keefe 1985, Blaikie & Brookfield 1987, Emel 1987). Like the earlier work on resource management by White and others, recent contributions from geographers emphasize both market and non-market institutions, the actual context of decision making and policy implementation rather than the higher-order abstractions of neoclassical economics, and social justice, human wellbeing, and ecological limits. The departure of some of this recent work from the human ecology approach to resource management is seen in the emphases on the political economy of resource allocation and development and the social construction of environmental ideas and practices.

Resource management theories can be grouped (albeit roughly) into three basic categories: neoclassical-economic, human-ecology, and political-economic (the latter includes the so-called political-ecology approach). Ecology and other physical sciences inform all approaches, although not uniformly. Other differences lie in the politics, social theories, methodologies, and research agendas of scholars working within these general theoretical frameworks. Our emphasis in the following sections is on the recent contributions of geographers to the political-economic area. In addition, we review the work of geographers who have considered critically the rôle of institutions and policies mitigating resource problems.

Critical views of neoclassical resource economics

Neoclassical economic theories of allocation and development are the dominant social science perspective in natural resource issues. Building on a welfare economic framework, these theories assume the objective of maximizing economic welfare from resource use. Renewable resources should be developed at a rate of maximum sustainable yield, and non-renewable or depletable resources at a rate that maximizes all future net benefits, with the future net benefits discounted at the appropriate rate. Pareto optimality (the allocation at which no person can be made better off without someone being made worse off) is the criterion by which the social welfare of the allocation or equilibrium is judged – even the inter-temporal equilibrium (the allocation of welfare streams between generations or into the future). Divergence between actual allocations, or rates of development, and their socially optimal criterion is attributed to failure of the market to internalize the values of resources and environmental services. To correct for this market failure, either private property rights must be made to capture in full all of the costs and benefits of production, or a non-market institution must intervene to assign limits and liabilities (e.g. charge for pollution emissions, tax or price to reduce use, reduce access to common property resources, or invest directly in ecosystems).

For the private market to achieve optimal resource allocation, several conditions must hold (after Davidson 1979 and Rees 1985):

(a) well organized forward markets exist for each date in the future;
(b) consumers know with actuarial certainty all their needs for resources at each date;
(c) consumers are able and willing to exercise all these future demands by currently entering into forward contracts for each date;
(d) entrepreneurs know with actuarial certainty the cost of production associated with production flows for each date;
(e) sellers can choose between an immediate contract at today's market prices and a forward contract at the market price associated with any future delivery date;
(f) entrepreneurs know with actuarial certainty the course of future interest rates;
(g) the social rate of discount equals the rate at which entrepreneurs discount future earnings and costs;
(h) no false trading occurs – no exchange or production at non-equilibrium prices;
(i) consumers and producers are economically rational beings;
(j) all parts of the economy are perfectly competitive, including the capital and labour markets;
(k) all goods and services are within the market system (there are no unpriced resources);
(l) all factors of production are perfectly mobile;
(m) the economy is free from government intervention.

Clearly, these conditions do not hold in reality and the abstract models based upon them are criticized by geographers and others, notably Davidson (1979),

Harris (1983), and Rees (1982, 1985). Nevertheless, many resource economists maintain that the market system as it exists (with most, if not all of the above conditions violated) can achieve an approximation of technological and allocative efficiency, and that observable sources of inefficiency can be corrected with appropriate government intervention (Freeman 1979, Pearce 1983).

The rich methodological and empirical tradition of what has come to be known (rather narrowly) as behavioural geography, stresses the importance of perceptions, attitudes, and values in motivating behaviour (White 1945, Kates 1962, Brookfield 1964, Kromm & White 1984). Behaviouralist resource geographers take issue in particular with neoclassical assumptions of complete knowledge, certainty of future events, and economic rationality. In so doing, they appropriately shift the focus of *homo economicus* to *homo socialis*. The behaviouralist view has not been supplanted, perhaps because of its theoretical ambiguity (Bunting & Gallant 1971, Lowenthal 1972, Wescoat 1987) or its methodological untidiness (Hewitt 1980), but it does broaden the discourse and inform neoclassical resource economics (e.g. see Nunn 1985, Clawson 1986). The typically positivistic orientation of behavioural resource geographers, and the failure of most to address the importance of political-economic and other institutional sources of causality (and thereby constraints), have in turn made the behaviouralist view the object of criticism (Harvey 1974, Hewitt 1986). A further criticism of the behavioural resource geography approach is that its contributors neglect to ask about the sources of values and beliefs affecting behaviour on, or in, the environment, or the processes through which belief systems are maintained (Emel & Roberts forthcoming). Nevertheless, the work by White (1945), Kates (1962), Burton, *et al.* (1978) and many of their followers clearly demonstrates the narrowness of traditional cost-benefit analysis, the restricted range of managerially perceived adjustments to resource development issues or hazardous events (usually engineered structures or other technical fixes), the importance of examining the potential responses (and costs) related to interventions in resource use systems (i.e. building in the floodplain following dam construction), and the failure of economic efficiency to capture or represent the full range of management goals.

More recent work from a political-economy perspective takes issue with nearly all neoclassical assumptions, particularly the ability and opportunity of consumers and producers to enter into market decisions, the perfect competitiveness of the market, the mobility of factors of production, the absence of unpriced goods and services, and the absence of political intervention. Harris (1983) is representative of the Marxist perspective wherein environmental problems are seen as a necessary consequence of economic development. These writers agree in stressing the inefficiency of the market in allocating natural resources, however they differ in the extent to which this inefficiency represents an anomaly easily rectified by institutional intervention. They emphasize the lack of political will to correct the inefficiencies of existing and inter-temporal allocations of resources and externalities such as pollution, soil erosion, and so forth. Case studies by Walker & Storper (1978), Blowers (1984), and Lowe & Goyder (1983) illustrate the dominance of economic interests relative to environmental protection efforts on the part of local and national environmental constituencies and regulatory bodies.

Rees's (1985) work on natural resources is probably the most comprehensive examination of the social, economic, and political dimensions of global natural use. It is also a pointed critique of the neoclassical and rationalist behavioural approaches. Avowedly eclectic, Rees argues that one must apply several perspectives, notably Marxist and other ideas stressing conflict of values, in order to explain 'the way natural resources, and the wealth or welfare derived from them, are distributed over space and time'. In offering her explanation of the processes and powers producing current resource and welfare allocations, she undertakes a thorough description and critique of the neoclassical economic view. Her analysis is particularly revealing in its denouncement of industry's (she uses the minerals industry in particular) failure to achieve even an approximation of economic efficiency, even though efficiency is often used as a justification for the free market system. Her analyses of distributive equity and economic development, and the additional problem of security threats to resource availability, clearly show the conflict inherent in resource management objectives, and the failure of neoclassical theory either adequately to explain or prescribe solutions.

Patterns of mineral exploration, exploitation, refining, and trade are explained by examining historical reasons for exploration bias, the implications of different types of investors involved in mineral search and production (private companies, multinational corporations, international agencies, or national governments), imperfections in the capital market, and types of risks (including political) involved in mineral search and production. Rees also argues that the intervention of the state in market operation has a long history. Intervention in trade to protect or encourage domestic mining and industrial interests has occurred at least since the 18th-century development of the British export trade, in part through protectionist measures. Through development of this explanation of minerals' distribution and production, she undercuts several neoclassical assumptions cited above.

In terms of flow resources such as water, firewood, and soil productivity, Rees argues there need be no scarcity in any absolute physical sense. Scarcity is rooted, instead, in the established socio-economic structures which deprive the poor of both effective demand and political influence. She cautions, however, that it is naïve to believe that reordering social relations automatically can or will solve resource scarcity problems. In example, she cites China's loss of an estimated 30 per cent of its arable land in the past two decades through soil erosion (citing Smil (1984)), and its deforestation problems which leave some 500 million people short of fuel for several months of the year (citing Rigdon (1983)). The problem is 'not just a question of reordering society to respond to the demands of the population, but of deciding which demands have priority over what timescale' (Rees 1985, p. 404). This problem of conflicting demands is also emphasized in Rees's analysis of renewable resource scarcity in advanced nations. She argues that problems cannot be reduced to crude trade-offs between aesthetic, waste assimilative, and productive uses of the environment (i.e. arctic alpines v. water supplies, wetland habitats v. food production, the aquatic status of Scandinavian lakes v. increased electricity charges).

While identifying, and refusing to simplify, the complexity and contradictions of resource/environment management, Rees does not equivocate on the enormity of the step from describing a desirable future to defining a pathway towards achieving it. She describes O'Riordan's (1976, p. 310) exhortation to

> individually and collectively seize the opportunities of the present situation to end the era of exploitation and enter a new age of humanitarian concern and cooperative endeavour with a driving desire to re-establish the old values of comfortable frugality and cheerful sharing,

as 'high-minded' and of 'little political relevance'. Instead, she offers a sceptical and pragmatic assessment of the real future resource concerns, qualifying her interpretation with the caveat that 'there is no absolute, objective reality'.

For Rees, the real resource problems are complex and cannot be explained by a 'naïve and simplistic' blaming of the capitalist system in general. She is particularly concerned that the role of resources in ameliorating or exacerbating global spatial inequalities be properly understood. She dismisses physical resource scarcity and geopolitically created scarcity as barriers to growth in advanced nations, but she also argues that there is small likelihood that trade in resources can reduce global inequalities in the absence of international institutions possessed of the power to enforce a new world economic order. The latter idea is dismissed as overly optimistic given the dominance of existing economic interests.

Rees is sceptical that any political-economic system, small-scale communitarian, centrally planned socialist, or representative market exchange, can necessarily resolve these conflicts between individuals demanding different goods and services from the resource base. She argues that a significant proportion of the population in advanced countries appears less materialistic now than in the recent past; but the forces of inertia are immense, deeply entrenched in systems of value, and in socio-economic and political structures.

Another dimension of irresolution lies in the engagement of all national economies in the global economic order. Even in countries where governments are willing and able to control their public industries, participation in the global economic system limits the extent to which they can remain competitive and, at the same time, fulfil equity or security objectives. Add to this the fact that a state has multiple other goals such as increasing national prestige, rewarding a political élite, curbing inflation, or avoiding balance-of-trade deficits, and it is not surprising that public policies to redress resource and environmental problems are only marginally effective. Yet, small shifts over a period of time have led to some rather dramatic transformations of institutions and decision criteria. The strength of the environmental movement is acknowledged in most countries by the creation of institutional structures designed to ensure that material goals are not pursued to the exclusion of environmental and social goals (O'Riordan 1981, Rees 1985). It is to these institutional interventions that we now turn.

Institutional interventions

Where resource use patterns appear out of balance or otherwise undesirable, it is common practice to turn to non-market institutions to solve the problem. Within resource geography, institutional analyses include many descriptions of specific institutions and policies, as well as critical assessments of performance along a number of criteria. In two particularly interesting areas of research, geographers have examined the rationalization of institutions such as property rights and administration systems, and the power relations that constrain institutions or cause contradictions in implementation of rationalized policy approaches.

Since White's (1961) initial work on the range of choice in resource use, geographers have characteristically taken a closer, more contextual look at what the economic rationalization of resources means. Analyses of water resource management efforts in the United States show that larger public entities have replaced smaller private entities during the past 100 years (Wescoat 1985), that formal rules and definitions have replaced discretionary decision making by the courts and administrative bodies, and that the types of inequalities engendered by formal versus discretionary systems can vary considerably. Using the critical approach of Kennedy (1976) and Unger (1976), Emel & Brooks (1988) argue that the forms (i.e. centralized versus local) of differentially rationalized property rights institutions exhibit different normative bases. As property rights in resources are rationalized, freedom and case-specific equity are traded for security and generalized equity. This increased rationalization also contributes to the further commodification of water resources (see, for example, Walker & Williams 1982, Emel 1987).

Others have examined the power relations surrounding specific efforts at rationalizing private property relations in resource use. In their work on ground-water management reform in Oklahoma, Roberts & Gros (1987) found High Plains irrigators able to block reforms that would reduce their water allocations (the benefits of which would accrue to society and future tax-payers and irrigators) because the largely urban support for such measures is diffuse and not easily mobilized. The voluntarism, locally controlled districts, absence of stringent enforcement mechanisms, and emphasis on advisory rather than regulatory capacity at the state level existing in some states of the High Plains reflect the power relations and stakes involved in resource use rather than an economically optimal depletion programme. Mitchell & King (1984) examine the conflict-laden issues of Canadian fisheries' management, stressing the particularly difficult trade-offs between fisheries' protection and job protection. Although they maintain an essentially pluralistic view of policy making, 'one which cultivates harmony and consultation in the industry', they argue that the best use of the fisheries resources raises fundamental questions as to

> how access and resource allocation are to be managed; what methods are to be used to rationalize excessive factor inputs and, intertwined with this, what rights are to be accorded to the resource users; what tradeoff path will be taken between desires for greater public revenues, higher incomes, more employment and lower consumer prices; and, faced with habitat degradation, how industrial development is to be successfully integrated with the fisheries (Mitchell & King 1984, p. 430).

Drawing on a wealth of empirical material, Rees (1985), Abs (1988), and others show that the search for ends–means rationality in policy making and implementation characteristic of neoclassical and other management approaches is likely to be thwarted in practice. Cost-benefit analysis, risk-benefit analysis, environmental impact assessment, programme planned budgeting, and hierarchical conceptualizations of management are all rationalist (see, for example, Mitchell (1979)). But policy is dynamic and is shaped by actions (even contradictory actions) at all levels in the decision hierarchy, from the legislature down to the individual resource user. Thus, regulatory techniques or administrative structures cannot be judged against an agreed upon, stationary policy target. Analysts must either adopt their own assessment criteria (i.e. economic efficiency, maximization of physical outputs, preservation, environmental stability, distributive equity, and so forth) or simplistically assume that specific policy ends can be identified from legislation, statutory duties, and stated management plans (Rees 1985).

Contributions of the human ecological approach to the institutional role in resource management are drawn largely from the ecological systems approach. Resource managers are encouraged to treat problems as social experiments, to see socio-economic development and environmental quality as unantagonistic concepts,[2] to generate a wide range of alternatives for accomplishing a management objective, acknowledge uncertainty, expect surprise, be sceptical of what we think we know, avoid doing the irreversible (at least in terms of the environment), integrate the environmental with economic and social understanding through every phase of the policy design process, to evaluate systems in terms of stability and resilience, and to design policy to meet criteria of resilience or robustness (Hollins 1978, Clark 1986, Walters 1986, Kates 1985a).

As Wescoat (1987) points out in his discussion of the political and moral bases of the range of choice research in resource geography, all sorts of 'isms' have been used to capture this theoretical position including: rationalism, managerialism, scientism, utilitarianism, decisionism, positivism, and behaviouralism. He argues that this particular approach (specifically referring to White's 'range of choice in use' concept) is pragmatic and that the political underpinning of this approach is what Habermas (1970) refers to as the explicit political commitment of the pragmatic model to democratic action. Wescoat advises critics of the human ecology approach that the debate would be more productive if they would refine their understanding of the distinctions between technocratic, pragmatic, and decisionistic models of rationality.

It is not difficult to see how the interdisciplinary forms of ecological, systems analysis, and natural and technological hazards assessment could be accused of engendering technocratic rationality. There is an overtone to these works which is suggestive of automatic decision possibilities given enough scientific analysis and strategic management. Because these concerns about conventional natural hazards work are discussed below, we do not pursue them here except to clarify the pragmatic approach that Wescoat attributes to the range of choice concept. The pragmatic approach stresses the interaction of scientific expert and politician. Experts do not replace or dominate politicians as in the technocratic model of science and politics; but neither does the politician make decisions informed only by ideology rather than scientific discussion as in the decisionistic model (albeit crudely defined). Rather, 'scientific experts advise decision-makers and politicians consult scientists in accordance with practical needs' (Habermas 1970, p. 67). The public mediates the 'transposition of technical and strategic recommendations into practice' and ensures that social interests determine what needs are gratified by technology (Habermas 1970, p. 68).

If indeed the 'pragmatic' label fits the human ecology approach, the lack of emphasis placed on overarching political or social theory is to be expected (see, for example, Rorty 1982). However, a few words about theory and practice in the complex arena of resource and environmental management are useful at this juncture.

The case for theory informing policy choice in resource management is clear. We make choices hoping to ameliorate some problem, and we must know how the system works in order to suggest an approach. This understanding must be theoretical and not just based on random hunch or an accumulation of practical knowledge if we are to anticipate the effects of our policy decision on human–environment relations. Nevertheless, much human ecological work on resource issues has been incrementalist, flaunting the adoption of integrated worldviews. This may be due less to a lack of a proper methodological base than to the pervasive idea that social theories are too infirm to form the basis for policymaking (Goodin 1982). While this is an idea championed by many conventional political scientists (for example, Lindblom & Cohen 1979, and Wildavsky 1979), it is also picked up by ecologists in their prescriptions for resource and environmental management (Holling 1978).

It is difficult to empathize with incrementalism in the abstract, especially when it has been advertised by its proponents as 'a way of getting along without theory when necessary' (Braybrooke & Lindblom 1963, p. 118). Yet, for resource managers who realize the inadequacy of data, models, and conceptualizations fully to characterize and understand complex biogeochemical systems, the adaptive management approach appears responsible. Actions must be taken; decisions must be made. This problem of levels of theoretical understanding is critical for discussing resource/environmental issues. Is a full-blown social theory necessary in order to advise the prohibition of PCB production?

The importance of all parts of this geographic literature to resource and environmental management lies in its emphasis on the actual social, ecological, political, and economic factors that give rise to problems, institutions, policies, and outcomes. Efforts at resource management cannot be explained by depersonalizing the actors, by treating the problems as if they emerged from an historical vacuum, or by oversimplifying the policy-making and implementation processes.

Political ecology: a new direction and synthesis

The political ecology approach to resource management as exemplified by Blaikie & Brookfield (1987), Redclift (1987), and Rees (1985) is somewhat similar to the political-economy work in hazards and disaster also reviewed in this chapter. Resource problems are approached by examining the social order in dynamic relationship with environment. Both social order and environment are not static as with other approaches. Conflict and contradiction in the spheres of production, consumption, and nature are fundamental. Also fundamental is the unity of environment and development.

Prior to reviewing the new synthesis of political economy and human ecology, we should look at the argument for eschewing any resource management practice. Pepper (1984) reviews and synthesizes much of the Marxist and neo-Marxist literature on resources and environment existing up until 1983. Although the book was written for students, as a review of the historical, philosophical, and ideological aspects of environmentalism, Pepper disparages both the ecocentrist's idealism and the technocentrist's scientific intervention schemes. He is careful to develop the multiplicity of views within the ecocentrist camp, but at best, Pepper regards the ecosocialist faction of ecocentrism as naïve. In particular, he criticizes the movement for its failure to incorporate the importance of class struggle and for its 'corresponding negation of the pre-eminence of the mode of production in influencing social consciousness about the man–nature relationship' (Pepper 1984, p. 210). At worst, he relies on Bookchin's (1979) vilification of some environmentalists as ecofascists whose lifeboat ideology scenarios (based on Hardin (1974)) lead to repression and totalitarian control. Although Pepper states that much of Bookchin's critique of radical ecologist Andre Gortz (1979) is personal abuse, he finds justifiable Bookchin's ridicule of Gorz's utopian scenarios which reflect the naïveté of Schumacher's *Blueprint (for survival)* (1973), or Callenbach's *Ecotopia*.

Pepper questions the whole basis of the ecocentric left as represented by Commoner and Gorz. While Bookchin considers them vulgar Marxists who see ecological problems in economically reductionist form, Pepper calls them to account for the pluralistic approach of the European Green movement engaged as it is in parliamentary reform. On the other hand, Pepper introduces Enzensberger's (1974) worry that the left, in uncovering the inadequacy of the ecologists' social analysis, may disregard all the ecologists' have to say. Enzensberger (1974, p. 23) argues that even though the ecologists are naïve and utopian, they realize that 'any possible future belongs to the realm of necessity and not that of freedom, and that every political theory and practice, including that of socialists, is confronted not with the problem of abundance but survival'.

Having suggested that management intervention is ineffective and possibly ecofascist, Pepper argues that education to raise consciousness is one answer. 'Teachers should "attack" broadly and continuously by pressing for democratisation of education institutions and processes; they must reject authoritarianism and be in the forefront of a move to create a "unified class consciousness" ' (Pepper 1984, p. 223). He stresses, however, that people will not change their values simply by being taught new ones. What is called for are new social and economic goals for communities, and new relationships among people. In the end, his formula is 'reform at the material base of society, concurrent with educational change'. Ecocentric thinkers and activists are encouraged to combine with trade unions and labour movements to work for alternatives to capitalism, and to 'help to ensure that the growth of such socialistic experiments will avoid the pitfalls of centralisation, bureaucracy and a crude materialist outlook which have so bedevilled other similar developments'.

The conclusions that currently existing systems of production and property rights must be radically changed in order actually to solve environmental problems are typical of Marxist (and some other critical) analyses of resource/environment issues (for other examples see Mumy 1974, Walker & Storper 1978). One problem, however, is that radical change is not imminent, whereas there are environmental problems such as the following (after Dryzek 1987): declining proven quantities of specific resources in relation to rate of resource use; declining energy return on investment for fuels and increasing energy costs for nonfuel resources (Cleveland *et al.* 1984); excessive topsoil loss in comparison to the regenerative capacity of the land; deforestation in relation to the remaining area of forest cover; continuing buildup of carbon dioxide in the atmosphere; drought in much of Africa; the need for increasing quantities of fertilizers and pesticides necessary to maintain crop yields; a rapidly increasing rate of plant and animal species extinction; high human population growth rates; increasing damage to forests and lakes from acid rain; specific environmental catastrophes such as Bhopal and Chernobyl; and an enlarging hole in the atmospheric ozone layer. As nearly everyone interested in problems of resources and environment agrees, empirical evidence alone cannot indicate the current severity of ecological problems or the trends in that severity. The political economists show the relationships between the science and ideology (Harvey 1974); the ecologists (pluralist and otherwise) show the technical uncertainty of our science. (Holling 1978, Kates 1985b); and political groups of all kinds use this uncertainty as argument for promoting or inhibiting all sorts of activities (see Abs 1988). Furthermore, as Dryzek (1987) points out, empirical evidence is impotent because improvements in one indicator may simply result from the export of difficulties to another area. Problem displacements, spatially, temporally, and in terms of technical substitutions which often cause similar or worse ecological problems, are in fact primary means of ameliorating environmental offences. For the most part, however, we can agree that we do have several ecological problems – even if we cannot agree as to their exact severity or the exact attribution (and relationship) of their social and natural causation. The immediate question remains what can we do *now*, knowing what we think we know (from Marxist and other critical theories, from ecological studies, and from the empirical studies of pluralists, behaviouralists, and politcial economists), while also attending to long-term social change?

Generally, Marxists have avoided decision-making models relative to environmental management for political and epistemological reasons. However, Blaikie & Brookfield (1987) consider the decision-making arena non-revisionist and find large areas of agreement between Marxist and behaviouralist positions. This is a particularly refreshing and open-minded approach given the seriousness of many environmental–social problems and the unacceptability of doing nothing to mitigate problems in the short-term. On the other hand, their resort to boxes enclosing the decision-making process with a 'political economy' exogenous to the model somehow brings us back to the explanatory inadequacies of the neoclassical and behaviouralist theories.

The outstanding example of the kind of work we have in mind is Blaikie's (1985) earlier book on soil erosion. Blaikie writes in opposition to the conventional view that soil erosion, especially in the Third World, is not a particularly important problem and that induced innovations by farmers, governments, and private sector research institutions can cope with whatever problems exist. In fact, he says, the extension of conservation policies over wide enough areas to have an appreciable effect has been so slow as to constitute a failure. Why has policy usually failed? In the classical (colonial) model, the problem of soil erosion is seen as environmental rather than socio-economic, the blame being laid on irrational land users and overpopulation, and the solution found in the involvement of peasants in market economies. While this model has been subjected to critical scrutiny, Blaikie claims some of its characteristics remain embedded in contemporary policies. A new approach is therefore necessary. Soil degradation and erosion are caused by the interaction between land use, the natural characteristics of land, and the erosive forces of water and wind. While not neglecting the physical parameters, Blaikie's (1985, p. 32) emphasis lies on the social element – that is, why certain land uses take place in terms of the political-economic context in which land users find themselves. Erosion occurs in a number of social contexts: peasant and pastoral groups employing family labour; peasant and pastoral groups working under exploitative class relations; centrally planned economies; and advanced capitalist economies. Erosion is also contingent on other variables like rural population density, the state, and technology. Hence, while the social relations of production under which land is used are key elements in the explanation of soil erosion, there are many contingent elements which also have to be analyzed in any concrete instance – it is probably impossible to attempt a single theory, and Blaikie (1985, p. 35) therefore attempts only to theorize 'substructures in a theory of soil erosion'.

This theorization involves two systems, the physical and the socio-economic, in integration. Theory has to be location-specific, place-based and conjunctural, yet also non-place-based relations (market and class relations) must be integrated into the analysis. Following a bottom-up approach, Blaikie begins with the smallest decision-making unit, usually the household, making land use decisions under constraints of asset holdings. Place-based decisions are mapped, generalized over time, and compared with spatial variations in the physical determinants of soil erosion, with feedback loops noted between the two systems. This mechanistic yet conceptually precise scheme then needs animating by political-economic analysis.

Focusing on the peasantries of underdeveloped countries, Blaikie proposes two spheres of political-economic relations which explain soil erosion: social relations of production at the level of the enterprise; and exchange and other relations at the level of the world economic system. In both spheres surplus is extracted from peasants: at the local level through wage labour or rents; at the international level through unfavourable terms of trade and low product prices. Capitalism can be periodized in terms of its relations with the peasantries of the world – that is, in terms of requirements (raw materials, land, labour power, markets) and the means used to render peasants malleable to these needs (force, state taxation, etc.). But the essential relation with soil erosion lies through surplus extraction: 'surpluses are extracted from cultivators who then in turn are forced to extract "surpluses" (in this case energy) from the environment (stored up fertility of the soil, forest resources, long-evolved and productive pastures, and so on), and this in time and under certain physical circumstances leads to degradation and/or erosion' (Blaikie 1985, p. 124). Through incorporation into the world economic system peasants are marginalized (in the sense of losing control over the structure and location of their lives), have changed relations of production, and are proletarianized: a significant aspect of this process is spatial marginalization in combination with peasant differentiation, which can lead to poor peasants crowding on to land of limited agricultural potential along with other desperate survival strategies. An important element is the overuse of common land among marginalized people. Political-economic process is then related to natural process: steep-sloped areas, with a higher propensity to soil erosion than flat areas and subject to economic and political peripheralization, suffer from similar processes of environmental deterioration. In general, therefore, small producers cause soil erosion because they are poor and desperate, and, in turn, soil erosion exacerbates that condition. A set of socio-economic conditions called underdevelopment lies at the centre of this poverty syndrome (Blaikie 1985, Ch. 7).

Blaikie is less successful in laying out a model of large enterprise mining of the soil. However, the circumstances encourging owners of capital to use up the natural content of the soil and then withdraw from an area can be laid out. And the general tendency for the natural resources and labour power of underdeveloped countries to be increasingly incorporated into the global economy can be specified. Control of natural resources by new classes removes the locus of decision making so that effective means of conservation often do not lie in local or national hands (Blaikie 1985, Ch. 8). Bringing this tenuous theory into relation with his earlier discussion of small producers, Blaikie (1985, Ch. 9) concludes that soil erosion in underdeveloped countries will not be substantially reduced unless it seriously threatens the capital accumulation possibilities of the dominant classes. As the impact of soil erosion is diffuse, patchy, and difficult to measure, and because powerful people can easily adjust, the degree of threat has to be substantial. Likewise, small producers cause soil erosion under circumstances of threat to livelihood which also makes co-operation difficult and state intervention ineffective. In addition, ideas about soil conservation are varied and do not directly reflect economic imperatives. Drawing these arguments together, Blaikie (1985, p. 149) proposes that strategic choices in soil conservation policy have to be consistent with a broader view of the direction of development and social change: on this there are different, politically-biased positions – for example, socialist utopianism, populism, rational policy making, and authoritarianism, each having a different perspective on soil conservation. The outlook for major success in conservation seems bleak, but a practical pessimist *can* make some suggestions first in the realm of ideas, for example focusing research on the implications of erosion for inequality and impoverishment and second in terms of practical projects, for example, rascal-proof systems of local management of watersheds. Blaikie (1985, p. 154) admits these suggestions may seem tame and reformist, but finds it better to end . . . in an honest whisper than a spurious bang'.

These recent attempts to identify the social and environmental causes of environmental degradation, resource depletion, and maldistribution of benefits and liabilities are honest in their emphasis on complexity, and their making of conditional and multiple hypotheses (Blaikie & Brookfield 1987). In large part, these efforts attribute resource and environmental problems to the contradictions of production (not simply capitalism), a conflict of values, or both. The significance of this recent work by Blaikie (1985), Blaikie & Brookfield (1987), Rees (1985), Redclift (1984, 1987) and others is in the development of linkages between resources, economies, institutions, individuals, and societies. These works do not contribute particularly to management-level issues in terms of offering specific approaches. Instead, the major emphasis is on the social, political, and economic origins of environmental problems and the consequences of resource depletion and deterioration. A second major emphasis, introduced by Redclift (1987), is on the way technology mediates the social relations with environment. He cautions against overestimating the potential of science and using technology as a way of distancing ourselves from contradictions of development for the environment, rather than resolving them. Instead we must be pro-active rather than reactive to production processes. We must make explicit needs and the historically determined interpretations of these needs. Finally, any attempt at resolution of resource issues must recognize the importance of the local knowledge of those actually making decisions, the political-economic conditions underlying that knowledge and constraining decision, and the many implications of state intervention.

In arguing that no necessary relationship exists between a specific form of political economy and ecological production, Rees (1985), Blaikie (1985), Blaikie & Brookfield (1987) and Redclift (1987) depart from Pepper's (1984) line of reasoning, but not significantly. All tend to agree that we must innovate and find forms of articulation in order to address the contradictions inherent in the development of the environment. This will not be possible without a multiplicity of differences, which we should not fear to deepen and develop – not in the spirit of being right in splendid isolation, but in order to foster both short-term and long-term initiatives and alliances capable of informing political, economic, and scientific practice on the threshold of the 21st century.

Natural hazards

When we turn from resource management to hazardous aspects of the human use of nature, the radical critique grows stronger – some would say more strident – and the theoretical alternative becomes more clearly linked with Marxist and neo-Marxist politics. As mentioned above, the leading school of natural hazards research in geography was initiated by White's (1945) work on human adjustment to floods in the early 1940s. Joined by Burton and Kates in the early 1960s, and later by their students, the White–Burton–Kates School is composed of a number of interlinked North American researchers centred on the University of Colorado, University of Toronto, Clark University, and (recently) Brown University. The White–Burton–Kates School of natural hazards is highly influential at the national and international level, is heavily and consistently funded, and has published extensively. In many ways it represents conventional geography's outstanding success story in the academic–governmental arena, even though its main researchers have long considered themselves interdisciplinary intellectuals rather than simply geographers. In 1978, Burton, Kates, and White published a summary of research work conducted under grants from the US National Science Foundation over the preceding decade. We summarize their views mainly as presented in that 1978 volume, before proceeding to critiques and the evolution of radical theoretical alternatives.

Environment as hazard

As with resource management, members of the hazards school spend most of their time working with themes which have practical implications. The lack of philosophical discourse makes placing their work in its intellectual context more of a matter of interpretation than citation. However it seems clear that the school emerged from a line of thought (human ecology) opposed to the environmental determinism of Huntington and Semple in the late 19th and early 20th centuries. Whereas determinism argues that history is naturally determined (Peet 1985), the human ecology school sees relations between people and environment more from the standpoint of human adjustment to environment, taking into account a variety of influencing factors, and avoiding the notion of direct determination by nature (Barrows 1923). Thus in White's (1945, p. 36) initial formulation the basic elements of human ecological analysis are floods as natural hazards, human occupancy of flood plains, and the adjustments of occupants to flood hazards. Or as Kates (1971) later phrased it, a natural hazard is an interaction of people and nature governed by the co-existent adjustment processes in the human use system and the natural events system. In Burton *et al.* (1978) these two events systems are conceptualized as functioning independently, with people transforming the environment into resources and negative resources (hazards) through their use of natural features (Fig. 3.1). Human response to hazards is related to people's perception of the phenomena and their awareness of adjustment opportunities. Adjustments and adaptations create and change the capacity of individuals, managerial units, and social systems to absorb the effects of extreme environmental fluctuations. Burton, Kates, and White visualize the range of choices available to individuals or agencies as a tree of alternatives involving increasingly active and complex adjustments. People choose the degree of risk they bear, and the adjustment they make, using subjective methods which Burton, Kates, and White describe in terms of a bounded rationality model similar to Simon's (1956) 'satisficing'. There are interpersonal and intercollective variations in the perception of hazards, knowledge of adjustments, and methods of evaluating decision criteria. So what causes natural hazards, they ask? Most people questioned about hazards in their locality view them as inexplicable, or as acts of supernatural forces. Scientists have different views, among which three perspectives can be discerned: (a) hazards are natural events needing scientific investigation, better technology, and improved warning systems; (b) technological adjustments aggravate natural problems; (c) social variables other than technology are significant but opinion is divided between neoclassical economists and radicals. Burton, Kates, and White find each perspective valuable, none by itself adequate, and favour an unspecified interaction between the three. Their more concrete arguments involve an increase in the global population concentrating people in hazard-prone areas, particularly in underdeveloped countries, with poorly chosen and inappropriate technologies aggravating the problem; this seems to be a mixture of neo-Malthusian and technological causation (see also Marston 1983, p. 340).

This kind of behavioural geographic theorization proved amenable to national and international sponsorship of funded research aimed at alleviating the consequences of natural disasters. However, supportive recognition by a majority of governmental, academic, and financial institutions was opposed by a minority reaction characterized by severe criticism from other academics, especially those of a radical persuasion. Immediately the Burton, Kates, and White book was criticized, in terms of its contribution to the broad set of ideas on similar topics in social theory as a whole (Torry 1979a), for virtually ignoring the vast anthropological and sociological literature and bypassing historical and epidemiological studies (Torry 1979b, p. 368). Their ideas on

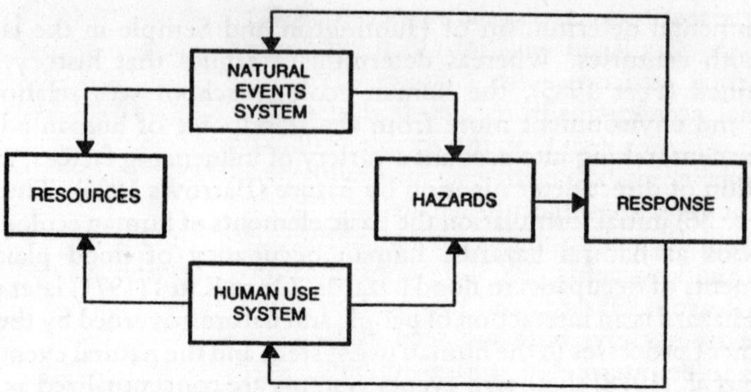

Figure 3.1 Natural hazards (after Burton, Kates & White).

individual and group adjustments to hazards were singled out for particular criticism: Walker's (1979) more radical critique found the model of response based on purposeful rationality inadequate, especially in that the burden of fault is heavily weighted towards the psychological propensities of the individual. While Burton, Kates, and White recognize that social conditions influence individual perception and response, Walker continued, there is no workable theory of social process, and the authors waver inconclusively between individual and social causes of irrational behaviour. Waddell (1977) earlier found the reason for the missing link between nature and the individual in White's (1974) disregarding of the Marxist literature which seeks to identify the social-structural causes of natural disasters. These severe criticisms drew only a defensive reply from Burton *et al.* (1981), whose work continued more or less as before (e.g. Burton & Kates 1986). As the debate quickly escalated into an alternative radical theory we will outline three critiques of the conventional hazards school in considerable detail.

Poverty of the technocratic view

It has to be stressed again that Burton, Kates, and White occupy a liberal position in a resource field dominated by neoclassical economies and engineering. They are very much aware of the criticisms both of the resource field in general, and their contribution in particular. Thus, White, Burton, and Kates devoted the first of a series of books on risks and hazards to a collection of studies critical of their own position: in Burton's (1983, p. vi) editorial words there was 'a wide-spread feeling of discontent and dissatisfaction "with the dominant view" in hazards research – the ship is leaking badly' producing the need for new elements 'to be incorporated into the eventual design of a new vessel'. Hewitt's (1983a) essay, which opened this book, uses a Weberian (bureaucratic) rather than Marxist (class) analytic to launch what is essentially an internal critique. The dominant view in social science, he argues, is part of a bureaucratic ethos which channels scientists into analytical approaches which reflect the positions taken by hierarchical organizations sponsoring research. The dominant view in hazards research is particularly suited to institutions carrying out technical work, with technology wedded to science of the most advanced kind; this makes it the creature of the most powerful, wealthy, and centralized institutions, the leading organizations of government, business, and culture. A privileged position then enables the dominant view to resist fundamental criticism – 'its changes have been chameleon-like exercises in superficial novelty-absorbing, co-opting or ignoring dissent at will (Hewitt 1983a, p. 4). As a result, an intractable problem for human societies is tackled by an archaic, inflexible strategy even while the truths and effectiveness of the dominant view are subject to debate.

Hewitt finds a close analogy between the bureaucratic–technocratic view of hazards and Foucault's (1965) description of the invention of madness by the Age of Reason. Hazards and madness are initiated in ways that seem uncontrollable by society; both are seen as judgements on human activity; both threaten to be interpreted as punishments for a disorderly and useless science. And just as poverty-stricken 'crazy' people were confined in the 18th century, so natural calamities are separated from the rest of human-environmental relations and social life for special research, using a language stressing the 'un'-ness of the problem – they are *un*scheduled events occurring among *un*aware people. Disasters happen because of the chance recurrence of natural extremes, with socio-economic factors or habitat conditions only modifying the details. The geography of risk is therefore synonymous with the distribution of such natural extremes – like earthquakes – and the natural features associated with them – faults. While disaster is attributed to nature in the dominant view, there *is* something society can do. That something is not everyday human activity – which makes the problem worse – but public policy backed by the most advanced geographical and managerial capability. Most research and financial expenditures thus fall into three areas:

(a) monitoring and scientifically understanding geophysical processes with the goal of prediction;
(b) planning and managerial activities to contain natural processes where possible or re-arrange human activities where not;
(c) emergency measures and quasi-military organizations for relief and rehabilitation.

The main areas of expertise are those of the physical sciences and engineering, with the social sciences studying crisis behaviour, or the places or groups living in disaster-prone areas, and increasingly confining themselves to the research motifs outlined above – how people appraise risks, respond to forecasts, cope with emergencies – thus reinforcing the geophysicalist and technological reductionism of the dominant view (Hewitt 1983a, pp. 5–24).

It is, of course, the case that natural hazards are connected with geophysical processes. But in Hewitt's alternative view uncertainty and vulnerability flow as much from ordinary life as from natural fluctuations. Recent disasters have occurred in settings characterized by extraordinary sociocultural change so that natural hazards are among life's certainties for people who must devote most risk-avoiding energies to the social uncertainties of everyday life. People's traditional means of dealing with natural problems have been weakened, if not destroyed, by modernization, especially alienation from land and nature. Moreover, relief and construction are disproportionately focused on restoring the infrastructure of the powerful institutions of society rather than on directly responding to the needs of the victims. For Hewitt, then, a new consensus is needed, less influenced by the pressures and interests of technocratic institutions, yet still able to influence these, so that change can be achieved pragmatically (Hewitt 1983a, pp. 24–8). This alternative view as expressed in Hewitt's own work involves achieving critical distance from conventional misconception of climatic hazards and agricultural development in Third World countries; being critical of market processes as they operate, especially in non-Western contexts; and in general seeing 'the form and level of risks from nature as a function not merely of spontaneous natural events but of human development' (Hewitt 1983, p. 199).

The poverty of theory

The dominant view in hazards research has also been criticized from a more explicitly Marxist perspective, most extensively by Watts (1983a, 1985). This critique begins with the theme explored by Hewitt: knowledge is shaped by the preoccupations brought to it; the world is interpreted within an historically conditioned imagination; theories and concepts cannot be taken for granted. But as even these few phrases, drawn from Watts (1983a, p. 231) indicate, the critique is cast in somewhat different terms, epistemological questions are pursued at greater length, and the need to re-situate theory in the wider body of (critical) social theory is more urgently expressed. Watts places geographical work on hazards under the more general heading of cultural ecology and subject to the criticisms launched against this field: it sees people and nature as discrete entities with the interaction between the two conceived along neo-Darwinian lines through the biological optic of organismic adaptation. In this view, maladaptation becomes a type of cybernetic malfunction, or results from mistaken perception, imperfect knowledge, or inflexible decision making – all remediable through the right set of policies. Instead, hazards, research and cultural ecology need placing in the context of political economy – specifically the study of how social formations reproduce themselves through the labour process. In this view, environmental relations are instances of the productive process, and adaptive strategies or coping responses are grounded in the social relations of production in concrete historical circumstances; calamities thereby yield valuable information about the stricken society (Watts 1983b, p. 26). For example, in West Africa, peasant households are inserted into the nexus of social and spatial relations. Relations between the household economy and commodity markets can deteriorate, forcing peasants to super-exploit land and labour to remain at the same level of subsistence. Surplus extraction is then transmitted to the environment as individual and communal lands are 'mined' to pay debts, rents, or transmit profit over space via unequal commodity exchange. Reproduction squeezes account for the abandonment or irrelevance of traditional adaptive strategies. And the existence of high risk conditions among the poor accounts for the persistence of high population fertility among farming households, when having children for the sake of family survival deepens only the pressure on fragile ecosystems. In general, the idea is to stimulate ecological questions in socio-spatial context, disequilibrium in the social system being translated into ecological disequilibrium. In this Marxist approach, explanatory emphasis is replaced on to issues like patterns of social differentiation and the process of capital accumulation and surplus extraction (Watts 1985, pp. 24–5, 30).

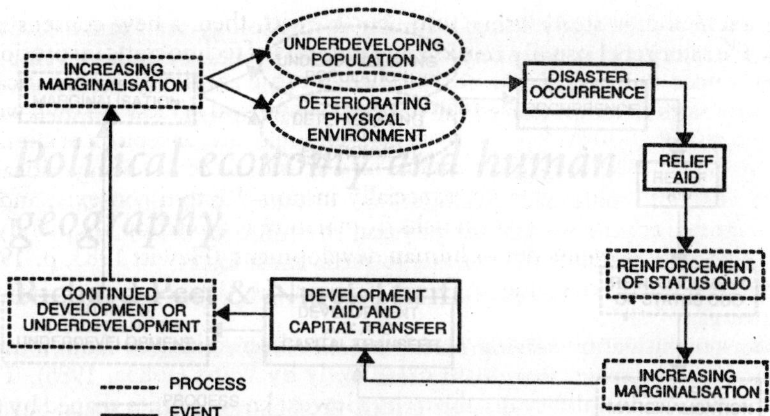

Figure 3.2 Marginalization and disaster in the Third World (after Susman, O'Keefe & Wisner).

Of poverty and the poor

An even blunter version of this radical view, presented more as theoretical alternative than direct critique, comes from Wisner *et al.* (1977). After reviewing the available data, they find that the Earth is becoming a more dangerous place in which to live, for the frequency of large-scale disasters is increasing, especially in underdeveloped countries. Causal theories blame: the poor, for reproducing themselves too frequently, or living in hazardous places; or nature, in terms, for example, of long-term climatic change (Wisner, *et al.* 1976). Instead, they propose, increasing numbers of people are becoming more vulnerable to physical events that have long recurred. Explanation has, therefore, to be sought in the social analysis of vulnerability. They argue that peasants understand their environments well and have extensive repertoires of survival adjustments. But traditional 'people's science' is distorted, sometimes to the point of destruction, during contact with the capitalist mode of production and as colonial and neocolonial governments denegrate peasant knowledge. Direct capital penetration means an increase in commodity export and a decrease in the land base used for peasant livelihood. This shrinks the range of peasant choice and produces decision pathology which, in turn, explains irrational or non-adaptive behaviour such as overgrazing. Greater vulnerability explains a significant increase in disasters. Concepts of naturalness should therefore be removed from the study of natural disasters – in this, the poorest Guatemalan peasants surviving the earthquake of 1976 and renaming it a 'classquake', knew more than the finest scientists (O'Keefe *et al.* 1976, p. 567).

The most innovative alternative conception emerging from this critical stance was the theory of marginalization (Fig. 3.2), which links vulnerability and disasters back into broader processes of social change. Part of the process of underdevelopment is the control and exploitation of indigenous resources by governing élites and outside interests. Also as part of the underdevelopment process, surpluses crucial to maintaining peasant economic flexibility are expropriated. Peasants are forced to look for livelihoods in hazardous areas or change their resource use in ways that exacerbate their vulnerability to disasters (for a detailed case study using a similar methodology see Franke & Chasin 1980). Disaster relief usually reinforces the same process of underdevelopment that produced vulnerability in the first place. This continues marginalization and encourages deterioration of the physical environment. The implications of this theory are:

(a) because of continued exploitation, disasters will increase as socio-economic conditions and the physical environment deteriorate;
(b) the poorest people will continue to suffer most losses;
(c) relief aid will continue to work against the weakest, most suffering groups;
(d) disaster mitigation relying on high technology reinforces underdevelopment and increases marginalization;
(e) to reduce vulnerability disaster efforts need placing within social development planning.

In brief, instead of inheriting the Earth, the poor are being eaten up by it (Susman *et al*. 1983).

A study of environmental bankruptcy in Africa (Timberlake 1986), advised by O'Keefe and Wisner, similarly puts causal emphasis on the social causes of natural disasters. Natural events, such as earthquakes and droughts are triggering mechanisms among people made vulnerable to disaster by the development of society: the geography of social relations thus determines the occurrence and extent of natural disasters. Thus an overemphasis on export crops for the global market leads farmers to overcultivate the declining areas devoted to food production leading to land degradation and vulnerability to famine. Famines may be triggered by drought, but whether drought becomes disaster depends on the previous management of the land, and this, in turn, depends on the social relations dictating management practice. In the case of the cash crop squeeze given above, relations with the global market, reinforced by development agencies such as the World Bank, are the underlying causes of natural disaster.

The political economy of natural disasters

As we can see, critiques of conventional hazards theory in the late 1970s and early 1980s quickly developed into alternative political-economy perspectives. The conditions for this theoretical development went beyond the progression, inherent in argumentative logic, from critique to counter-theory. A rising death toll from disasters in the Third World and an intensification of media interest and public awareness focused research on the causes of such problems. By the 1980s the political-economy perspective had been accepted as occasionally worthy of research funding, recognition through publication, even incorporation into eclectic theories of natural events. This perspective is still emerging. But we can gain insight into its structure and course of movement through an extensive summary and analysis of one of its leading versions: Watts's structural analysis of famine in northern Nigeria.

For Watts, the ancient and persistent human experience of famine is simultaneously biological and social: its aetiology may be as economic as it is environmental, its effects as much political as physiological. Famine refers to 'a societal crisis induced by the dissolution of the accustomed availability of, and access to, staple foods on a scale sufficient to cause starvation among a significant number of individuals' (Watts 1983c, p. 13). In analyzing disasters like famines, Watts (1983c, p. 14) finds that 'both hazards research and human ecology have suffered from neurotic obsession with individual rationality, a profound ahistoricism and not least a neglect of political economic structure'. Recently attempts have been made to overcome these limitations, for example, by looking at the changing structure of permanence as local systems become parts of global networks. While akin to political economy, this work usually fails to specify the structure of the entire productive system. Loose terms like subsistence system and cash economy are used and, to the extent production is considered, analytical priority is lent to energy flows rather than to access to, and control of, productive assets – i.e. the social relations of production. Work on risk and risk aversion is useful, especially Scott's (1976) concept of the moral economy of peasants based in survival and simple reproduction. But this too is deficient in its theory of the larger society: risk must be grounded in the form and quality of the social relations in which peasants participate. Such a view would suggest: (a) that subsistence crises are symptoms of the structural ability of the socio-economic system to cope with unusually harsh ecological conditions; (b) hence that crises enable us to view the darkest corners of social systems; and (c) that appreciating that risks are mediated by socio-economic structures shows that development and modernization have not solved age-old problems and, in some cases, have aggravated them. Watts, therefore, suggests that we see famines as parts of the general history of subsistence crises. Analyzing famine demands a careful deconstruction of the structure of society and its historically specific systems of production – such a project might carry the title 'The social production of famine'.

Turning to Marxism for analytical guidance, Watts finds two lines of thinking on peasant societies, such as those of northern Nigeria. The first poses peasants as semi-proletarians released for wage labour as capitalism destroys pre-capitalist relations of production. The second sees capitalism subordinating peasant social formations through a dialectical process of preservation – dissolution. The latter allows for specificity and contingency, as with uneven patterns of incorporation into international capitalism, emergence of hybrid forms of production, and regionally specific transition processes. Following the second, articulation of modes of production view, Watts begins his empirical research with a detailed historical/structural account of the pre-capitalist mode of production in northern Nigeria: the Sokoto Caliphate, forged from thirty pre-existing emirates in the *jihad* (holy war) of 1806. Production took the form of an upland millet-sorghum complex, together with irrigated gardens, hunting and gathering, livestock ranching, and artisanal manufacturing emphasizing dyed cloth for local and export markets. The fundamental unit of production was the household (*gida*) or, in a more protracted form (including clients and slaves) the *gandu*, with larger work groups forming as co-operative means of overcoming labour bottlenecks in the agricultural cycle. This largely self-sufficient world of the household was integrated into a state structure through the expropriation of surplus in the form of taxes, a process sustained by Muslim ideological apparatuses grounded in the belief that all land belonged to the community vested in the emir as head of state. While Watts distances

Table 3.1 Subsistence security and resource structures in 19th-century Hausaland.

Agronomic or domestic level	Community level	Regional or state level
Agronomic risk aversion Intercropping (crop mixtures) Crop rotation (moisture preservation) Crop experimentation (short-maturing millets, etc.) Exploitation of local environment (famine foods) Secondary resources (dry season crafts) Domestic self-help and support	Interfamily insurance (risk sharing) Extended kin groups (*gandu*) Reciprocity (gift exchange, mutual support) Élite redistribution to the poor Storage, ritual sanctions Anti-famine institutions Patron–clientage Communal work groups (*gayva*)	Regional and ecological interdependence between desert edge and savannas Local and regional trade in foodstuffs from surfeit to deficit regions Role of state: (a) central granaries based on grain tithe; (b) state relief and tax modification

Source: based on Watts 1983c, pp. 110–11.

himself from Althusserian structuralism, believing it to have degenerated into abstract, sterile taxonomies, his description remains similar to the version of the Asiatic mode of production outlined by this literature (Taylor 1979, pp. 182–4).

Watts also draws on French ecological history in the form of conjunctural studies of the impact of climate on history emphasizing long-term *structural* adaptations to climatic changes. Whereas in ecological anthropology structural adaptation is interpreted organismically, Watts interprets social reproduction in the face of recursive stress (such as drought) in political-economic terms; that is, as a metabolic *inner*action, within the unity of society and nature, but also by the social relations of the labour process. Watts further draws on Scott's (1976) concept of moral economy in which subsistence security is projected on to the screen of peasant rationality in the form of behavioural conservatism: the safety-first maxim. Synthesizing these ideas with a mode of production approach, Watts achieves a broad conceptualization of pre-capitalist Hausaland in which the tensions, constraints, and controls of peasant practices emerge from the political, economic, and ethical–legal system as a whole: in particular, the moral economy of peasants stems from social productive relations seen to possess ethical qualities. Thus Hausa farmers traditionally possessed an adaptive flexibility to accommodate climatic risks at three levels: the household, the community, and the region (Table 3.1). Famines occurred in pre-colonial Hausaland and caused socio-economic dislocation. But Watts argues that droughts were expected, precipitating a logical chain of events – there was an indigenous relief system involving a social map of expectations.

Watts then proceeds to show how the traditional economy was transformed after 1900. He interprets the British colonial social formation as an articulation between merchant capitalism and persisting non-capitalist forms of production. Merchant capital does not itself have to organize production on a capitalist basis to extract surplus value. Instead it incorporates commodity (groundnuts and cotton) trade into the international circuits dominated by Europe, leaving production to reorganize itself – although the colonial state, acting in the interest of merchant capital, does play a crucial interventionary role. In northern Nigeria a mixed production system emerged, growing both export crops and food. However, export cropping made profound inroads into the food production system. Indirectly, use-value society changed into one oriented towards exchange-value, the culture of reproduction was likewise transformed, and with it the social security of the household. The *gandu* (extended household) lost its pre-eminent position in favour of the nuclear household, a less secure reproduction structure. *Gandu* landholdings were fragmented into small plots scarcely sufficient for subsistence needs. Large parts of craft production, which previously had employed off-season labour, were destroyed by competition from cheap European wares. Wage labour on railroads and in mines drew workers from intensive cereal production, causing an increasing dependence on foodstuffs controlled by (indigenous) merchant capital which was less reliable at times of famine because prices rose. The system of indirect colonial rule intensified the power of the traditional aristocracy and changed the form of its exercise, allowing tax extortion and corruption as ways of increasing surplus extraction. Rural indebtedness, part of the commodity-producing system, reached crisis proportions. Meanwhile, the volatile prices of groundnuts and cotton insured that peasant incomes varied within much wider margins than in the pre-colonial period. Watts also stresses the role of a colonial state which, because of an imperial demand for self-sufficiency, had a primordial interest in direct taxation, while its efforts at relief at times of crisis were ineffective due to incompetency and a misconception of the problem. This, in turn, affected the moral economy by speeding up the change to a money economy and regularizing previously flexible tax payments. In general, therefore, commodity production under merchant capitalism disrupted the traditional relations of Hausa production without improving its technology:

> for large sections of the Hausa peasantry, the margin of subsistence security was progressively undermined by a retarded capitalism; the bases of the moral economy were eroded in a significant way . . . colonialism dissolved many of the response systems that served to buffer households from the vagaries of a harsh and variable semiarid environment . . . . The patterns of change . . . not only altered the extent of hunger in a statistical sense but changed its very etiology (Watts 1983c, p. 226).

Thus, in analyzing the effects of the 1969–74 drought on the people of the Nigerian Sahel, Watts emphasizes an historical political-economy analysis of the resilience of food systems in relation to environmental and economic perturbations. Analytical stress is laid not on the psychological attributes of peasants, who traditionally had a series of effective response strategies, but on the present precariousness of the rural poor, caught in a simple reproduction squeeze. In other words, peasant households are not intrinsically pathological;

instead they are now constrained in their ability to respond to threats, disturbances, and perturbations (Watts 1983c, p. 465).

This extended review of Watts, makes clear the difference between the old models and the new. Theoretical emphasis is replaced on to the social structures originating individual responses, these structures are interpreted in social relational terms, the historical dimension is more deeply appreciated and appropriated, and the links with radical social theory explored and developed as sources of a more generalized comprehension. It goes without saying that the politics which emerges from such a retheorization of natural hazards is entirely different – liberal modification gives way to social transformation.

Conclusions

From the early critiques of resource management and natural hazards we find alternative political-economic theories evolving in the late 1970s and 1980s. A new view of social relations to the environment began to appear, rooted eventually in Marxist political economy, but also incorporating ideas drawn from a wide variety of critical perspectives. Even during the few years of the mid-1980s we find political-economic explanation shifting in response to internal criticisms. Comparing the three works discussed in detail in this chapter, Watts (1983c), Blaikie (1985), and Rees (1985), we find Watts drawing heavily on mode of production theory, embedding social practices with regard to nature, specifically famine avoidance, in the whole way of life of a people. At the time of his early work Watts revelled in structuralist explanation, seeing diversity as complex manifestations of common elements in the social relation to nature. Blaikie also emphasizes the social relations by which surplus is extracted from producers causing them, in desperation, to mine the soil, but does not construct an elaborate structural conception of these relations, preferring an emphasis on geographical context and contingency. The latter theme is clearly reflected as well in the natural resource work of Rees. The various approaches have yet to be reconciled in political ecology as in geography as a whole. But there are promising signs. Thus Bassett (1988, pp. 469–70) concludes that peasant–herder conflicts over land in the Ivory Coast are 'largely conjunctural and thus difficult to theorize' yet 'these conflicts contain structural features related to the larger political economy which can be theorized without reducing the complexity of the situation in a crude, deterministic way.' We join with him in believing that political ecology, a model linking society, political economy, and environment, already offers rich theoretical insights into some of the most fundamental concerns of human geography. Indeed, the conclusion of this chapter is, that on reviewing the work of the last two decades; we find abundant evidence of intellectual maturation, the development of an alternative perspective, healthy levels of internal disagreement, and an increasing engagement with what should be the central concern of geographers of all persuasions: the contradictory relations between people and Earth.

Notes

1 O'Riordan (1972, p. 19) suggests that resource management 'should be visualized as a conscious process of decision involving judgment, preference and commitment, whereby certain desired resource outputs are sought from certain perceived resource combinations through the choice among various managerial, technical and administrative alternatives'.

2 Holling (1978, p. 18) describes his 'adaptive environmental assessment and management' approach as 'not only absolutely compatible with the dynamic concepts of development and the rational use of natural resources, but it also tends to promote the generation of self-reliant and endogenous approaches to the environmental problems – approaches appropriate to local conditions, needs, and socioeconomic structures'.

References

Abs, S. L. 1988. The limits of 'rational' planning: a study of the politics of coastal mineral extraction. *Geoforum* **19**, 227–44.

Barrows, H. H. 1923. Geography as human ecology. *Annals of the Association of American Geographers* **12**, 1–14.

Bassett, T. J. 1988. The political ecology of peasant–herder conflicts in the northern Ivory coast. *Annals of the Association of American Geographers* **78**, 653–72.

Blaikie, P. 1985. *The political economy of soil erosion in developing countries*. New York: Longman.

Blaikie, P. & H. Brookfield 1987. *Land degradation and society*. London: Methuen.

Blowers, A. 1984. The triumph of material interests: geography, pollution and the environment. *Political Geography Quarterly* **3** 49–68.

Bookchin, M. 1979. Ecology and revolutionary thought. *Antipode* **11**, 21–32.

Braybrooke, D. & C. E. Lindblom 1963. *A strategy of decision*. New York: Free Press.

Brookfield, H. 1964. Questions on the human frontiers of geography. *Economic Geography* **40**, 283–303.

Bunting, T. E. & V. C. Gallant 1971. The environment grab bag. In *Geographical inter-university resource management seminar*, 19–43. Waterloo, Ont.: Waterloo Lutheran University, Department of Geography.

Burton, I. 1983. Foreword. In *Interpretations of calamity*, K. Hewitt (ed.). Boston: Allen & Unwin.

Burton I., R. W. Kates & G. F. White 1978. *Environment as hazard*. New York: Oxford University Press.

Burton, I. & R. W. Kates (eds) 1986. *Geography, resources and environment*. 2 vols. Chicago: University of Chicago Press.

Burton, I., R. W. Kates & G. F. White 1981. The future of harzards research: a reply to William I. Torry. *Canadian Geographer* **25**, 286–9.

Clark, W. C. 1986. Environmental history, Editorial in *Environment* **28**, i.

Clawson, M. 1986. Integrative concepts in natural resource development and policy. In *Geography resources, and environment: themes from the work of Gilbert F. White*, R. W. Kates & I. Burton (eds), 69–82, vol. 2. Chicago: University of Chicago Press.

Cleveland, C. J., R. Costanza, C. Hall & R. Kaufmann 1984. Energy and the U.S. Economy: a biophysical perspective. *Science* **225**, 890–7.

Davidson, P. 1979. Natural resources. In *A Guide to Post-Keynesian economics*, A. S. Eichner (ed.), 151–64. White Plains, NY: M. E. Sharpe.

Dryzek, J. S. 1987. *Rational ecology: environmental and political economy*. Oxford and New York: Basil Blackwell.

Emel, J. 1987. Groundwater rights: definition and transfer. *Natural Resource Journal* **27**, 653–73.

Emel, J. & E. Brooks 1988. Changes in form and function of property rights institutions under threatened resource scarcity. *Annals of the Association of American Geographers* **78**, 241–52.

Emel, J. & R. Roberts 1989. Ideologies of property: property rights in Texas and New Mexico groundwater resources, *Economic Geography* **65**.

Enzensberger, H. M. 1974. A critique of political ecology. *New Left Review* **84**, 3–26

Foucault, M. 1965. *Madness and civilization*. New York: Mentor Books.

Franke, W. & B. H. Chasin 1980. *Seeds of famine: ecological destruction and the development dilemma in the West African Sahel*. Montclair, NJ: Allanheld Osmun.

Freeman, A. M. 1979. *The benefits of environmental improvement: theory and practice*. Baltimore, NJ: Johns Hopkins University Press.

Goodin, R. E. 1982. *Political theory and public policy*. Chicago and London: University of Chicago Press.

Gorz, A. 1979. *Ecology and politics*. Boston: South End Press.

Habermas, J. 1970. *Toward a rational society*. Boston: Beacon Press.

Hardin, G. 1974. Living on a lifeboat. *BioScience* **24**, 561–8.

Harris, A. 1983. Radical economics and natural resources. *International Journal of Environmental Studies* **21**, 45–53.

Harvey, D. 1974. Population, resources and the ideology of science. *Economic Geography* **50**, 256–77.

Hewitt, K. 1980. Book review: The environment as hazard, by I. Burton, R. W. Kates, and G. F. White. *Annals of the Association of American Geographers* **70**, 306–11.

Hewitt, K. 1983a. The idea of calamity in a technocratic age. In *Interpretations of calamity*, K. Hewitt (ed.). Boston: Allen & Unwin.

Hewitt, K. 1983b. Climatic hazards and agriculture development: some aspects of the problem in the Indo-Pakistan subcontinent. In *Interpretations of calamity*, K. Hewitt (ed.), 181–201. Boston: Allen & Unwin.

Hewitt, K. 1986. The idea of hazard in a technocratic age and natural hazards research. In *Natural resources and people*, K. Dahlberg & J. W. Bennett (eds), 303–42. Boulder, Col.: Westview Press.

Hollins, D. S. (ed.) 1978. *Adaptive environmental assessment and management*. Chichester: Wiley.

Kasperson, R. E. (ed.) 1983. *Equity issues in radioactive waste management*. Cambridge, Mass.: Oelgeschlager, Gunn & Hain.

Kasperson, R. E., G. Berk, D. Pijawka, A. B. Sharaf & J. Wood 1980. Public opposition to nuclear energy: retrospect and prospect. *Science, Technology and Human Values* **5**, 11–23.

Kates, R. W. 1962. *Hazard and choice perception in flood plain management*. Research Paper 78. Department of Geography, University of Chicago, Chicago.

Kates, R. W. 1971. Natural hazards in human ecological perspective: hypotheses and models. *Economic Geography* **47**, 438–51.

Kates, R. W. 1985a. Hazard assessment: art, science, and ideology, In *Perilous progress: managing the hazards of technology*, R. W. Kates, C. Hohenemser, & J. X. Kasperson (eds), 251–64. Boulder, Col.: Westview Press.

Kates, R. W. 1985b. Success, strain, and surprise. *Issues in Science and Technology* **2** 46–58.

Kates, R. W., C. Hohenemser & J. X. Kasperson (eds) 1985. *Perilous progress: managing the hazards of technology*. Boulder, Col.: Westview Press.

Kennedy, D. 1976. Form and substance in private law adjudication. *Harvard Law Review* **89**, 1685–778.

Kromm, D. E. & S. E. White 1984. Adjustment preferences to groundwater depletion in the American High Plains. *Geoforum* **15**, 271–84.

Lindblom, C. E. & D. K. Cohen 1979. *Usable knowledge*. New Haven, Conn.: Yale University Press.

Lowe, P. D. & J. Goyder 1983. *Environmental groups in politics*. Resource Management Series, vol. 6. London: Allen & Unwin.
Lowenthal, D. 1972. Research in environmental perception and behaviour: perspectives on current problems. *Environment and Behavior* **4**, 333–42.
Marston, S. A. 1983. Natural hazards research: towards a political economy perspective. *Political Geography Quarterly* **2**, 339–48.
Mitchell, B. 1979. *Geography and resource analysis*. London: Longman.
Mitchell, B. & P. King 1984. Resource conflict, policy change and practice in Canadian fisheries management. *Geoforum* **15**, 419–32.
Mumy, G. E. 1974. Law, private property, and environment: a provocation to a debate on solving the problem of environmental pollution. *Maryland Law Reform* **4**, 69–76.
Nunn, S. C. 1985. The political economy of institutional change: a distribution criterion for acceptance of groundwater rules. *Natural Resources Journal* **25**, 867–92.
O'Keefe, P., K. Westgate & B. Wisner 1976. Taking the naturalness out of natural disasters. *Nature* **260**, 566–7.
O'Riordan, T. 1976. *Environmentalism*. London: Pion.
O'Riordan, T. 1981. *Environmentalism*, 2nd edn. London: Pion.
Pearce, D. 1983. Are environmental problems a challenge to economic science? In *An Annotated Reader in Environmental Planning and Management*, T. O'Riordan & R. K. Turner (eds), 253–64. Oxford: Pergamon Press.
Peet, R. 1984. The social origins of environmental determinism. *Annals of the Association of American Geographers* **75**, 309–33.
Pepper, D. 1984. *The roots of modern environmentalism*. London: Croom Helm.
Redclift, M. 1984. *Development and the environmental crisis: red or green alternatives?* London: Methuen.
Redclift, M. 1987. *Sustainable development: exploring the contradictions*. London: Methuen.
Rees, J. 1982. Profligacy and scarcity: an analysis of water management in Australia. *Geoforum* **13**, 289–300.
Rees, J. 1985. *Natural resources: allocation, economics and policy*. London: Methuen.
Rigdon, S. 1983. Resource policy in communist states. In *Scarce natural resources: the challenge to public policy*, S. Welch & R. Miewald (eds), 65–84. Beverly Hills, CA.: Sage Publications.
Roberts, R. & S. Gros 1987. The politics of ground water management reform in Oklahoma: implications for the Eastern Ogallala. *Groundwater* **25**, 535–44.
Rorty, R. 1982. *Consequences of pragmatism*. Minneapolis: University of Minnesota Press.
Schumacher, E. 1973. Small is beautiful. London: Blond & Briggs.
Scott, J. 1976. *The moral economy of the peasant*. New Haven, Conn.: Yale University Press.
Simon, H. A. 1956. Rational choice and the structure of the environment. *Psychological Review* **63**, 129–38.
Smil, V. 1984. *The bad earth*. Armonk, New York: M. E. Sharpe.
Soussan, J. & P. O'Keefe 1985. Research policy and review 5: biomass energy problems and policies in Asia. *Environment and Planning A* **17**, 1293–301.
Susman, P., P. O'Keefe & B. Wisner 1983. Global disasters, a radical interpretation. In *Interpretations of calamity*, K. Hewitt (ed.), 263–83. Boston: Allen & Unwin.
Taylor, J. 1979. *From modernization to modes of production*. New York: Humanities Press.
Timberlake, L. 1986. *Africa in crisis: the causes, the curses of environmental bankruptcy*. Philadelphia, PA: New Society Publishers.
Torry, W. J. 1979a. Anthropological studies in hazardous environments: past trends and new horizons. *Current Anthropology* **20**, 517–40.
Torry, W. I. 1979. Hazards, hazes and holes: a critique of the *Environment as hazard* and general reflections on disaster research. *Canadian Geographer* **23**, 368–83.
Unger, R. M. 1976. *Law in modern society: a criticism of social theory*. New York: Free Press.

Weddell, E. 1977. The hazards of scientism: a review article. *Human Ecology* **5**, 69–76.
Walker, R. 1979. Review of *The environment as hazard*. *Geographical Review* **69**, 113–14.
Walker, R. & M. Storper 1978. Erosion of the Clean Air Act of 1970: a study of the failure of government regulation planning. *Environmental Affairs* **7**, 189–257.
Walker, R. & M. Williams 1982. Water from power: water supply and regional growth in the Santa Clara Valley. *Economic Geography* **58**, 95–119.
Walters, C. 1986. *Adaptive management of renewable resources*. New York: Macmillan.
Watts, M. 1983a. On the poverty of theory: natural hazards research in context. In *Interpretation of calamity*, K. Hewitt (ed.), 23–62. Boston: Allen & Unwin.
Watts. M. 1983b. Hazards and crises: a political economy of drought and famine in Northern Nigeria. *Antipode* **15**, 24–34.
Watts, M. 1983c. *Silent violence: food, famine and peasantry in Northern Nigeria*. Berkeley, CA: University of California Press.
Watts, M. J. 1985. Social theory and environmental degradation. In *Desert development: man and technology in sparselands*, Y. Gradus (ed.), 14–32. Dordrecht: Reidel.
Wescoat, J. L. 1984. Integrated water development: water use and conservation practice in western Colorado. Research Paper 210. Chicago: University of Chicago, Department of Geography.
Wescoat, J. L. 1985. On water conservation and reform of the prior appropriation doctrine in Colorado. *Economic Geography* **61**, 3024.
Wescoat, J. L. 1986. Expanding the range of choice in water management: an evaluation of policy approaches. *Natural Resources Forum* **10**, 239–54.
Wescoat, J. L. 1987. The 'practical range of choice' in water resources geography. *Progress in Human Geography* **11**, 41–59.
White, G. F. 1945. Human adjustment to floods. Research Paper No. 29. Chicago, IL: University of Chicago, Department of Geography.
White, G. F. 1961. The choice of use in resources management. *Natural Resources Journal* **10**, 20–40.
White, G. F. 1969. *Strategies of American water management*. Ann Arbor, MI: University of Michigan Press.
White, G. F. 1974. *Natural hazards: local, national, global*. New York: Oxford University Press.
Wildavsky, A. 1979. *Speaking truth to power*. Boston: Little Brown.
Wisner, B., P. O'Keefe & K. Westgate 1977. Global systems and local disasters: the untapped power of people's science. *Disasters* **1**, 47–57.
Wisner, B., K. Westgate & P. O'Keefe 1976. Poverty and disaster. *New Society* **9**, 546–8.

4 *The challenge for environmentalism*

Timothy O'Riordan

Twenty years ago, environmentalists were portrayed as Jeremiahs, purveyors of doom always ready to spoil the occasion. Talk of famine and global destruction backed by unproveable computer predictions poured forth. As the world economy apparently survived the OPEC oil price squeeze and the subsequent anxieties about commodity scarcity, so the worst fears of the anxious were alleviated. The world entered the 1980s in a new mood of optimism and market-orientated politics where squeezing the world's poor became part of the process of continued wealth creation. It was also assumed that the impoverished would somehow remain manageable in terms of civil order and public health, and that business could become even better than usual.

Economies began to grow. It was conveniently forgotten that most US wealth was borrowed, that the Third World debt was far out of all proportion to the capacity of borrowing nations to repay, and that on all fronts short-term economic gains were being won at the expense of huge, but latent, environmental losses. In the late 1980s the accountants began to draw up the whole balance sheet. Third World debt write off will probably cost the Western economies as much as the post-OPEC costs. Global environmental damage is beginning to make headline news – loss of tropical forest cover, widespread drying of the savannah margins, regional pollution of inland seas and oceans, atmospheric contamination on a vast scale in the form of increased acidity and greenhouse gas warming, and growing alarm over the distribution of toxic chemicals in consumer goods and waste discharges.

The most dramatic outcome of all this will lie in perturbation of global climates. This in turn will affect food availability and bring great hardship, causing the poor to damage further the healthy metabolism of ecosystems, most notably those of forests, estuaries, and grasslands. We do not really know precisely what rôle these ecosystems play in regulating the viability of life on Earth, but many distinguished scientists and commentators are expressing sufficient alarm to cause some politicians to take stock. To date these politicians do not include those representing the powerful group of seven nations that effectively control the world economy. As the scientific evidence becomes remorseless in its predictions, and as the links between poverty, environmental damage, and civil insurrection grow more firmly established, so environmentalism will have run the first lap of its course. It will become embedded in the political culture as a permanent force with which to reckon, and to which adjustments will have to be made if the human race as we know it is to survive.

The changing meaning of environmentalism

This introductory section is designed to provide a simplified summary of the arguments that follow. It is also written as a bridge between the preceeding critical analysis of nature and resources by Emel and Peet, and the subsequent more political interpretations of environmentalism.

There is a generally held view amongst scholars of environmental history that attitudes rarely provide a guide to actions towards nature. All societies exhibit some sort of schizophrenia in their feelings about the natural world (see especially Glacken 1967, Tuan 1974). The duality consists of a nature-as-usufruct view, and a nature-as-nurture perspective. The former line of argument visualizes nature as malleable and manipulable through the ingenuity of scientific understanding and technological application. What matters is managerial wizardry, which is claimed to be able to eliminate all impediments to meeting the objective of improving both the lot of nature as well as the wellbeing of the human race.

Greens tend to see an interventionist perspective as the enemy, on the grounds that arrogance built on ignorance breeds disaster not only for humans but also for life on Earth in general. The intervenors genuinely believe themselves to be environmentalists. Their purpose is to improve the world by conscious planning and management, so that nature can be better off as much as the human race is upgraded. This perspective is not confined to Western-style capitalism: centrally planned economies built on state capitalism reveal many of the same symptoms. The guiding incentive is optimism and expansionism: only the structure of ownership differs. To the intervenor, nature is inherently exploitable, and resilient.

The nature-as-usufruct view dates back to the Hebrews and the Greeks (see Passmore 1974), cultures that had to struggle against a harsh environment to make a living. Nature had to be overcome if life was to progress. Only human ingenuity created fertility and progress. Consequently the environment came to be seen as a metaphor for triumph over struggle, for dominant forms of social management over the weak, and for the production of capital and resource surplus as an essential prerequisite to the class domination that was necessary to allow society to progress. Nature-as-usufruct was an excuse for persistent exploitation of the weak, whether the weak was the natural world or lesser mortals (see especially Leiss 1972).

This is still the dominant view in all industrial societies. Indeed the 'I–thou' perspective, according to Bennett (1976), is a relatively rare phenomenon shared by marginal cultures, in marginal environments, with marginal technologies. The North American Indians, the Australian aborigines, and the Canadian Inuit all recognized that nature was potentially hostile and parsimonious. They never developed a social organization or a technology that would allow them to exploit the natural world consciously. But inadvertently they left their mark, notably via fire, a major ecological hazard, and through excessive killing of game, apparently during times of both relative abundance and climatic stress. Such societies were ecologically dominant even though they may not have thought so. But their propensity for excessive depletion of environmental resources was mediated by complicated social institutions that rewarded sharing and reciprocity and punished profligacy that had no apparent social meaning. So their natural relationships were in a state of uneasy equilibrium, prone to population crashes and explosions, and always close to the edge of scarcity.

The nature-as-nurture view arose as an antidote to the nature-as-usufruct perspective. In essence, ever since it broke clear of pure subsistence economics, which was a very long time ago, human society has always recognized its capability to destroy its environs as greater than its ability to restore the damage within a manageable period of adjustment. So the nurture line seeks to place humankind in its ecological setting, simply as one of the sentient species in a world that may have no equal anywhere in the universe. Originally nurture was almost akin to sanctity, but Passmore and others (see Black 1970) have endeavoured to show that the nurture philosophy is set in a more survivalist and utilitarian mould, namely that the wellbeing of humans must only be possible in a world where nature has its rightful place as a democratic partner in evolution. The nurture view is more cautious, pessimistic, and critical. It is a counter to the excesses of zeal that transport the intervenors into new realms of exploitation and domination. For some, nurture is very Earth-centred, expressing a faith in the vitality of the globe set in a history and a future with a span of tens of billions of years. For others, nurture is still a profoundly anthropocentric concept, since the survival of humankind depends on survival of the Earth. We cherish the Earth in order to save ourselves.

At the heart of all this debate are two fundamental issues. One is the inevitability of exploitation; humans cannot live on Earth without exploiting its natural resources, that is by taking more than they return. This is the essence of entropy, the steady increase in chaos and energy dissipation that makes it more and more costly in time and effort to maintain the organs of social stability. Admittedly, this is a very difficult thesis to prove, but it has an intuitive plausibility. Eventually we may begin to spend more effort in maintaining order than we generate in new energy and wealth. At that point, a steady state may have imposed itself.

Exploitation is a function of class domination, surplus accumulation, avarice, and an unwillingness to give up comforts even when others are suffering as a result. Exploitation is rooted in both private and state capitalism. The sustainable economy is a myth and an elusive goal, that provides a convenient excuse for an endless search (see Redclift 1987).

The other fundamental issue is the paradox that there is hope for a better future. That hope may also be chimerical. But it is real in the hearts and minds of almost all people, irrespective of their condition. One fascinating aspect of humans in their concern for moral virtue and enlightenment even when they see precious little evidence of these in their day-to-day existence. The environmental movement today, as it has always been, is the tension between exploitation and sustentation, between arrogance and compassion, between lack of feeling and hope. That tension must never be relaxed; to do so would be fatal for life as we know it.

Environmentalism and green politics

Environmentalism is an awkward word. Its heptasyllabic cumbersomeness reflects its conceptual heavyhandedness; it is neither easy to define nor to visualize. It has a will-o'-the-wisp-like character that allows the opportunist or the lazy thinker to interpret it as they wish. One very real danger is to equate it with greenness or green politics, which have different meanings. Environmentalism is a collage of values and views of the world, a general patterning of predispositions, being first and foremost a social movement, though one with political overtones. Being green is a subset of environmentalism. Greenness applies to demands for fundamental reform in specific policy areas such as wilderness and wildlife protection, pollution control, conservation of resource use (including recycling), and appropriate management so as to replenish renewable resources. It is a philosophy that embraces Earth-centredness, a sense of altruistic communalism, non-violence, and a concept of time that is almost timeless. Being green is more than calling for reform: it is a striving for a special kind of life-style, which as yet is chosen by only a tiny fraction of contemporary Western society.

In the past decade green politics has moved from the fringe of voluntary environmental pressure groups and green parties into mainstream party politics, manifesting itself in debates over defence, energy production, transport, settlement and places of work, housing, social services, and the protection of wild and semi-wild habitats and scenic landscapes. Green politics has benefited from the charismatic qualities of a small band of media-responsive people, notably Hazel Henderson in the United States, Petra Kelly in West Germany, Bruce Lalonde in France, and Jonathon Porritt in the United Kingdom. Petra Kelly outlines her characteristically robust green case in a foreword to Jonathan Porritt's book *Seeing green* (Porritt 1984, p. x).

> We are all people of the old world trying to create a new one. We must constantly be asking ourselves whether we now want to continue to support the status quo, seeking to cope with crisis *after* the event by using outdated and ineffectual methods of crisis management . . . [die Grünen] is a party capable of deciding between morality and power, a party which will be prepared to counter repression with creative disobedience.

Porritt has, in turn, tried to set out a green manifesto in his book. He chastizes the main British political parties for wearing green arm bands on their growth generating sleeves. Greenness, he says, is the politics of ecology and life interests, not the politics of exploitation and class interests (Porritt 1984, p. 238). His approach is dominated by a concern for peace and non-violence, total disarmament, growth redirected towards basic social and material needs, and the establishment of self-managing communities geared to meeting essential requirements through equality of opportunity. These are the conditions, he maintains, for Earth-respectfulness and people-caring; the prerequisites for patterns of evolution that greens prefer to the term growth.

Porritt's green manifesto received only a lukewarm reception even among his sympathizers. He was unable to unravel the many contradictions between environmental perspectives and green positions. This was particularly noticeable with regard to his position on the character of central–local government relations and the ability of the individual to be enabled to run his or her own affairs. His tendency towards an authoritarian tone also unsettled a number of sympathetic commentators. Spretnak & Capra (1986) have assessed the position of green politics over much of Europe and North America. They sense that greenness will come of age only when the tide of social outlook flows in quite a different direction. That has yet to happen. Green politics is about creating the conditions that will channel the tide; environmentalism is about the many cross-currents within the complex patterns of tidal forces that constitute modern social values.

Green politics can be exploited for all kinds of politically expedient ends. Despite complaints from green parties that 'we are really different' and 'we were here first', the mainstream European political parties are now moving into green arenas with predatory purpose. Green politics are, therefore, party politics. Environmentalism is an uneasy amalgam of beliefs and prejudices which are political in the sense that they shape values and allegiances, and tilt slightly the prevailing distribution of power, away from capital and the established interests of manipulators, towards the nurturers introduced earlier.

The challenge for modern environmentalism is, therefore, to overcome a paradox. On the one hand, environmentalism is becoming subsumed within the political struggle for green votes: in that narrow sense it is succeeding. On the other hand, environmentalism as a mosaic of contested positions could be splintered into competing segments so that its more powerful underlying social critique is lost. Green activist Rudolf Bahro summarizes this dilemma from a green perspective:

> Every rejected proposal of ours that contains the *whole* message is worth a hundred times more than an accepted one that just sets about correcting the symptoms without intervening in the suicidal logic of the overall process (quoted in Spretnak & Capra 1986, p. xvi).

Since environmentalism has always made headway at the rhetorical level, these schismic forces could weaken its conceptual cohesiveness, and subsequently its reformist power.

In some respects environmental rhetoric is almost becoming too successful. The speed with which the international community responded to the calls to reduce ozone-depleting chloro-fluoro-carbons (CFCs) in aerosol propellants and foam injectors seemed to prove that environmental concern can result in effective political action. But one should be circumspect. To begin with, CFCs are relatively simple to replace. Chemical companies and foam users can make adjustments, maintain profit margins, indeed capitalize on their green image. Thus CFCs can be cut by 50 per cent by 1999. However, CFCs remain in the stratosphere for 30 years, so the remaining 50 per cent that will still be emitted, will be very damaging to ozone. In addition, other atmospheric warming gases are present in much greater abundance, notably carbon dioxide and methane, both of which are far less easy to control. Carbon dioxide (CO_2) stems from fossil fuel burning and forest depletion, while methane (CH_4) emanates from cattle and rice paddies, both offshoots of the self same forest depletion.

So rhetoric must always be matched by the scientific underpinnings to

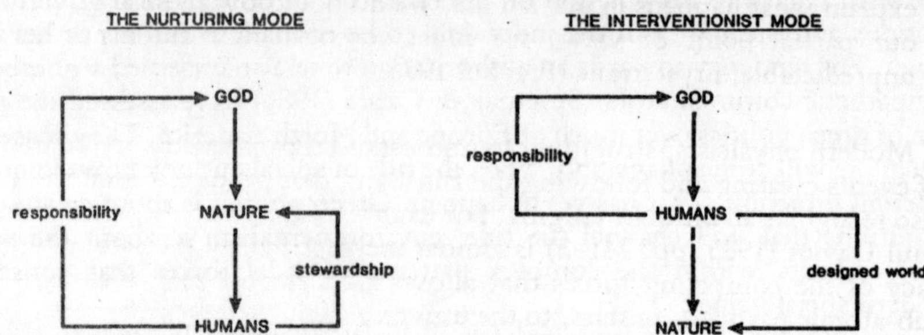

Figure 4.1 Relationships to nature.

environmental problems and a more realistic appraisal of the scale of the response that will be required. That response lies right at the very heart of the contemporary politics, and this is the ultimate challenge for modern environmentalism.

The modern meaning of environmentalism

Scholars, analysts, and activists have conceptualized environmentalism as a constructive tension between two major worldviews. What follows is based on the writings of Glacken (1967), Passmore (1974), Sandbach (1980), O'Riordan (1981), Capra (1982), Cotgrove (1982), Milbrath (1984), and Pepper (1984). All these writers have identified, to a greater or lesser degree, a distinction between on the one side a *conservative* and *nurturing* view of society–nature relationships, where nature provides a metaphor for morality (how to behave) and a guide to rules of conduct (why we must behave so), and on the other side a *radical* or *manipulative* perspective in which human ingenuity and the spirit of competition dictate the terms of morality and conduct (Fig. 4.1).

Responsibility to a cosmic force

In both modes there is a God. The God metaphor stands for the force of creation, a superhuman and unimaginable phenomenon lying beyond human intelligence and consciousness. God is the personification of the unknown, the unknowable and the ultimately mysterious. This is important, because, again, in both conceptions humans are supposed to have a sense of responsibility to the wonder of creation when seeking guidance about how to act on Earth. The loop of responsibility becomes the vital restraining mechanism in avoiding the inherent destructive powers of society's excesses. Charles Frankel (1976) concluded that nature ought to provide a moral brake on any tendency to believe that knowledge of environmental and social processes is so complete that there is no room for doubt:

> Indeed the appeal to 'Nature' may well be a useful reminder that human purposes fade, and that the sacred truths of an era are usually only collective follies. It also reminds us that, although there are laws, presumably, that

explain what happens in human life, we do not know these laws, and, from our partial point of view, we must accept nature as in part random, unpredictable, mysterious (Frankel 1976, pp. 111–12).

Modern physicists, struggling to understand the extraordinary combination of events creating and following the Big Bang that produced time and matter, also recognize an ultimate enigma. For example, the noted theoretical physicist Paul Davies (1985, pp. 241–3) is almost mesmerized by the astounding intricacy of the competing forces that allows all forms of matter to exist, from sub-atomic particles, to stars, to the universe itself. Apparently even infinitesimal changes in the fundamental constants that govern the laws of physical phenomena would long ago have destroyed matter and even life as we currently know it:

> It is tempting to believe, that a complex universe will emerge only if the laws of physics are very close to what they are. If physics is the product of design, the universe must have a purpose, and the evidence of modern physics suggests strongly to me that the purpose includes us (Davis 1985, p. 241).

Of course, these views are by no means shared by everyone. But the heartland of traditional environmentalist philosophy has always held that the wonder of creation was a restraining influence on any tendency to global destruction, whether the motive was manipulative or nurturing, and that environmentalism had an affinity with the cosmic scale. This is important, because, as we shall see, part of the contemporary critique is that society–nature relationships have become agnostic in their attitude towards any concept of creation as mystery. In the void are exploitative structures of power and corporate superiority that threaten the stability of global environmental processes (for what is still, in my opinion, the best analysis, see Ensenberger 1974).

The nurturing mode

The twin modes depicted in Figure 4.1 display an important shift in emphasis in the story of the creation and in the role and morality of human beings. In the nurturing mode, God created the Earth, then human beings. This is the original Hebrew version of the Genesis myth as adopted by the more conservative Greek and Roman traditions and as subsequently taken up by the American Transcendentalists and the European Romantics of the 19th century (Pepper 1984, pp. 68–90 is especially clear on this evolution). Human morality was therefore shaped by the right of nature and by nature's seeming imperatives.

As long as humans regarded themselves as part of the cosmic life force that is Earth-in-Universe, this containment of morality posed no difficulty. The task for human beings, therefore, was to tend the Earth, avoid wastefulness and excess, recognize the spiritual component in all non-human existence, and, above all, to revere the creative force through acts of homage and environmental responsibility. The Greenpeace philosophy, widely circulated in Greenpeace literature, captures this perspective:

> Ecology teaches us that humankind is not the centre of life on the planet. Ecology has taught us that the whole earth is part of our 'body' and that we must learn to respect it as we respect ourselves.

This line of thought was shared by pre-colonial peoples on many continents. In 1854 Chief Seathl (Seattle), said the following to a treacherous United States government that had continually broken its promises to his people over land rights and peaceful co-existence:

> If men spit upon the ground they spit upon themselves. This we know, that the Earth does not belong to man, man belongs to the Earth. . . . Man did not weave the web of life, he is merely a strand in it. Whatever he does to the web, he does to himself (quoted in Lovelock 1988, p. 64).

Significantly, Church (1988) reveals that much of the original text of this speech has been misleadingly altered for quasi-religious reasons. Seathl was pessimistic about the rights of his people and the protective powers of the Great Spirit. Indeed, his kind of philosophy is difficult to put into practice in an interdependent world. It is the tension between what ought to be and what is that drives modern environmentalism into new political realms.

The manipulative mode

The manipulative mode interprets the pattern of creation somewhat differently. God created human beings first, then the Earth. The task of humans is to transform the untutored Earth into a 'designed garden' whereby nature and society are both improved. Through irrigation for example, the harshness and marginality of desert ecosystems are softened, and people can be enabled to create wealth out of the new resources provided. By removing wild forest and replanting with an appropriate mix of species, the loss of forest wildlife habitat can be transformed into a different wildlife complex. Wealth *and* beauty can be enjoyed in the new, human-engineered, forest. It is mostly a matter of careful forethought and sound ecological design: intervention through manipulation is supposed to make everyone better off.

This view, first espoused by the radical Hebrew and Greek traditions, was eagerly adopted by the Elizabethans, in the light of the first glimmers of the bright new age of science and technology. It has its modern counterpart in proposals for sustainable utilization, which will be discussed below. The interventionist mode is the cake-and-eat-it philosophy: carefully handled, nature can provide – and be improved upon so that it can provide – even more.

Modern environmentalist conceptions

These two fundamental modes within environmentalism translate into a slightly more complicated picture today. This is illustrated in Table 4.1, which is again an amalgam of ideas drawn from the writings of the authors already cited. The manipulative mode is given the jargon title of technocentrism and the nurturing mode that of ecocentrism. Technocentrism finds allies in Galbraith's idea of the technostructure and Roszak's conceptualization of techno-

Table 4.1 European perspectives on environmental politics and resource management: contemporary trends in environmentalism.

Ecocentrism		Technocentrism	
Gaianism	*Communalism*	*Accommodation*	*Intervention*
Faith in the rights of nature and of the essential need for co-evolution of human and natural ethics	Faith in the co-operative capabilities of societies to establish self-reliant communities based on renewable resource use and appropriate technologies	Faith in the adaptability of institutions and approaches to assessment and evaluation to accommodate to environmental demands	Faith in the application of science, market forces, and managerial ingenuity
'Green' supporters; radical philosophers	Radical socialists; committed youth; radical-liberal politicians; intellectual environmentalists	Middle-ranking executives; environmental scientists; white-collar trade unions; liberal-socialist politicans	Business and finance managers; skilled workers; self-employed; right-wing politicians; career-focused youth
0.1–3% of various opinion surveys	5–10% of various opinion surveys	55–70% of various opinion surveys	10–35% of various opinion surveys
Demand for redistribution of power towards a decentralized, federated economy with more emphasis on informal economic and social transactions and the pursuit of participatory justice		Belief in the retention of the status quo in the existing structure of political power, but a demand for more responsiveness and accountability in political, regulatory, planning, and educational institutions	

cracy. Both writers visualize wholly contrived managerial systems where nothing is designed to be left to chance. Ecocentrism equates with greenness, the heartland of being green. Environmentalism seeks to embrace both worldviews: indeed it is the constant interaction between these positions that gives environmentalism its special dynamic qualities. It is on this crucial point that environmentalism can be distinguished from greenness.

Each main strand of thought is further subdivided into two credos or patterns of faith that drive an internally consistent structure of beliefs. A given individual may not hold a pure 'column' of any faith. But people tend to fit into particular patterns of worldview at various times in their lives. Table 4.1, therefore, portrays ideal types of pure versions, rarely found in actuality.

Interventionism

On the far right of technocentrism is a pattern of beliefs supporting the limitless capacity of people, when freed to seek their full potential, to exploit the Earth to improve public wellbeing and transform ecosystems. The collection of essays edited by Simon & Kahn (1984) best exemplifies this philosophy. Written by businessmen and academics who support the unfettering of governmental controls on resource development and individual enterprise, these essays propound the thesis

> that the nature of the physical world permits continued improvement in humankind's economic lot in the long run, indefinitely. . . . Of course, there are always newly arising local problems [but] the resilience in a well-functioning economic and social system enables us to overcome such problems, and the solutions usually leave us better off than if the problem had never arisen (Simon & Kahn 1984, p. 3).

The blots to progress are caused by

> the view that resource and environmental trends point towards deterioration rather than towards improvement, that there are physical limits that will increasingly act as a brake upon progress, and that nuclear energy is more dangerous than energy from other sources. These views lead to calls for subsidies and price controls, as well as government ownership and management of resource production, and government allocation of the resources that are produced. . . . We wish that there were grounds to believe that a shift in thinking will take place on these matters, but we do not find basis for firm hope (Simon & Kahn 1984, p. 4).

Exponents of this view tend to be found among the core of the capitalist class – the business and finance managers of major corporations and institutions, skilled workers and the high tech end of the trade union movement, the self-employed generally, conservative politicians, and the new breed of career-orientated youth, who see their future in business management or the leading professions of law, accountancy, and engineering. Research based on cross-national public surveys suggests that about one quarter of the population subscribes to this view, with an overall range of 10–35 per cent (Milbrath 1984, pp. 55–6). Milbrath defines these people as either 'The Rearguard' or 'The Establishment', making little distinction between the two.

Interventionists still essentially control the levers of political and economic power in all countries, totalitarian, socialist, or capitalist. Those responsible for governing tend to have a deep faith in the ability of humanly devised systems to conquer adversity. No one in power likes to accept limits on growth or an incapacity to advance the lot of all people. Interventionists may be a numerical minority but, in terms of effective influence, they remain very powerful.

It is important to realize that interventionists, as with all groups represented in Table 4.1, see themselves as environmentalists. This is particularly relevant

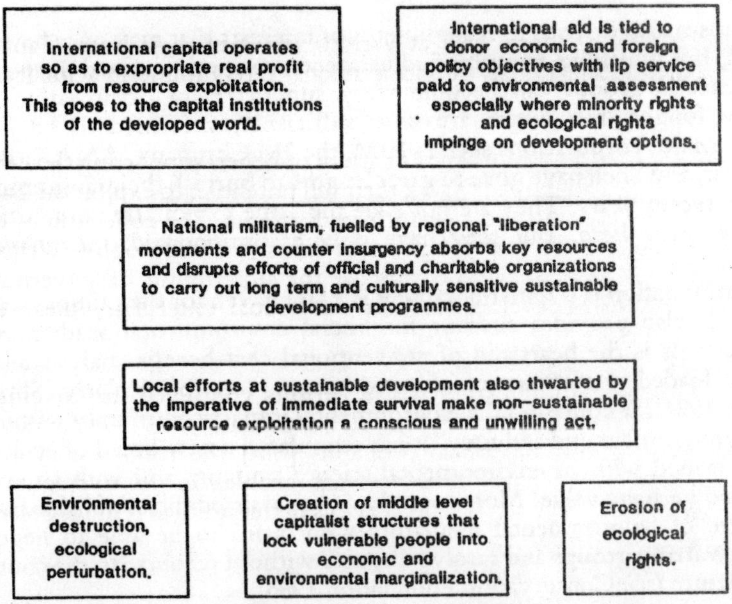

Figure 4.2 The causes of failure for sustainable development.

for the distinction between greenness and environmentalism. Interventionists believe that they can upgrade the quality of existence for all the world's people so long as the right entrepreneurial conditions hold. The quality of life is just as important for them as for the green advocate. The difference lies in the emphasis given to the meaning of that term and the method of achieving the objective. Interventionists see environmental considerations as *incidental* to economic and social advance; green proponents see such considerations as central to their concerns and as the prime objective. Moreover, green advocates fundamentally reject that it is possible to survive through interventionist practices: the Earth cannot absorb the effects of development and people will rebel through 'creative disobedience'. Environmentalism has, as yet, been unable to confront these powerful contradictions.

Accommodation

The accommodative column is a comfortable position. This is the arena of modest reform, tinkering at the margins, adjusting to the demands of environmental groups. Accommodation is a survival strategy. It is designed to retain the status quo of power and influence without giving too much away. It is prepared, if necessary, to make concessions, often quite significant ones, to reflect the changing public mood, and bend with political realities.

Accommodation is also visualized as being socially responsible. Accommodators adjust because they wish to show themselves as sensitive to social and environmental concerns. This is good neighbour politics – the ability to extend a hand to grasp a potential opponent, yet remain in control. Accommodation is also an exercise in corporatism, in the sense that accommodation provides a means whereby powerful institutions maintain their grip on the order of things.

Corporatism also extends to realignments of interests that may once have been opposed, but which, as a result of adjustments, can seek a united purpose. So environmental groups can become wary supporters of organizations they formerly fought, but which are now sufficiently reformed to be judged friendly. Major corporations such as IBM, the 3M Company, AMAX, British Petroleum, and Shell have gone to great lengths to burnish their environmental image in recent years. They are not *quite* the same exploitative organizations that they once were. But they have only accommodated; they have not reformed.

Accommodation is popular because it is a safe haven for the cautious and the anxious. It also provides succour for liberal environmental academics and consultants. It is the heartland of conventional cost-benefit analysis and the ethically loaded variations of that technique (see particularly Shrader–Frechette 1984). It nourishes the environmental impact community within and outside government and industry. It has stimulated a new breed of ecological planner, armed with an environmental science training, and with an eye for beauty and heritage value. More recently, it has assisted the establishment of a new cadre of environmental mediators who claim to be able to negotiate between warring groups and resolve disputes without recourse to the courts or the legislature (see Conservation Foundation 1986).

Hence accommodation is the whirlpool of contemporary environmentalism into which much intellectual debris is sucked. It is the preferred position for most middle-ranking executives and administrative officials. It is the choice of the majority of white-collar unions and the bulk of service professionals (in education, medicine, law, and planning). It is the convenient location for the liberal–socialist group of party politicians anxious to capture the green vote without alienating the establishment. It is a comfortable arena for the aid and development professionals and advisers who wish to see a Third World sufficiently prosperous as not to drain the aid budget or become overwhelmed by regional militarism, yet not so prosperous as to threaten trading relationships and northern hemispheric industrial survival.

Accommodation accounts for the environmental worldview of about half the populations of the developed nations. By definition, however, accommodation is a moving target. The techniques of cost-benefit analysis and project appraisal which are used in this perspective, not to mention pollution abatement and mediation–compensation strategies, have evolved almost out of all recognition from their origins in the late 1950s and early 1960s. Accommodation is the good part of the paradox within which environmentalism appears successful (see OECD 1985).

But accommodation is still a manipulative and technocentric position. It survives because it has led to superficially attractive reforms – not just in scientific methodology but also in institutional change. This is especially true of a growing responsiveness to the calls for more openness, participation, and accountability. Within the British constitutional system, for instance, statutory openness can be found in Canada, Australia, and New Zealand. While a Freedom of Information Act is still a long way off in the United Kingdom, some form of improved access to environmental information is now on the horizon (Freedom of Information Campaign 1986). Regulatory agencies, once excessively, indeed paranoically, secretive are now emerging from the shadows. Again, they still have a long way to go before they satisfy the demands of special interest groups, but right-to-know legislation in pollution controls and especially health and safety is now a central tenet of the European Community and a growing force in North America and Australasia (Baram 1986). This should have an important impact on employer–employee relations, and a statutory effect on regulatory styles. Environmental regulation is likely to become more participatory and mediative. This may slow down the process of regulation, but vastly improve its influence and effectiveness (Otway & Peltu 1985, J. C. Consultants 1986).

Accommodation is also appearing in a new and fascinating form: in the interconnection between green consumerism and green capitalism. Green consumerism reflects the rise in the environmentally conscious purchaser, who feels more comfortable buying products that are either inherently healthy or not painful or damaging to animals or ecosystems in their manufacture. Thus tropical hardwood companies now extol the virtues of selling timber from second growth, well managed plantations that encourage local employment. Hamburger chains strive to rid themselves of the tropical forest–cattle ranch connection and also produce CFC-free foam containers. Some cosmetic companies now produce natural products which are not tested on animals (see Elkington 1988).

Similarly, companies are seeking new products that are environmentally acceptable, they make profits on new environmental pollution abatement technologies, and they are investing in old buildings on derelict land to make them habitable and profitable. Green capitalism is a dreadful contradiction in terms, but it survives in these days of accommodation (see Elkington & Burke 1987).

Communalism

Nevertheless, from the ecocentric viewpoint, accommodation is still an insufficient response, and at worst a false strategy. Ecocentrists look for a different kind and character of society and political relationships. They are inherently radical and reformist, despite their conservative traditions concerning nature. They tend to see in the nature metaphor a symbol of a new communalism, based on federated political structures, economically self-contained communities, and much more effective collective and individual power at the level of the household.

The concept of household is central to the ecocentric line. It means a cohesive unit of activity, involving social, economic, and psychological transactions which merge around the application of collective self-sufficiency. In the communalist mode economic relationships are intimately connected with social relationships and feelings of belonging, sharing, caring, and surviving. There is now a vast range of literature on this topic, much of which is ephemeral, abstract, and, as yet, politically naïve. Ekins (1986) has put together a collection of essays that encapsulates the mood of this movement, yet reveals its intellectual poverty and practical weaknesses.

Communalism extends from the anarchist traditions of the last century, notably the writings of Kropotkin, Morris, and Ebenezer Howard (see Miller 1984 for a useful review; and Galtung 1986 for a more contemporary perspective). Its modern champion was Fritz Schumacher (1973), the original source for a variety of movements usually termed the new economics. Communalism addresses established socialist principles of sharing and caring. Its adherents look to reforms in basic minimum wages, the provision of essential social services through both communal and state-run enterprises, changing the status of women and other, erstwhile minority groups (see, notably, Caldecott & Leyland 1983, Merchant 1983), and in the scope for the informal economy. The last is an amalgam of the 'black' economy where transactions are paid in cash but not taxed, and the 'barter' economy where services and commodities are produced in the household for domestic consumption.

Communalism feeds on idealism, in faith in the inherent co-operative character of humankind, and in the ability of co-operative people to realize that they can achieve their ends more safely and expediently through co-operation rather than conflict. They are encouraged by the lively debate currently emerging amongst geneticists, namely that genetic selection is advanced more through some forms of interactive co-operation than through pure selfishness. They are also encouraged by the modest, but not insignificant, flowering of networks of community organizations, established on a self-help basis, which are struggling to demonstrate by example that this ethos is both practicable and enjoyable (Robertson 1985, Dauncey 1986). They are further excited by the talk, now fashionable within such establishment circles as the OECD, of community investment banks and local economic initiatives, through which small entrepreneurs may obtain start-up funds and advice to enable them to undertake constructive economic activity in their local communities.

These are very early days; but the tide of opinion is shifting from accommodation towards communalism. The gap between the two positions is by no means as large as it was. Widespread youth unemployment and vast regional disparities in social wellbeing in a number of countries and economic opportunity have concentrated even the more conservative economic minds on means of stimulating local economic initiatives. Part of the wider repercussions of the green movement is the linkage between economic revival and environmental rehabilitation. Where environmentalism may make its most important contribution is in developing a credible programme of economic and political reform which releases individuals from the fetters of deprivation centring on degraded physical environments, hopelessley inadequate social service provision, poor job prospects in the conventionally defined sense of employment, and declining personal health. Still to be grasped is the possibility of operationalizing a renaissance of spirit, mind, body, community, and environment. This is both the charter of the communalists and the penumbra which the accommodators are beginning to explore. The challenge for environmentalism is to mobilize a coalition of accommodation and communalism. Yet this is precisely the arena where the awkward paradox of success and failure is most evident.

Gaianism

Gaianism is a relatively new word in the environmental dictionary. Gaia was the Greek name for the Goddess of the Earth, the nurturing mother figure from whom all sustenance derived. Gaia's mythical daughter was Themis, the Goddess of Justice. The Greeks equated justice with the retribution of environmental systems when abused. As the American historian Donald Hughes put it:

> It is because the Earth has her own law, a natural law in the original sense of these words, deeper than human enactments and beyond repeal. . . . Who treats her well receives blessings; who treats her ill suffers privation, for she gives with evenhanded measure. Earth forgives but only to a certain point, only until the balance tips and then it is too late (Hughes 1983, p. 56).

Obedience to natural laws and evenhandedness of retribution are central ideas in the Greek interpretation of Gaia.

But there is another view of Gaia. This is ascribed to the geochemist James Lovelock (1979, 1988) who seeks to put the notion of a cybernetic biosphere on a scientific footing. Lovelock believes that the geochemistry of the Earth is manipulated by living organisms so that natural perturbations in atmospheric, aquatic, and marine chemistry are smoothed out to retain, at least for the geochemically forseeable future, a complicated, steady state of life-sustaining conditions. Gaia is the label attached to what is essentially a randomly occurring, multi-centred homeostatic process of quite extraordinary scale and complexity. It is as though life on Earth has sustained the very conditions necessary for its survival for nearly 300 million years. This contrasts markedly with the view, held by evolutionary biologists, that life varies by separate random processes, and that the physical environment and biological development are independent of each other.

The Lovelock thesis attributes no intelligence to Gaia. Gaia is simply a biochemical mechanism that happens to have resulted in the biosphere as we know it. The Gaian message is profound. Gaia is transcendental to human beings: Gaia would occur whether human kind survived or not: Gaia has neither conscience nor compassion. But Gaia is not indestructible. Lovelock chides the interventionists:

> Each time we significantly alter part of some natural process of regulation or introduce some new source of energy or information, we are increasing the probability that one of these changes will weaken the stability of the entire system, by cutting down the variety of response (Lovelock 1979, p. 145).

Lovelock is also aware that humankind has the capacity to alter the homeostatic relationships that constitute Gaia, notably by disturbing the great tropical ecosystems of the life-rich estuaries, the savannah margins, and the tropical forests. He is also concerned that high tech agriculture could damage soil ecosystems to the point where remedial treatment might become prohibitively costly.

Lovelock's main message is, however, an optimistic one. He visualizes humankind as a collectively adaptable organism with an immense capacity for learning. If we are truly Gaian, we can pull back from the brink and devise appropriate responses to protect us from our folly. People, in society, are part of the homeostatic response that safeguards Gaia.

The true Gaianist is therefore caught in an interesting dilemma. On the one hand there is the temptation to cry doom and bemoan the destruction of crucial ecosystems and with them their immense capacity for stabilization. This tends to be the position adopted both by scientists (see especially Myers 1985) and thoughtful institutional analysts (see especially World Resources Institute

1986). They genuinely believe that humankind has a choice between destroying Gaian systems or becoming consciously part of Lovelock's democratic entity.

On the other hand, the true Gaianist actually sees people as contributing integrally to the intrinsic processes of homeostasis that are Gaia – so there is no special need to worry. The dilemma can be resolved by being active in collectively ordained causes. For example, the Chipko Andolan 'hug-the-trees' movements of Nepal and northern India are sometimes cited as a form of Gaian homeostasis: village women stopped their menfolk from removing the trees by non-violent protest. The movement has spread throughout the Himalayas with important repercussions for community politics and official forestry policy. Chipko has by no means halted Himalayan deforestation, but it has at least alerted world opinion to a very serious problem. Chipko also linked ecology to politics. As Bandyopadhyay & Shiva (1987) observe in a thoughtful article;

> Chipko's search for a strategy of survival has global implications. What Chipko is trying to conserve is not merely local forest resources but the entire life-support system, and with it the option for human survival. Ghandi's mobilisation for a new society, where neither man nor nature is exploited and destroyed, was the beginning of this cultural response to the threat to human survival. Chipko's agenda is the carrying forward of that vision against the heavier onus of the contemporary crisis (Bandyopadhyay & Shiva 1987, p. 34).

The difficulty is how to ensure that essential adjustments can be made in time without unnecessary human and ecological suffering.

Gaianism manifests itself in other ways. The mystics and the romantics seek to exploit it as a metaphor for a co-evolutionary natural ethic where human beings are no different from any other sentient being. This view establishes a solidarity not just with all organisms presently alive, but with all living matter, past, present, and future. One manifestation of this view, predating the Gaia thesis is the deep ecology movement (see especially Fox 1984, Devall & Sessons 1985, Tobias 1985), emerging from the writings of philosophers in Scandinavia, California, and Australia. According to its adherents, deep ecology provides a justification of the sacredness and quality of life and demotes the value of humans to an equivalence with all other forms of life. Thus, self-realization becomes an exercise in recognizing inherent solidarity with the complex of life forces that is nature. Deep ecology is not a widely accepted movement, but it does exhibit itself in elements of the Animal Rights Campaign and the anti-vivisection movement. In politically pragmatic terms, however, this facet of Gaianism is unlikely to become a major force (see also Grey 1986).

Gaianism also manifests itself in radical green movements, notably the anarchist, the anti-corporatist, and those rejecting so-called material values. This is a curious political arena, for it links greenness to socialistic and communistic beliefs which exhibit tendencies toward collectively inspired authoritarianism and state-organized communalism. The most extreme proponents of this view believe in radical action to break down corporatist structures and destroy the link between the state and the military–industrial complex. The less extreme concentrate on countering oppression of all minorities, including nature, which they visualize as innocence betrayed. They tend to support and initiate citizens' action groups aimed at opening the minds of the bourgeoisie to their condition and to the predicament of the globe. They tend to be oppressively self-righteous and often fail to advance their cause.

Overview

Figure 4.1 and Table 4.1 present a picture of contradictions and tensions dominated by a failure to agree over cause, symptom, and action. The peculiar difficulty facing the environmental movement is how radical to become in approaching the matter of reform. The accommodators believe they can push towards the centre and achieve limited objectives in the short term, while awaiting structural changes in the economy. They visualize a fundamental shift towards a more information-based style of economic transaction and more robot-run manufacturing and clerical work, releasing vast numbers of people into constructive self-employment and job sharing, with a more diffuse distinction between leisure, social activity, and work. They fudge the issue of how to persuade any government to adopt the central credo of their position, namely a basic (untaxed) income for all (equivalent to a negative income tax), and a social tax on land based on value in use.

The radicals within the ecocentric tradition are impatient for a transfer of power and for means to restrict the exploitative and oppressive activities of many national and multinational corporations, the military, and the petty despots who syphon off vast amounts of wealth from the really poor – always at the expense of Gaian stabilizing forces. The radicals survey a progressive wasteland which cannot be salvaged because it is still strewn with mines – the detonators of which are triggered by the corporatist structures and interventionist values they love to hate.

Interpreting and carrying out sustainable development

The refuge of the environmentally perplexed is sustainable development, namely wealth creation based on renewability and replenishment rather than exploitation. The trouble is that this is essentially a contradiction in terms for a modern capitalist culture. Development that does not destabilize environmental gyroscopes cannot produce real improvements in standard of living for a growing population without massive redistribution of wealth and power. At the heart of the sustainability debate, therefore, is the essence of global communalism, namely an ecological basis to economies and local self-reliance. That means not just a comprehensive shift in power, but also wholesale changes in institutional alignments. Agriculture and forestry ministries, grazing and irrigation departments, minerals, regulators and energy suppliers – all would need to be transformed if they were to meet the demands of sustainability. This was a topic that, understandably yet frustratingly, the Brundtland Commission (1987) failed to address.

Sustainability is becoming accepted as the mediating term that bridges the gap between developers and environmentalists. Its beguiling simplicity and apparently self-evident meaning have obscured its inherent ambiguity. A distinction must be made between sustainability and sustainable utilization.

Sustainable utilization is the official term adopted by the international conservation community (IUCN 1980) to denote a rate of resource take which equals the rate of renewal, restoration, or replenishment. In agriculture, the farmer derives fertility from the soil equal to the ability of the soil to supply nutrition. The fisherman and forester draw from sea or forest resources equivalent to their refurbishment. It is possible to increase the natural yield by manipulation, but the basic principles of replenishable extraction still apply. Implicit in this narrow definition are three precepts:

(a) *Knowability*. The amount, rate of renewal, and other characteristics of replenishment are understood and calculable.
(b) *Homeostasis*. Renewable resource systems operate broadly around equilibria, or can be manipulated to approximate steady states following human intervention: homeostasis is believed to be a preferential state of nature.
(c) *Ecosophy*. The act of drawing upon a renewable resource, even below the threshold of allowable take, has no implication for the capacity for survival of life forms that operate within the resource complex that is being exploited.

None of these precepts is unchallengeable. Knowledge is imperfect, and few ecological models are particularly helpful in pinpointing the correct point of optimal take. Buffer margins to allow for uncertainty and to protect the interests of other ecological values have to be imposed. Natural systems do not normally operate around short-term equilibria, so artificial manipulation of equilibria will have to be conducted with caution.

Sustainability is a much broader phenomenon, embracing ethical norms within the Gaianist tradition, including taking into account the rights of future generations of all living matter. Sustainability deals with structures and arrangements that ensure that sustainable utilization actually takes place. Sustainability is, therefore, a reformist notion in the radical tradition of opening up institutions of economic investment and resource development to a far greater sense of Gaian accountability.

Sustainability is based on a neo-Marxist ecological theory of value. This line of reasoning is well explained by Redclift (1984, pp. 7–19). Marx believed that labour was the unit of profitability, and that capitalists exploited labour by drawing profit beyond what was paid to labour in the process of resource transformation. The modern interpretation of this is that an element of profit now comes from the uncompensated exploitation of marginalized peoples and environmental systems, neither of which have the ability, or even the understanding of their predicament to improve their lot.

Figure 4.2 illustrates the pressures for non-sustainable resource draw and the self-perpetuating imperatives of international capitalism, international aid, national militarism and counter insurgency, and cultural conflict. This combination of forces renders impossible any realistic hope of achieving sustainable ultilization in much of the Third World.

Making this more specific, Figure 4.3 illustrates the links between capital movement, the seed–fertilizer–pesticide axis of major multinational chemical corporations, desertification, and indebtedness, all of which lead to marginalization of the vulnerable who intensify the degradation of the soil in a desperate effort to survive. A total of $10 000 million has been spent by the major chemical companies in acquisition of seed manufacturers and distributors

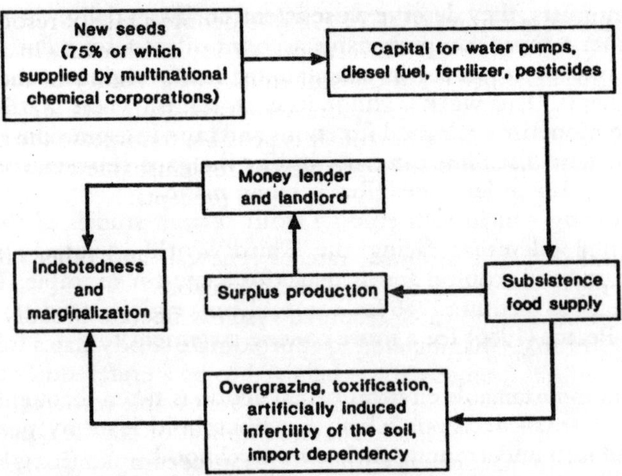

Figure 4.3 Environmental exploitation in Third World agriculture.

(Goldsmith & Hildyard 1988, p. 213). There is now a very close link between high yielding varieties and all forms of chemical application, from fertilizers to fungicides; the one cannot exist without the other. Third World food production, aided and abetted by international chemical capital, is locked into expensive chemical dependency. This destroys the viability of marginal farmers and increases the power of the landowner and debt-collector. The solution requires co-ordination between agencies responsible for agriculture, soil management, forestry, aid schemes, and industrial regeneration, integrating their efforts at the regional and local level through extensive use of **animateurs**, locally based extension agents acting as catalysts for local self-help schemes.

Blaikie (1984) provides a disturbing case study from Nepal of the processes through which exploited peasants become politically and economically marginalized to the point where, even to survive, they must destroy the soil resource that is their livelihood. Oppression, neglect, and militarism combine to deny access to basic resources for existence. Indebtedness increases so the poor become impoverished from both ends: they cannot make enough to survive and are increasingly dependent on those who exploit them to remain alive. They are the truly vulnerable; the hazard lies in social institutions, not acts of nature. Reform cannot be achieved by natural manipulation, by means of soil conservation or replanting. Reform must be achieved through the transformation of social relationships. This is the true nature of the ecological crisis, a general point which is further developed across a range of case studies by Blaikie & Brookfield (1987).

Sustainability and ecological economics

Sustainable development may still be a distant dream, but it is at least spawning a new breed of economist. Since economists command attention, even ecological economists, they deserve close attention. Essentially, they are trying to piece together a more comprehensive account of resource value, based on the ecological role of resources in maintaining local, regional, and global life-support systems. This work is still in its early stages. There is still too strong a tendency to monetize ecological functions and human aspirations: it is doubtful whether the new discipline can truly bridge the gaps between culture ecology and economy. But at least the effort is being made.

The following conclusions emerge from current studies of the three great environmental dilemmas facing the Third World – namely soil erosion, desertification, and tropical forest depletion (see, for example, Brown 1986, World Resources Institute 1986 for comprehensive global studies; and Warford (1986) and Pearce (1986) for a more concise statement):

(a) Most non-sustainable environmental action is taken through the accumulation of small decisions taken at a household level by people who are trapped into undermining their own livelihood.
(b) Such actions are essentially uncontrollable unless the structural conditions inducing poverty and desperation are altered.
(c) Middlemen who exploit the desperation of the poverty-stricken and the landless exploit any propensity to accumulate capital by expropriating surplus through extortion and debt-creation.
(d) Militarism, and especially civil war, now commonplace in poor Third World countries, strike against any successful approach to sustainable development by drawing capital into arms, removing able-bodied labour into warfare, and physically destroying the vital infrastructure of rural development. It is unlikely that any long-term agricultural programme built on sustainability principles can remain unscathed.
(e) International aid is not geared to sustainable development at the microscale. Aid is linked to established political structures and, to a degree, is dependent on recipient government support. Recent studies of World Bank aid, even programmes with an allegedly specific environmental component, indicate socially divisive and environmentally destructive outcomes.

Criticism of the World Bank has been most furiously developed by the *Ecologist*, co-ordinated by its editor Edward Goldsmith. Five issues of the magazine have been devoted to this cause (vol. 14, nos 5 and 6 (1984); vol. 15, nos 1 and 2 (1985); vol. 16, nos 2 and 3 (1986); vol. 17 no. 7 (1987). Goldsmith sums up his views:

> Environmental degradation in the Third World is the . . . inevitable consequence of present development policies, and Third World people are poor, because they have been impoverished by previous development, because they have been robbed by developers of their means of sustenance, and are now condemned to scratching an ever more marginal existence from land that ever more closely resembles the surface of the moon (Goldsmith 1985, p. 7).

This reads like unsubstantiated rhetoric. But World Bank ecologist Robert Goodland (quoted in World Resources Institute 1986, p. 199) admits that World Bank-financed major water projects have caused massive damage and that there is an acute shortage of environmental specialists on the Bank staff. The United States Treasury, also quoted in the World Resource Institute (1986), suspects

> that the problems encountered in the environmental aspects of projects may be an instance of the over emphasis [on quantity rather than quality]. If environmental considerations threaten expeditious project processing, the environment is assigned low priority and is left to be dealt with later.

The World Resources Institute report indicates that progress is being made, but at an agonizingly slow pace.

Despite numerous studies of the need for global sustainability, and even more reports on global ecological disturbance, no serious attempts at institutional reform are in progress. Various studies and analyses paint a sombre picture (see annual volumes of *World Resources*, published by the World Resources Institute, the *State of the World*, published by Worldwatch, and the *Earth Report*, published by Mitchell Beazley). Up to two thirds of the world's population are affected by the impoverishment of natural wealth and wellbeing caused by non-sustainable resource exploitation. Those in the most vulnerable positions, deprived even of the basic necessities of food, shelter, healthcare, and education, are steadily losing their entitlement. In their desperate struggle to survive, they are destroying the only asset that can give them life – the earth system that sustains them, even if at a sub-marginal level. Nearly all Third World countries squander their natural resource assets by selling them at a loss and buying expensive imported goods. Repetto (1986) calculates that the Indonesian government has 'lost' $90 billion of its potential gross national product by selling itself short in this way. Government attempts at holding down fuel and food prices actually add to the exploitation and wasteful utilization of these desperately precious resources.

The problem lies not just with despots and international capitalism. Nor does land reform or the opening of the market in peasant agriculture provide an accessible solution. The solution, essentially unattainable, lies in unfettering the myriad social, political, and economic restrictions that perpetuate a peculiar brand of environmental capitalism in the Third World, changing cost/benefit analyses from mechanistic measures to culture-linked sustainability audits, and linking the fundamental principles of basic needs entitlement to locally based ecodevelopment. In short, sustainable development is essentially about the provision of basic needs for all in a form that is ecologically and culturally acceptable (Fig. 4.4).

Economists such as Pearce (1988) and Warford (1986) advocate the introduction of national resource audits (a taking into account of the value of their natural assets to a nation), based on sustainable utilization principles. They also favour the imposition of depletion taxes to ensure that there are no unnecessary distortions in the prices of key commodities essential to basic needs survival. Such dramatically interventionist measures would be administratively burdensome if operated from the centre. The aim must be to adopt the communalist advocacy of collective self-reliance. Here is where more decentralized, federated structures, based on clusters of 'households' (each with their own entitlement of natural resources and incentives to create replenishable stocks of

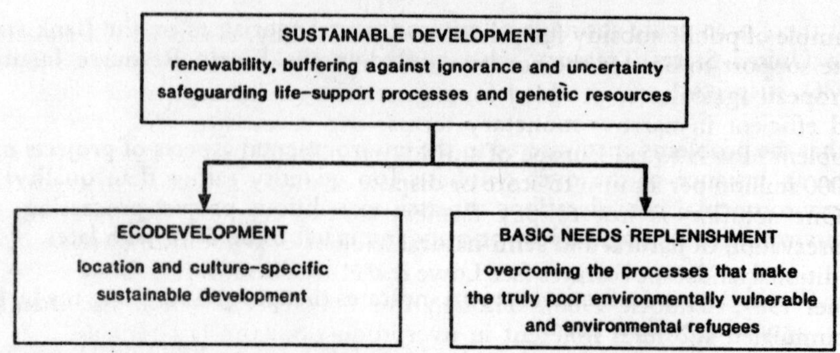

Figure 4.4 Sustainable development.

new resources), could provide a solution. Clearly, any such action would have to come about by small stages, operating on the basis of informed advice and creative experimentation. One can be hopeful, but not excessively optimistic.

Sustainability and European environmentalism

Sustainability has been gratefully grasped by accommodators in the post-industrializing world. Environmentalism at last can be visualized as a force for good, a device for having the wealth-creating cake and eating it. Sir Arthur Norman, Chairman of the De La Rue conglomerate and a powerful industrial voice in green politics defines sustainability as

> the bringing about of a *productive partnership* between conservation and development interests, to promote policies and programmes which encourage *profitable and sustainable* economic activity whilst ensuring that any environmental effects are kept within limits *acceptable* to the population at large (Norman 1985, p. 5; emphasis added).

This line of reasoning has been adopted by William Waldegrave (1987), formerly the United Kingdom Minister of State at the Department of the Environment, a man highly regarded for his green beliefs. Waldegrave recognizes that green politics are an electoral asset; he also believes that environmental repair is a wealth-producing and job-creating activity. He looks with envy at the West German economy, where over DM10 billion have been invested in automobile exhaust control technology and the reduction of sulphurous and nitrogenous emissions from fossil fuel-burning plants. Further, it is estimated that over 300 000 new jobs have been created by France's environmental protection programme since 1980, and OECD economists (OECD 1985) see as much as 2 per cent of gross domestic product being generated by creative environmental management.

Naturally, the greens spurn all those optimistic statistics as ludicrous. Paying for the consequences of non-sustainable exploitation can hardly be seriously considered as a *benefit*. Nevertheless, this is the contemporary rhetoric of Western interpretations of environmental sustainability. It is yet another example of public subsidy for private and public folly. A further illustration of state support to bolster the consequences of state-encouraged private folly is European agriculture, prodded for a generation to become more productive and efficient in narrow monetary terms, and succeeding all too well. The problem now is to rid Europe of surplus agricultural production costing over £7000 million per annum to store or dispose of through subsidized exports.

One solution is not to pay farmers to overproduce, but to invest in conservation of natural and semi-natural habitats or woodland, or to maintain traditional landscape features (see Lowe *et al.* 1986, Countryside Policy Review Panel 1987, Baldock 1988). Though this may cost a nation less than the accumulated subsidies inherent in overproduction, the fact remains that the state is intervening to protect an industry that has expanded beyond its effective rôle in a modern economy. Landscape gardening is unlikely to be an attractive proposition for the progressive farmer, yet it is improbable that the European Community can sustain its present commitment to agricultural subsidies for another decade. Sustainability will become a device for state support for an industry that is out of environmental economic balance. The better solution is to diversify farm income away from agricultural production into recreation, education, and local servicing as well as into non-farm entrepreneurial enterprise. This may be possible in a rapidly changing economic scene where computer-aided self-employed business ought to be able to flourish. But it will require a fairly dramatic adjustment in agricultural thinking, and even more imagination in the guidance of settlement and buildings in the countryside.

Also relevant to the future of European environmentalism will be the steady coalescence of a number of radical causes, ranging from consumer protection, public health and healthy foods, feminism and ecofeminism, peace movements and associated civil disobedience, animal rights, and a more general interest in cosmic solidarity. This coalescence will not take place in an organized fashion, and many disparities will remain. But it is possible that a more coherent environmentalism will emerge on the radical left. In the light of the relative success of the accommodation–sustainability centristic coalition, this may emerge as the future force of radical greenness. It is not clear how far such a coalescence will result in an effective green movement. To date the signs are not good; far too many greens prefer to be true to their own pure positions (the fundamentalists) than sacrifice principles in order to gain power (the realists). In all probability greenness will remain on the radical fringe. This will especially be the case if the centrists organize an effective programme of reform focusing on:

(a) greater use of environmental appraisal before any policy, programme, or project is contemplated;
(b) the adoption of environmental audits, making use of the flows of natural systems as measures of input and output;
(c) more attention to decision analysis through which uncertainties can be handled by probability arrays, and weights can be attached to particular outcomes;
(d) introduction of more formal mechanisms for involving relevant and representative sections of the public as part of a process of environmental mediation and bargaining;

(e) more consistent application of environmental trade-off deals through which individuals and communities are compensated for particular losses of amenity and local environmental benefits in such a way that, in the long run, they are not demonstrably worse off in environmental wellbeing.

These measures are already beginning to be instituted. More exciting and permanent developments are in the offing. But their success will not placate the green fringe. The bias of environmentalism will shift from the right of centre (see Fig. 4.2) to the left of centre over the next few years, but peace will never reign. For environmentalism to survive, it must always experience an internal as well as an external struggle.

References

Baldock, D. (ed.) 1988. *Removing land from agriculture: implications for farming and the environment*. London: Council for the Protection of Rural England.

Bandyopadhyay, B. & V. Shiva 1987. Chipko: rekindling India's forest culture. *Ecologist* **17**, 26–34.

Baram, M. 1986. *Risk communication and the law*. Boston: Boston University, Department of Law and Public Health.

Bennett, J. W. 1976. *The ecological transition: cultural anthropology and human adaptation*. Oxford: Pergamon Press.

Black, J. 1970. *The dominion of man: the search for ecological responsibiliy*. Edinburgh: John Black.

Blaikie, P. 1984. *The political economy of soil erosion*. Harlow, Essex: Longman.

Blaikie, P. & H. Brookfield 1987. *Land degradation and society*. London: Methuen.

Brown, L. C. 1986. *The state of the world 1985*. London and New York: Norton.

Brundtland Commission 1987. *Our common future*. Oxford: Oxford University Press.

Caldecott, L. & S. Leland (eds) 1983. *Reclaim the earth: women speak for life on earth*. London: Womens' Press.

Capra, F. 1982. *The turning point*. London: Wildwood House.

Church, C. 1988. Great Chief sends modified word. *ECOS* **4**, 40–1, also FoE, 26–28 Underwood St, London N17JQ, England.

Conservation Foundation 1986. *Resolving environmental disputes: a decade of experience*. Washington, DC: Conservation Foundation.

Cotgrove, S. 1982. *Catastrophe or cornucopia: the environment, politics and the future*. Chichester: Wiley.

Countryside Policy Review Panel 1987. *New opportunities for the countryside*. Cheltenham: Countryside Commission.

Dauncey, G. 1986. A new local economic order. In *the living economy*, Paul Ekins (ed.), 264–72. London: Routledge & Kegan Paul.

Davies, P. 1985. *Superforce: a search for a grand unified theory of nature*. London: Unwin Paperbacks.

Devall, W. & G. Sessons 1985. *Deep ecology: living as if nature mattered*. Salt Lake City: Peregrine Smith Books.

Ekins, P. (ed.) 1986. *The living economy: a new economics in the making*. London and New York: Routledge & Kegan Paul.

Elkington, J. 1988. *Green pages*. London: Hodder & Stoughton.

Ellington, J. & T. Burke 1987. *The green capitalists: industry's search for environmental excellence*. London: Gollancz.

Ensenberger, H. 1974. A critique of political ecology. *New Left Review* **84**, 3–32.

Fox, W. 1984. Deep ecology: a new philosophy for our time? *Ecologist* **14**, 194–232.

Frankel, C. 1976. The rights of nature. In *When values conflict*, L. H. Tribe, C. S. Shelling & J. Voss (eds), 93–114. New York: Wiley.
Freedom of Information Campaign 1986. *Newsletter*. 3 Endsleigh St, London WC2, England.
Galtung, J. 1986. Towards a new economics: on the theory and practice of self reliance. In *The living economy*, P. Ekins (ed.), 97–109. London: Routledge & Kegan Paul.
Glacken, C. 1967. *Traces on a Rhodian shore*. Berkeley, CA: University of California Press.
Goldsmith, E. & N. Hildyard (eds) 1988. *The Earth report: monitoring the battle for our environment*. London: Mitchell Beazley.
Grey, W. 1986. A critique of deep ecology. *Journal of Applied Philosophy* **3**, 211–15.
Hughes, D. 1983. Gaia: an ancient view of our planet. *Ecologist* **13**, 54–60.
International Union for the Conservation of Nature (IUCN) 1980. *World conservation strategy*. Geneva: IUCN.
J. C. Consultants 1986. *Risk assessment for hazardous installations*. Oxford: Pergamon.
Leiss, W. 1972. *Man's domination of nature*. New York: George Braziller.
Lovelock, J. 1979. *Gaia: a new look at life on earth*. Oxford: Oxford University Press.
Lovelock, J. 1988. Man and Gaia. In *The earth report*, E. Goldsmith & N. Hildyard (eds). London: Mitchell Beazley.
Lowe, P., G. Cox, M. MacEwen, T. O'Riordan, & M. Winter 1986. *Cultivating conflict: the politics of agriculture, forestry and conservation*. Farnborough, Hants.: Gower.
McNeely, J. & D. Pitts 1985. *Culture and conservation*. London: Croom Helm.
Merchant, C. 1983. *The death of nature: women, ecology and the scientific revolution*. New York: Harper & Row.
Milbrath, C. 1984. *Environmentalists: vanguard for a new society*. Buffalo, NY: State University of New York Press.
Miller, D. 1984. *Anarchism*. London: Dent.
Myers, N. 1985. *The Gaia atlas of planet management*. London: Pan.
Norman, A. 1985. Introduction. In *Sustainable development in our industrial economy*, 1–5. London: Centre for Environment and Economic Development.
Organization for Economic Cooperation and Development (OECD) 1985. *Environment and economics*. Paris: OECD.
O'Riordan, T. 1981. *Environmentalism*. London: Pion–Methuen.
Otway, H. & M. Peltu 1985. *Regulating industrial risks*. London: Butterworth.
Passmore, J. 1974. *Man's responsibility for nature*. London: Duckworth.
Pearce, D. W. 1986. The sustainable use of natural resources in developing countries. In *Sustainable development*, R. K. Turner (ed.), 211–38. London: Bellhaven Press.
Pepper, D. 1984. *The roots of modern environmentalism*. London: Croom Helm.
Porritt, J. 1984. *Seeing green: the politics of ecology explained*. Oxford: Basil Blackwell.
Redclift, M. 1984. *Development and the environmental crisis: red or green alternatives?* London: Methuen.
Redclift, M. 1987. *Sustainable development: exploring the contradictions*. London: Methuen.
Repetto, R. 1986. *Natural resource accounting in a resource based economy: an Indonesian case study*. New York: World Resources Institute.
Robertson, J. 1985. *Future work: jobs, self employment and leisure after the industrial age*. Farnborough, Hants.: Gower.
Sandbach, F. 1980. *Environment, ideology and policy*. Oxford: Basil Blackwell.
Schumacher, E. 1973. *Small is beautiful*. London: Blond & Briggs.
Shrader–Frechette, C. 1984. *Science policy, ethics and economic methodology*. Dordrecht, North Holland: Reidel.
Simon, J. & H. Kahn 1984. *The resourceful earth: a response to global 2000*. Oxford: Basil Blackwell.
Spretnak, C. & F. Capra 1986. *Green politics: the global promise*. Santa Fe, NM: Bear.
Tobias, M. (ed.) 1985. *Deep ecology*. San Diego, CA: Avant Books.

Tuan, Y. F. 1974. *Topophilia: a study of environmental perception, attitudes and values*. Englewood Cliffs, NJ: Prentice-Hall.
Waldegrave, W. 1987. Introduction. *ESRC Newsletter*, January, 5.
Warford, J. J. 1986. *Natural resource management and economic development*. Washington, DC: World Bank.
World Resources Institute 1986. *World resources*. New York: Basic Books.

Part III
NEW MODELS OF UNEVEN DEVELOPMENT AND REGIONAL CHANGE

Introduction

Richard Peet

Geography has long been fascinated with the specificity of place. Originally, regional specificity was conceived as arising directly from the local natural environment, with the frictions of space preserving naturally derived ways of life. Then location theory became the new focus of geographic interest in the late 1950s, replacing the study of regions. By comparison with *natural* determination, location theory explained regional differences in terms of the *social* allocation of activities in space. The years of the late 1950s also marked the transition from geography as qualitative description, to geography as quantititative analysis, explicitly based in neoclassical economic theory. This transition, or as some phrased it 'quantitative revolution', was marked by intense intellectual excitement and a sense of discovery. Geography was regaining its status as a social science which it had lost with the decline of environmental determinism in the 1920s. But as Schoenberger, Smith, and Lovering point out in Part III, shades of disillusionment began to close over the prospect of unlimited theoretical development. Why? The true test of theory is its ability to explain reality. Yet the world began to change dramatically in the late 1960s and early 1970s, in ways opaque to the standard, increasingly conventional, neoclassical location theories. Sharp and discountinuous shifts occurred in global space opposed to neoclassical theory's equilibrium view of the world. The direction taken by regional development varied greatly from that predicted by location theories emphasizing transport cost minimization. Location theory was profoundly unable to structure the analytical mind in correspondence with these, the historical and geographical dynamics of international capitalism. The result was the extensive critiques of the 1970s (e.g. Massey 1973) and subsequent attempts at theoretical replacement in the later 1970s and 1980s.

In Part III Schoenberger explores three avenues of departure from traditional location theory – recent theory stresses the role of technological change, focuses on the structure and organization of firms and, more generally, emphasizes location as an aspect of social production as a whole. This last emphasis was part of the development of a self-consciously Marxist theory of location, focused on the labour process, or more generally on the regime of capital accumulation, as a mediating theoretical layer between general theories of capitalism as an entire way of life and more specific sectoral and regional studies. In this approach, the geographically varying social relations between capital and labour are isolated for particular attention. Schoenberger concludes that the creativity of class actors makes theory construction a long, arduous, and as yet only partially completed task.

Smith also begins with a summary of several critiques of neoclassical location theory – in this case his critical emphasis lies on its ahistorical and equilibrium-centred character, its fetishization of space, and its legitimation function as a 'manual of policy for corporate executives'. For Smith, however, an important route to the new radical location theory lay through dependency theory, its derivatives, and various other theories of global production and uneven development. The results in geographic theory were the concept of spatial divisions of labour and labour theories of location, which Smith continues to find preferable to a perceived return to geographical eclecticism via empirical studies of localities. Concluding on this note, Smith outlines some future research themes which could claim the interest of theorists and empiricists alike: a focus on the production of space; an interest in questions of scale; and the making of a new regional geography.

In much of the urban and regional research literature, therefore, the political-economy perspective is now so established as almost to be taken for granted. But the perspective has been adopted more slowly in the study of agriculture and rural space (Marsden *et al.* 1986). Yet the need for such a view is enhanced by contemporary processes of the concentration of agricultural capital (growth of agribusiness), the penetration of agriculture by industrial and finance capital, and the increased adoption of high-technological solutions to agricultural crises. In such an environment, earlier emphases on locality, community, and family have proven unsatisfactory while, similarly, the standard, neoclassically based agricultural economics, with its narrowly technocratic predictive models, is likewise insufficient. Things began to change in the late 1970s and early 1980s as work using the political-economy approach sought to provide a macrostructural analysis of agricultural social relations removed from the idiographic community studies of the past (Buttell & Newby 1980). Below, Cloke reports on similar work in and around rural geography. He sceptically remarks that even much of the work employing a structural approach pays lip service to it rather than reconstructing its methods and critical concerns. Rural geographers still have problems differentiating their object of study – rurality and rural space. There are problems also with borrowing from critical rural sociology – for example, a focusing on agriculture to the detriment of other elements of rural social reproduction. Rural geographers should integrate their work with broader analyses of capital restructuring. And they need also to focus on the state as a central organ in rural policy and planning. Cloke therefore outlines a tentative research agenda aimed at producing a materialist conception of rural society based on theories of the unevenness of capitalist development, the impacts of restructuring, and the role of the state – a far cry from the rural geographies of the past.

The taken-for-granted nature of the political-economic perspective in general does not mean accepting a unilinear sequence of evolving ideas. In urban and regional research in particular, political economy has involved 'internal and sympathetic' critiques (Sayer 1985), debates over such fundamental issues as the necessity of uneven development under capitalism (Browett 1984, Smith 1986), and the growth of several new variants of political economy. Lovering contributes to this formative process by reviewing the debate over economic restructuring. In the discourse of radical social science, this term means qualitative change in the relations between the components of the capitalist economy – in spatial terms it means transformation in the economic roles played by regional societies in the global capitalist system. He traces the upsurge in interest in restructuring to the changing political context, especially the erosion of the regional industrial base of traditional socialist politics. The restructuring approach began as a critique of neoclassical location theory and proceeded through spatial versions of the Marxist theory of uneven development. Marxist theories may be criticized in terms of the degree of their removal of analytical categories from empirical categories, so that solutions become difficult to find. An analysis is needed, Lovering continues, that starts from capital accumulation but can also encompass regional and local specificity and incorporate the non-economic dimensions of life. The early restructuring literature complemented abstract theories of capital accumulation with a set of medium-range theories of the labour process, technology, the state, etc., to arrive at impacts on localities. But for Lovering, this attempt at restructuring the theory of restructuring proved unsuccessful, at least in its spatial divisions of labour form, while the original base in Marxist structuralism also lost favour. As a result early conceptions of the system as a whole were replaced by later conceptions of people living in places, involved in diverse relationships or, more generally, cultural rather than economic formations. The new openness of the restructuring approach means that quite different theoretical directions can be incorporated – for example, postmodernist and locality research. In summary, restructuring is not a theory but a package of concepts, tools, and claims. Lovering proceeds to unpack this package from a realist–localities perspective, providing a research programme combining non-reductionist methodologies with a preliminary set of analytical tools.

Likewise Corbridge, surveying the geography of development, finds increasing criticism of the economism, rationalism, and theoretical arbitrariness of Marxist theory, and signs of a post-Marxist development geography emerging which has assimilated these criticisms. Corbridge provides an introduction to some of the themes of this post-Marxist discourse. A materialist ontology, a stress on economic inequality, and an emphasis on human agents acting under conditions not of their choosing are the aspects that continue Marxism. The 'post' dimension is characterized by methodological eclecticism, anti-economism, a degree of scepticism about the labour theory of value, non-functionalist accounts of power, and a politics more open to ideas of feasible reform based on concepts of moral justice. (It should be pointed out that Corbridge's earlier (1986) critique of radical development geography has been counter-critiqued (Watts 1988, Peet 1988), and his suggestions of a move to post-Marxism (in the sense of 'beyond' or 'after') represent only one response to a continuing debate which has yet to be resolved.) Corbridge concludes with a most useful outline of the directions more empirical and theoretically diverse Marxist development studies might take – examining the various forms taken by a changing world system, especially its regional and local forms, taking seriously the constitutive roles of class, gender, and ethnicity, stressing contest, struggle, and the role of the state in the periphery, and re-examining the possibilities *and* contradictions of socialism in Third World countries.

Where does this leave the political economy of global space? In a state of rapid but diffuse growth, scattered over several academic disciplines, shaded by several political orientations, and characterized still by debate over fundamental methodological and empirical issues. There remain several different models of global development even within the structuralist perspective. For example, Graham *et al.* (1988) follow a mode of production structuralism emphasizing the changing relations of capitalist production which produce variants, termed competitive, monopoly and global capitalism, each with a typical geography of production. However, as several authors in Part III concur, the 1980s witnessed an increasing dominance of concepts and terms derived from the French Regulation School. This school of thought originated in the critique of Althusserian structuralism in the middle to late 1970s (Lipietz 1985). Different members of the School show an uneasy distancing from Althusserian Marxism. For Aglietta (1979), writing originally in the mid-1970s, social systems reproduce the determinant relations which assure their integrity and cohesion; so long as these fundamental relations are not challenged, social systems develop continuously. But crises rupture the continuous reproduction of social relations as corrective mechanisms break down.

> In that event a direct threat is posed to the reproduction of the invariant element, and hence to the system itself. When this happens, the system reacts as a totality to plug the gap by modifying the form of regulation. A change of regime takes place (Aglietta 1979, p. 20).

The difference between this conception and Althusser's structuralism resides in Aglietta's recognizing different regimes of accumulation in the history of the capitalist mode of production. Moreover, these regimes do not automatically materialize but need regulation – that is coercive political forces, social institutions, and ideologies – to assure the necessary cohesion of the strategies and expectations of human agents. For Aglietta, as for Althusser, social institutions are produced, transformed, and renewed by class struggle, but for Aglietta the process of struggle cannot be assigned limits nor be confined by determinism.

> In a situation of historical crisis, all that a theory of regulation can do is note the conditions that make certain directions of evolution impossible, and detect the meaning of the actual transformation under way. Thereafter, however, the future remains open . . . history is initiatory (Aglietta 1979, pp. 67–8).

We can act in history but not calculate it.

Looking back on this formulation critically, Lipietz (1985, p. xvii) later argued that in Aglietta's work '"regulation" seemed to bend historically, in an almost functional manner, to embrace the new exigencies of regimes of accumulation'. His own version (Lipietz 1986, 1987) reformulates the left stream of what has now become a politically and theoretically heterogeneous Regulation School. Marxists have over-schematized, over-generalized, and dogmatized their thinking, he believes. There is a history of human variety to be discovered, the objective subject creating history through struggle. Like Aglietta, Lipietz focuses on the study of regularities in human relations – thus the concept mode of production marks a certain system of relations, contradictions, and crises. Within a mode of production, regimes of accumulation describe stabilizations in the allocation of the social product between consumption and accumulation, imply correspondence between the conditions of production and reproduction, and involve particular linkages between modes of production. Regulation consists of the norms, habits, laws, and networks which make human behaviour approximately consistent with schemas of social reproduction. Regulation does not have the function of making regimes of accumulation work. Rather, a regime and a mode of regulation are stabilized together because they temporarily ensure crisis-free social reproduction.

This is not the most easily comprehensible analytical position! In practice the several phases of capitalist development are usually identified by examining transformations in the labour process – that is, changes in the organization of work. Under capitalism this means essentially looking at the relations between workers and machines. Using the terminology of Marx in *Capital*, vol. 1, and Aglietta's extensions of these terms, Dunford & Perrons (1983, Ch. 9) identify four major developments in the capitalist labour process: manufacture, dominant between 1780 and 1870, but continuing in some industries until today; machinofacture beginning with the Industrial Revolution but dominant from 1870 to 1940; scientific management (Taylorism) and Fordism beginning in the late 19th century but dominant from 1940 to the 1980s; and neo-Fordism beginning in the 1970s and quickly expanding thereafter. Changes in the organization of work are linked to changes in the sectoral structure of the economy, in the methods of wage determination and modes of consumption, in the monetary system, and in spatial relations. The process of social transformation takes the form of a succession of major crises, as possibilities for economic growth are exhausted by a regime, or a mode of regulation becomes inappropriate for a regime of accumulation.

There is general agreement on the application of Gramsci's term Fordism to the postwar 'intensive regime of accumulation, focused on mass consumption' (Lipietz 1986, p. 26). Based in assembly line methods, the deskilling of labour, the social contract and Keynesian welfare state regulation, the Fordist regime was spatially associated with the dominance of the Euro-American Manufacturing Belt. This system began to come apart in the late 1960s and early 1970s, new forms of production have emerged, and the mode of social regulation has been modified (Scott & Cooke 1988). While many commentators agree that socio-economic transformation is presently underway (e.g. Piore & Sabel 1984, Lash & Urry 1987), the use of several different characterizations (neo-Fordism, post-Fordism, flexible accumulation) indicates considerable disagreement on the nature of the emerging regime. Using the term flexible production system Scott (1988) argues that the typically rigid mass production processes of Fordism are rapidly giving way to changeable, computer-enhanced processes, situated within a system of malleable external linkages and labourmarket relations. The new regime of flexible accumulation, he argues, is founded on three ensembles of industrial sectors: revivified artisanal and design-intensive industries mainly producing for final consumption; high-technology industries and their surrounding input suppliers and dependent subcontractors; and services, especially to businesses. Scott (1988a, 1988b, 1988c, Storper & Scott 1986) emphasizes horizontal and vertical disintegration in production and external economies which enhance the flexibility of the resulting industrial complexes. Many of the new producers deliberately seek alternative locational environments uncontaminated by Fordism, such as enclaves within older manufacturing regions (revitalized craft industries in inner cities, high-technology complexes in the far suburbs) and new areas on the margins of capitalist industrialization (the artisinal industries of the Third Italy, the technopoles of Western Europe, the high-technology Sunbelt of the United States). In general, Scott (1988a, p. 183) argues, the conflicting and confusing cross-currents in the space economies of North America and western Europe spring from

> the copresence of an aging regime of Fordist accumulation alongside an ascending regime of flexible accumulation, giving rise in turn to an intricate pattern of old and new industrial spaces implicated in a widening international division of labour. Whatever the future evolutionary path of this system may be, it is evident that the landscape of capitalist production is today drastically different from what it was even a couple of decades ago.

In similar vein, Harvey (1987, 1988) argues that cultural and intellectual life have been radically transformed in ways that parallel the political economic changes in regimes of accumulation. By the early 1970s, he says, modernism had lost all substance of social critique and had become closely linked to a Fordist regime of accumulation characterized by rationality, functionality, and efficiency. In the urban context, postmodernism in planning and development now plays an active rôle in promoting new cultural attitudes and practices consistent with the new regime of flexible accumulation.

However, before an impression of harmony is created, let it also be said that the concept of a new regime of flexible accumulation has its critics. Approaching the matter from the perspective of old industrial regions, Hudson (1988) suggests that Fordism never established more than a tenuous hold on these areas, and that recent changes involve only a selective reproduction of pre-Fordist and Fordist methods of production. Murray (1987) looks at the Emilia–Romagna area in the Third Italy, claiming that the optimistic picture of resurrected, dignified craft work in small firms characterized by the absence of capital–labour conflicts is an illusion – for him, this area is not representative of a post-Fordist regime of flexible specialization. And Polert (1988) deconstructs the term flexibility and finds it part of a futurology discourse which obscures complex and contradictory processes in the organization of work; by asserting a sea-change in structure, it fuses description, prediction, and prescription into a self-fulfilling prophecy.

Nevertheless, it seems probable that extensions of the regulationist concept of regimes of accumulation will continue as a leading edge of the new political economy of space. Certainly, the geography of global capitalism is a growth area in terms of academic interest (e.g. Cooke 1986, Scott & Storper 1986, Clark *et al.* 1986, Taylor & Thrift 1982, 1986, Peet 1987, Gottdiener 1988, Henderson & Castells 1988, Sayer & Morgan 1988, Wolch & Dear 1989). But it remains extraordinarily diffuse. A regulationist conception of global space might integrate concepts such as Massey's (1984) spatial structures of production, or the literature on the new industrial countries (Foster-Carter 1985, Hamilton 1986, Hart-Landsberg 1987, Bienefeld 1988) into Lipietz's (1986) regional categories. Likewise it is probably insightful to retheorize the spaces of deindustrialization in regime-of-accumulation terms (e.g. Dunford & Perrons, 1983, Moulaert & Swyngedouw 1987). The growing literature on services

(Daniels 1982, Gershuny & Miles 1983, Bressand & Distler 1985, Daniels 1986a, 1986b), which includes a significant spatial component (Marshall 1988, Stanback & Noyelle 1982, Stanback *et al.* 1981, Ch. 5, Urry 1987, Walker 1985) can similarly be recast in a more critical regulationist discourse (Petit 1986, pp. 37–43).

Corbridge and others may see all this as a movement to 'post-Marxism'. But the central intent of regulationist reasoning, especially as practised in geography by Harvey, Scott, Moulaert, Swyngedouw and others, seems more accurately to be characterized as asking macro-theoretical questions of the logic of capitalism in its manifold regional forms. This represents a return to what was never more than a *neo*structuralist political economy of space, but using an intermediate-level theoretical technology. Themes from the past and modifications from the present are combined into a theory for the future. This is an area where political economic geography plays a formative rôle.

References

Aglietta, M. 1979. *A theory of capitalist regulation: the U.S. experience*. London: New Left Books.

Bienefeld, M. 1988. The significance of the newly industrializing countries for the development debate. *Studies in Political Economy* **25** (Spring), 7–39.

Browett, J. 1984. On the necessity and inevitability of uneven spatial development under capitalism. *International Journal of Urban and Regional Research* **8**, 155–76.

Browett, J. 1986. Industrialization in the global periphery: the significance of the newly industrializing countries of East and Southeast Asia. *Environment and Planning D, Society and Space* **4**, 401–18.

Bressand, A. & C. Distler 1985. *Le Prochain Mode*. Paris: Seuil.

Buttell, F. H. L. & H. Newby (eds) 1980. *The rural sociology of advanced societies*. Montclair, NJ: Allanheld & Osmun

Clark, G., M. Gertler & J. Whiteman 1986. *Regional dynamics: studies in adjustment theory*. Boston: Allen & Unwin.

Cooke, P. (ed.) 1986. *Global restructuring, local response*. London: Economic and Social Research Council.

Corbridge, S. 1986. *Capitalist world development: a critique of radical development geography*. London: Macmillan.

Daniels, P. 1982. *Service industries: growth and location*. Cambridge: Cambridge University Press.

Daniels, P. 1986a. The geography of services. *Progress in Human Geography* **10**, 436–44.

Daniels, P. 1986b. Producer services and the post-industrial space economy. In *The geography of de-industrialization*, R. Martin & B. Rowthorn (eds), 291–321. London: Macmillan.

Dunford, M. 1988. *Capital, the state and regional development*. London: Pion.

Dunford, M. & D. Perrons 1983. *The arena of capital*. New York: St Martin's Press.

Foster-Carter, A. 1985. Korea and dependency theory. *Monthly Review* (October), 27–32.

Gershuny, J. I. & J. D. Miles 1983. *The new service economy*. New York: Praeger.

Gottdiener, M. (ed.) 1988. *Modern capitalism and spatial development*. New York: St Martin's Press.

Graham, J., K. Gibson, R. Horvath & D. Shakow 1988. Restructuring in U.S. manufacturing: the decline of monopoly capitalism. *Annals of the Association of American Geographers* **78**, 473–90.

Hamilton, C. 1986. *Capitalist industrialization in Korea*. Boulder, Col.: Westview Press.

Hart-Landsberg, M. 1987. South Korea: the fraudulent miracle. *Monthly Review* (December), 27–40.
Harvey, D. 1987. Flexible accumulation through urbanization: reflections on 'Post Modernism' in the American city. *Antipode* **19**, 260–86.
Harvey, D. 1988. The geographical and geopolitical consequences of the transition from Fordist to flexible accumulation. In *America's New Market Geography: Nation, Region and Metropolis*, G. Sternlieb & J. W. Hughes (eds). New Brunswick: Center for Urban Policy Research, Rutgers University.
Harvey, D. & A. Scott 1988. The practice of human geography, theory and specificity in the transition from Fordism to flexible accumulation. In *Remodelling Geography*, W. Macmillan (ed.). Oxford: Basil Blackwell.
Henderson, J. & M. Castells (eds) 1988. *Global restructuring and territorial development.* London: Sage Publications.
Holland, D. & J. Carnalho 1985. The changing mode of production in American agriculture: emerging conflicts in agriculture's role in the reproduction of advanced capitalism. *Review of Radical Political Economics* **17**, 1–27.
Hudson, R. 1989. Labour-market changes and new forms of work in old industrial regions: may be flexibility for some but not flexible accumulation. *Environment and Planning D, Society and Space* **7**, 5-30.
Lash, S. & J. Urry 1987. *The end of organized capitalism*. Cambridge: Polity Press.
Lipietz, A. 1985. *The enchanted world.* London: Verso.
Lipietz, A. 1986. New tendencies in the international division of labor: regimes of accumulation and modes of regulation. In *Production, work, territory*, A. J. Scott & M. Storper (eds), 16–40. Boston: Allen & Unwin.
Lipietz, A. 1987. *Mirages and miracles.* London: Verso.
Marsden, T., R. Munton, S. Whatmore & J. Little 1986. Towards a political economy of capitalist agriculture: a British perspective. *International Journal of Urban and Regional Research* **10**, 498–521.
Marshall, J. N. (ed.) 1988. *Services and uneven development*. Oxford: Oxford University Press.
Massey, D. 1973. Towards a critique of industrial location theory. *Antipode* **5**, 33–9.
Massey, D. 1984. *Spatial divisions of labour: social structures and the geography of production.* London: Macmillan.
Morgan, K. & A. Sayer 1988. *Microcircuits of capital.* Cambridge: Polity Press.
Moulaert F. & E. Swyngedouw 1987. Regional development and the geography of the flexible production system. Manuscript. Villeneuve d'Ascq. France: Cerie, University of Lille.
Murray, F. 1987. Flexible specialization in the 'Third Italy'. *Capital and Class* **33**, 84–95.
Nusbauer, J. (ed.) 1987. *Services in the global market*. Boston: Kluwer Academic Publishers.
Peet, R. (ed.) 1987. *International capitalism and industrial restructuring.* Boston: Allen & Unwin.
Peet, R. 1988. Review of S. Corbridge, *Capitalist world development. Economic Geography* **64**, 190–2.
Petit, P. 1986. *Slow growth and the service economy*. London: Frances Pinter.
Piore M. & C. Sabel 1984. *The second industrial divide.* New York: Basic Books.
Polert, A. 1988. Dismantling flexibility. *Capital and Class* **34** (Spring), 42–75.
Sayer, R. A. 1985. Industry and space: a sympathetic critique of radical research. *Environment and Planning D, Society and Space* **3**, 3–29.
Scott, A. 1988a. Flexible production systems and regional development: the rise of new industrial spaces in North America and Western Europe. *International Journal of Urban and Regional Research* **12**, 171–86.
Scott, A. 1988b. *Metropolis: from the division of labor to urban form*. Berkeley, CA: University of California Press.

Scott, A. 1988c. *New industrial spaces*. London: Pion.

Scott, A. & P. Cooke 1988. The new geography and sociology of production (special issue). *Environment and Planning D, Society and Space* 6, 241–370.

Scott, A. & M. Storper (eds) 1986. *Production, work, territory: the geographical anatomy of industrial capitalism*. Boston: Allen & Unwin.

Smith, N. 1986. On the necessity of uneven development. *International Journal of Urban and Regional Research* **10**, 87–103.

Stanback T. M. & T. J. Noyelle 1982. *Cities in transition*. Torowa, NJ: Allanheld & Osmun.

Stanback, T. M., P. J. Bearse, T. J. Noyelle & R. Karasek 1981. *Services: the new economy*. Torowa, NJ: Allanheld & Osmun.

Storper, M. 1987. The new industrial geography, 1985–1986. *Urban Geography* **8**, 585–98.

Storper, M. & A. Scott 1986. Production, work, territory: contemporary realities and theoretical tasks. In *Production, work, territory: The geographical anatomy of industrial capitalism*, A. J. Scott and M. Storper (eds), 3–15. Boston: Allen & Unwin.

Taylor M. & N. Thrift (eds) 1982. *The geography of multinationals*. London: Croom Helm.

Taylor, M. & N. Thrift (eds) 1986. *Multinationals and the restructuring of the world economy*. Beckenham: Croom Helm.

Urry, J. 1987. Some social and spatial aspects of services. *Environment and Planning D, Society and Space* **5**, 5–26.

Walker, R. A. 1985. Is there a service economy? The changing capitalist division of labour. *Science and Society* **49**, 42–83.

Watts, M. 1988. Deconstructing determinism: Marxisms, development theory and a comradely critique of *Capitalist world development: a critique of radical development geography*. *Antipode* **20**, 142–68.

Wolch, J. & M. Dear (eds) 1989. *The power of geography*. London and Winchester, Mass.: Unwin Hyman.

5 *New models of regional change*

Erica Schoenberger

Introduction

Regional development patterns in advanced capitalist countries have changed dramatically in the last 20 years. Traditional industrial regions mushroomed in the postwar period with the growth of major manufacturing sectors such as steel and automobiles, only to plunge into crisis. Massive layoffs in these industries were not compensated, in terms of the number of jobs, wage levels, occupational structure, or job security, by the rise of new economic sectors such as electronics or services. Meanwhile, some formerly lagging or peripheral regions suddenly 'took off', seemingly of their own accord.[1] This pattern of shifting regional growth and decline has occurred at a number of scales, including the international level with the rise of the so-called newly industrializing countries (Weinstein & Firestone 1978, Perry & Watkins 1978, Blackaby 1979, Carney *et al.* 1980, Bluestone & Harrison 1982, Massey & Meegan 1982, Tabb & Sawers 1984, Massey 1984, Markusen 1985, Martin & Rowthorn 1986, Clark *et al.* 1986, Scott & Storper 1986, Peet 1987). The unexpected reversal of regional growth patterns, and the realization that regional growth and decline were linked phenomena – 'two sides of the same coin' – revealed the inadequacy of traditional spatial theory (Gertler 1987).

Theories of regional growth and development must account for these changes – not only why and how they have occurred, but why at particular times in particular places. In attempting to come to grips with these phenomena, recent research on the causes of regional growth and decline has departed from earlier theoretical traditions in significant ways.

Traditional models of regional growth were largely derived from non-spatial economic theory. Their aim was to insert space into the analysis without reconstructing the basis of the analysis. Thus, in a conventional neoclassical approach, capital and labour are treated as undifferentiated factors of production, flowing across space in response to differences in marginal rates of return (profit and wage rates).[2] Given the standard assumptions of identical production functions, perfect competition, information and factor mobility, capital theoretically will flow to areas where wages are low – due to a relative abundance of labour – and the return to capital high due to its relative scarcity. This is the basis for the expectation that regional growth patterns will converge over time: an expectation that did not hold in practice (Borts 1960, Borts & Stein 1968).

The rise of the Sunbelt and the decline of the Rustbelt in the United States seem, at first glance, to fit the model quite well. A region characterized by low wages and a low level of investment ends up as the recipient of capital flows from a high-wage, capital-abundant region. Two problems immediately arise, however. First, the newly growing region had exhibited precisely the same characteristics for over a century. An explanation of the *timing* of the shift is lacking. Second, the *form* investment took cannot be explained by this approach. The growth of the Sunbelt is linked to the rise of entirely new industries (e.g. electronics and services), the resuscitation of traditional resource-based sectors (energy, petrochemicals, and mining), and the decentralization of manufacturing from the core. Finally, though the model posits regional *convergence* over time, the growth of one region apparently occurred at the expense of another region which was plunged into decline.

Alternatively, some conventional regional growth models were principally derived from Keynesian theory, with its emphasis on disequilibrium and the rôle of demand in explaining economic growth (Richardson 1973, Holland 1976, Cooke 1983). Export base models posited regional growth as a function of external demand for a region's exports and, hence, the region's resource endowment (Perloff *et al.* 1960, North 1974a, 1974b, Tiebout 1974a, 1974b). Cumulative causation models suggested that already growing regions tend to perpetuate and even enlarge their advantages over lagging regions (Myrdal 1957, Hirschman 1958, Kaldor 1970, Richardson 1973).[3] Interindustry linkages and agglomeration economies further reinforced divergent urban and regional growth paths (Perroux 1950, Hoover & Vernon 1959, Pred 1966).

Keynesian models, however, also are unable fully to explain the timing and form of regional growth and decline. Implicitly, growing regions enjoy an initial comparative advantage over competing regions in these models. It is relatively simple to analyze cases in which comparative advantage is derived from natural resource endowments. But the growth of new manufacturing or service-based regions where natural resources play little or no part suggests that comparative advantage must be understood as an historically created phenomenon, not a natural attribute of particular places (Harvey 1982).

These problems in previous theory point towards the theoretical and methodological issues confronted by the new models of regional change. Three main avenues of departure from traditional theory characterize more recent efforts. First, theory has sought explicitly to analyze the role of technological change in transforming spatial patterns of development, whether through the rise of new sectors or the restructuring of traditional industries. Second, the structure and organization of firms and industries, the kinds of linkages generated in terms of flows of authority, information, and goods, and associated spatial effects, have received considerable attention. In effect, the rediscovery of the corporation as an active agent of change marked a significant break with traditional regional theory. Third and finally, Marxist geographers have carried the themes of technological change and corporate structure further into an analysis of capital, labour, and the geography of the social production process as a whole. This chapter is organized around these three themes.

Technological change and regional growth

Transformations in transportation and communications technologies have long been a concern of geographers due to their evident implications for the importance of distance in the functioning of the space-economy. Indeed, these considerations remain highly relevant in an era of increasingly rapid transport and globally integrated telecommunications networks that, among other things, have greatly enhanced the spatial reach of large corporations. However, traditional theory remained essentially silent on the issue of technological change in products and production processes. By contrast, these forms of technological change have become a prime focus of attention in recent work.

New products, new industries, new growth

Technological change may, in the first instance, be embodied in new products and new industries. In the context of regional development, these industries are of interest in their own right, especially as their growth rates outstrip, often to a spectacular degree, the growth of mature sectors. This fact alone helps explain the already voluminous, but by no means conclusive, literature on the locational proclivities of high-tech firms (Malecki 1979, 1980, 1983, 1984, Oakey *et al.* 1980, US Congress 1982, Armington *et al.* 1983, Office of Technology Assessment 1984, Oakey 1985a, 1985b, Thwaites & Oakey 1985, Hall & Markusen 1985, Castells 1985, Glasmeier 1985, Cooke 1986, Markusen *et al.* 1986, Rees 1986, Sayer 1986, Schoenberger 1986, Storper & Scott 1988). Rapid growth also accounts for the mad scramble by local and state economic development authorities to lure high-tech firms to their areas (Luger 1984, Vaughan & Pollard 1986). The essential assumption is that innovation and entrepreneurship foster regional growth or, to put it another way, those regions fortunate enough to possess a favourable environment for innovation will prosper while others will not (Thomas 1985).

Despite this attention, high tech or innovativeness as analytical categories have resisted effective theorization. By and large, these studies have been empirical investigations of the location-specific factors that appear to be associated with the proliferation of innovative firms emphasizing, for example, the presence of major research universities, low levels of unionization, or the sort of cultural and environmental amenities thought to be attractive to geographically mobile scientific and professional workers. Some doubt about the relevance of these factors as explanatory variables has emerged, however (Markusen *et al.* 1986, Malecki 1986, Armington 1986, Storper & Scott 1989).

The prototype models of this pattern of regional growth are Silicon Valley in California and Route 128 around Boston in the United States. Both are characterized by dense networks of inter-firm transactions, the proliferation of small, new entrepreneurial firms, and continued product innovation thought to guarantee self-reinforcing and self-propelled regional growth paths. However, the historical specificity of these high-tech centres and the evident importance of agglomeration economies which underlie a pattern of spatial concentration rather than dispersal of high-level technical activities (e.g. research and development) suggest that this particular model will not be easily generalized (Saxenian 1984, 1985, Massey 1985, Gertler 1987).

Storper & Scott (1988) discuss the social and economic preconditions for development on the model of Silicon Valley and observe that these preconditions existed in a much larger number of places than actually went on to develop in this way. They propose the concept of a window of opportunity that was briefly open in the early days of high technology industrial development when the actual locational dynamics were still essentially undetermined. Once the seeds of this type of development took root in a small number of places, this window of opportunity closed as the forces of agglomeration tended to concentrate development in the early sites. The concept of a window of opportunity is not meant to suggest that the actual dynamics are unknowable – rather, that they were historically highly contingent and cannot be satisfactorily theorized.

It has been hypothesized that new industrial complexes will tend to arise in new areas unfettered by the social traditions and infrastructural rigidities characteristic of mature industrial regions (Storper & Scott 1988). The close connection between defence spending and the early development of such sectors as electronics and computers along with aircraft and missile production has also been linked to locational concentration in the US Sunbelt region (Markusen & Bloch 1985, Markusen 1986). The central idea is that new industries will benefit from the ability to create a social and infrastructural landscape suited to the particular needs of the sector in question. For this reason, formerly peripheral areas become strong candidates for new growth. Which peripheral areas will ultimately be selected is still open to question. High-tech complexes in Silicon Valley and Orange County, California developed on the immediate periphery of existing industrial agglomerations, while new sectoral growth in the Route 128 area overlays a declining traditional manufacturing base. Thus, even the concept of 'peripheral' has a certain ambiguity in this discussion.

Aside from the specific features of new sectors themselves, technological change in the form of new products/new industries is also important for its linkages with other sectors and activities that can, under certain circumstances, spark a generalized wave of growth in the economy (Schumpeter 1939, 1961, Mandel 1978, Mensch 1979, Freeman *et al.* 1983, Hall 1985). These intersectoral effects occur in dfferent ways. The age of the automobile called forth not only massive investments in the production of vehicles, but tremendous collateral effects through linkages to supplier and complementary industries. Moreover, it was associated with equally massive long-term investments in roads, suburban housing, and other aspects of a restructured built environment. Finally, the geographical flexibility afforded by trucking influenced the location patterns of other industries and, through altering relative distances among places, the growth prospects of different regions. While growth in the macro-economy has a significant influence on the development prospects of individual regions, the precise spatial impacts of such innovation waves are difficult to specify. The fate of growth pole policies, which attempted to translate a Schumpeterian vision of technical and industrial change into a mechanism for regional development, provides a useful lesson in the difficulties of fixing spatially the complex processes of technological transformation according to a coherent theoretical logic (Perroux 1970, Hansen 1970, Darwent 1974, Thomas 1975).

Technology, process, and spatial change

The intense focus in the literature on product innovation should not obscure the fundamental importance of technological change in the process of production and its implications for the spatial allocation of economic activity. The age of the microchip, indeed, may have its most significant impact in the production processes of other industries rather than in the direct creation of employment.

A number of themes arise in this connection, most of which will be developed below. They centre on the relationship of technological change to the demand for labour. Progressive mechanization of production processes, for example, will reduce the level of employment associated with a given level of investment. Indeed, it has been argued that the crisis of traditional manufacturing regions described above is less a problem of *investment*, which has continued in these areas, than a problem of *employment* consequent on the labour-saving bias of investments that have been made (Varaiya & Wiseman 1981, Gertler 1984a, 1987). Under these circumstances, sustaining or expanding employment growth requires a progressively greater growth of output and the effective demand necessary to absorb it. Much, then, depends on conditions in the macro-economy (Freeman *et al.* 1983).

In practice, more attention has been paid to the relationship between technological change and the *kind* of labour demanded, specifically on the issue of skills. It has been widely argued that technical change is associated with the *deskilling* of labour in production (Braverman 1974, Noble 1978, 1984). As labour demand becomes more homogeneous, firms are less constrained to locate near supplies of skilled and experienced industrial labour. This is seen to be one of the essential factors underlying the increased geographic mobility of capital in search of a lower-cost, non-unionized workforce, and thus for the shift of investment from the industrial core to peripheral regions (Frobel *et al.* 1980, Bluestone & Harrison 1982, Peet 1983).

Others have argued that there is no determinate relationship between technical change, the demand for labour, and the spatial allocation of production (Walker 1985, Storper 1985b). New techniques may require new skills (e.g. electronics technicians or computer programmers) (Durand *et al.* 1984, Shaiken 1984). Alternatively, if automation displaces sufficient amounts of labour, the importance of labour costs and labour control in the location calculus of the firm may be diminished.

In this sense, several possible industrial location and regional development outcomes ensue. Firms may remain in the same location but reduce employment or shift to a different local source of workers if local labourmarkets permit the replacement of, for example, skilled male workers by unskilled women or immigrants. In this case, the original regional economy may continue to grow, but the *nature* of the growth in terms of occupational and income structures may change significantly. Alternatively, the firm may change location to gain access to a supply of relatively docile workers in peripheral regions where the prospects of unionization and other forms of worker resistance appear slight (Bluestone & Harrison 1982, Massey 1984, Peet 1983). Where this phenomenon is sufficiently widespread, quite dramatic shifts in the patterns of regional development and stagnation may be seen as investment is directed away from traditional industrial regions toward erstwhile peripheral areas.

As this suggests, the restless search for solutions to problems of competitiveness and profitability gives rise to a range of technical and spatial strategies that act sometimes as alternatives, sometimes as complements, to one another. The weakness of approaches that attempt to tie a *particular* technological tendency to a counterpart spatial pattern can be exemplified by an examination of one of the more widely adopted paradigms of technical and spatial change – the product cycle.

The product cycle

Product cycle theory attempted explicitly to tie notions of technological change in products *and* production processes to changing locational patterns and, by extension, changing patterns of regional growth (Vernon 1966). The model continues to exert considerable influence (Berry 1972, Norton & Rees 1979, Rees 1979, Erickson & Leinbach 1979, Park & Wheeler 1983, Rees & Stafford 1986, Moriarty 1986, Gross & Weinstein 1986, Flynn 1986).

In essence, the model envisages a set of parallel evolutionary paths along the axes of product technology, production process, the demand for labour, competition, and location. Once past the introductory phase, product technology is fixed and the production process becomes increasingly mechanized and standardized. The demand for labour accordingly shifts from skilled to unskilled, creating the possibility for decentralization of production away from concentrations of skilled labour. At the same time, the technological lead which allowed the innovating firm to be relatively insensitive to cost pressures erodes as new competitors enter the field. Increased price competition puts firms under considerable pressure to decentralize production in order to reduce labour costs. In the product's old age, as the market stagnates and declines, these pressures are intensified and production is likely to shift offshore to low-wage developing nations with output exported back to the core. Crucially, relocation of production is *permitted* by standardization of product and production process and *impelled* by increased price competition.

The model has a certain plausibility as a descriptive device. From it one may deduce certain propositions concerning the prospects for regional development and its character in different phases of the cycle. Benefits accrue to the region hosting the early stage of the cycle but, by implication, places are engaged in what amounts to a life-and-death struggle to achieve or defend their positions as innovators. A region at the tail end of the product cycle finds its fate as a branch-plant economy based on low-skilled and low-wage production – a form of regional growth, but not particularly desirable or stable as alternative lower-wage areas can always be found. The prospects for regions that are neither innovative nor sufficiently poor to offer the lowest-wage labourforce are decidedly grim.

The product cycle model has been extensively criticized in recent work (Gertler 1984b, 1987, Markusen 1985, Walker 1985, Storper 1985a, Sayer 1985a, Clark *et al.* 1986). The question of which regions are likely to be innovative is especially difficult to resolve. In his original formulation dealing with the international economy Vernon (1966) suggested that high-income countries would be most capable of supporting technical innovation. But the case is less obvious when applied to different regions within a generally high-income national economy.

Further, the constituent dynamic elements of the model – product technology, production process, demand for labour, and competition – do not always follow the indicated evolutionary paths. It is difficult to justify the proposition that product technology is essentially fixed following the innovative/introductory stage – competition may continue for some time to be based more on product technology, differentiation, performance, and quality than on price (Sayer 1985a, Schoenberger 1985, 1986).

Similarly, one can challenge the assumption that production inevitably evolves into standardized, mass production (Piore & Sabel 1984, Storper 1985a, Walker 1988a).[4] Moreover, it is not clear that standardization and mechanization *per se* entirely remove skilled labour inputs (Piore & Sabel 1984; Duran *et al*. 1984, Shaiken 1984). Finally, firms are often able to stave off the proliferation of new competitors. To the extent that oligopolistic industrial structures can be maintained through the usual barriers to entry, the nature of competition in a sector may deviate significantly from that posited in the model – as may the spatial allocation of production (Markusen 1985).[5]

In this sense, although the product cycle model usefully identifies critical factors in the structuring and restructuring of industry and place, it treats them in an excessively mechanistic manner. While it seems to fit some industries at some points in time, others elude its scope.

Technology and region: a reconsideration

The impetus to draw the question of technological change into the discussion of regional development is certainly valid. It is incontestable that technological transformation is intimately linked with other processes of social, economic, and spatial change. For this reason, the question of technology provides a useful vantage point from which to analyze these other processes. Problems arise, however, when technology itself is taken as the fundamental starting point for regional theory. The origins of technological changes are themselves highly diverse. There is little reason to assume that technological change proceeds along a smooth and unidirectional trajectory in terms of the nature of products and production processes, the composition of labour demand, or locational requirements. As Walker (1988a) suggests, the path of technological change may be viewed more accurately as kinked and branching, impelled down many different channels.

A key point, then, is that the nature of technological change does not uniquely determine locational and regional development outcomes. Firms face a changing set of location possibilities and constraints influenced, but not wholly controlled by, changes in technique. Regional theory must *encompass* the question of technological change – it cannot be derived from it.

The organization of firms and industries: linkages, oligopoly, and regional development

In our conception of the modern corporation a number of characteristics come to mind: large size, assembly line, mass-production techniques, multidivisional and multilocational structures and, in general, more complex and more geographically diverse and extensive input–output flows within and across firms. In short, the nature of the central agent in the structuring and restructuring of economic and spatial relationships has changed dramatically over the last hundred years, and with it, the forces underlying the development prospects of particular regions. The recognition of this fact spurred the development of a new geography of enterprise focusing on the characteristics of firms and industries in order to explain the spatial organization of production. Attention turned to the rôle of corporate decision making and flows of goods, information, and profits between and within firms (Pred 1966, 1977, Krumme 1969, Hamilton 1974, Taylor 1975, McDermott & Keeble 1978, Hamilton & Linge 1979, Watts 1980, 1981).

To be sure, the small, single-location, single-product firm continues to exist, although it is sometimes viewed, mistakenly, as an archaic hanger-on from an earlier, more truly competitive era. Often, a more or less exclusive focus on the workings of large firms is justified by the argument that they control the commanding heights of the economy in terms of shares of assets, output, and employment: the analytical target is at the same time a strategic one (cf. Markusen 1985). Recently, however, a renewed interest has emerged in the internal operations of, and the interrelationships among, firms of all sizes (Sabel 1982, Taylor & Thrift 1982a, 1982b, 1983, Holmes 1986, Scott 1986a, 1986b).

Organizational and spatial structures of large firms

A number of factors influence the spatial structure of the large, complex, multilocational firm. In the first instance, the control structures of these firms are necessarily hierarchical. As one moves up the hierarchy, successive layers of management have increasingly broader responsibilities and horizons, ranging from the day-to-day operational management of individual plants, to strategic planning for the firm as a whole (Chandler 1962, Hymer 1972, Dicken 1976).

Second, with increasing scale and complexity, an increasingly refined and extensive technical division of labour within the firm crystallizes into a specialized array of functions, characterized by different work processes and factor demand structures. For example, management and high-level administrative operations, research and development, different segments of the physical production process, and lower-level clerical and intra-firm service operations all require different types of labour. This creates the possibility of locating the various functional activities within the firm in accordance with geographically highly differentiated sources of labour (Perrons 1981, Storper & Walker 1983, 1984, Massey 1984). The sheer scale and capital resources of the large firm promotes the realization of this possibility, as does the fact that material linkages internalized within the firm are relatively stable and predictable, permitting the co-ordination of geographically dispersed elements of the production process at a tolerable cost (Scott 1986a).

In contrast to the old spatial division of labour, which generated regional specialization on a *sectoral* basis, the foundation is laid for a new spatial pattern in which regions tend to have different *functional* specializations, influencing the character of regional development (Hymer 1972, Dicken 1976, 1986, Hansen 1979, Moulaert & Salinas 1983, Massey 1984).

Schematically, this is conceptualized in terms of regions arrayed in a functional hierarchy roughly paralleling that of the corporation (Hymer 1972, Lipietz 1982, 1986). High-level corporate control activities are concentrated in a small number of major metropolitan regions with occupational structures skewed toward white-collar professional and clerical support categories. A similarly small number of specialized research and development complexes may also be generated with growth patterns characterized by a bias towards high-level occupational and income structures (Saxenian 1984, 1985, Massey 1985, Malecki 1986). Segments of the production process still reliant on skilled labour inputs gravitate to already industrialized areas offering appropriate labour supplies, while the most standardized, least skill-intensive activities are freed to seek out low-cost labour in peripheral areas. Regional growth patterns in terms of occupational and income distribution are dramatically different in each of these cases.

There has been a tendency to treat this schema as an empirical tendency which can be generalized to all sectors, so that regions come to have strictly defined roles in the hierarchy of production and control (e.g. Frobel *et al.* 1980, Susman 1984). Indeed, some of the simpler versions of the new international division of labour (NIDL) model implicitly incorporate many of the same, overly mechanical assumptions of the product cycle concerning the evolution of products, production processes, the demand for labour, and location.

A number of writers, however, have argued for a more cautious interpretation (Jenkins 1984, Massey 1984, Sayer 1985a, Lipietz 1986, Herold & Kozlov 1987, Schoenberger 1988b). As Massey observes, this particular schema is only one of a number of possible spatial structures of production. Different kinds of products and production processes may be associated with rather different spatial divisions of labour (see also Storper & Walker 1984). In this light, Massey's insistence on the plural – spatial *divisions* of labour – is appropriate. Many of the themes introduced here are taken up later. It is nevertheless useful to add some observations concerning the relationship between corporate hierarchies and the character and dynamics of regional growth.

One of these concerns the way growth is transmitted through a space economy. The meaning of intra- and inter-regional economic relationships is different as these are mediated through geographically diverse but organizationally integrated corporate systems (Hamilton 1974, Dicken 1976, Pred 1977, Massey 1984). Growth impulses, in effect, are transmitted spatially through the corporate hierarchy, with interdependencies and feedback effects contained largely within the corporate structure itself. In this view, first expressed in the language of systems analysis, the primary regional growth dynamic is sought in the structure of intra-firm relationships.

A major concern is the spatial pattern of input output flows within and among firms. The implantation of, say, a branch assembly plant of a multilocational corporation induces a certain amount of regional growth in that new jobs are created and incomes generated. But the branch plant may not induce the development of local supplier or distribution networks. It is quite possible that the branch plant will have largely intra-firm but extra-regional linkages, drastically reducing the multiplier effects anticipated by traditional theory (Dicken 1976, Watts 1981, Martinelli 1986). True, the magnitude of the multiplier was always uncertain (Tiebout 1974a). But the view of the corporation as an integrated, spatially extensive system of flows implies that local multiplier effects may be severely circumscribed. Certainly, the failures of growth pole policies hinged to a great extent on this phenomenon (Hansen 1970, Darwent 1974, Holland 1976).

By the same token, the significance of agglomeration economies appears increasingly confined to a few elements of the corporate structure (Holland 1976, Scott 1983b). Headquarters cluster in a few large cities to permit intensive personal interactions among executives, bankers, lawyers, accountants, and other high-level professionals. Research and development complexes are sustained by cross-flows of information and personnel among firms and research institutions (Malecki 1986). Segments of the production process relying on highly skilled labour benefit from access to a well developed industrial labour market and a diversified supply of material inputs.

In contrast, the decentralized branch plant is relatively indifferent to such external economies. Because transactions are highly internalized within the firm, and because production is standardized, it does not require easy access to a local network of goods and services providers, or a specialized labour pool. This is the basis for a peculiarly distorted pattern of growth for regions favoured as branch plant sites.

In addition, given that corporate control centres are few in number and spatially concentrated, an oft-raised implication is that corporate control and decision making remain external to most regions experiencing growth through the implantation of other elements of the corporate structure (Hymer 1972, Westaway 1974, Dicken 1976, Pred 1977, Clark *et al.* 1986). Several problems may arise as a consequence. Most regions, even those that are growing, lack both a set of strategic functions and the high-level occupations and incomes associated with them. Another consequence, perhaps more serious for the long-term growth prospects of the region, is that locally generated profits are transmitted to the external headquarters and may be reinvested elsewhere (Watts 1981, Bluestone & Harrison 1982).

On the other hand, the extent to which individual regions have ever fully controlled their economic destinies should not be exaggerated. Regions have always been subject to the vagaries of national and international economic systems. Even locally integrated, single-plant firms serving regional markets are controlled by a class (the owners of capital) rather than the region *per se* (Massey 1984). There may be benefits from integration within the structure of a large corporation. The number of jobs generated by a major corporate branch plant may be considerably higher than that produced by local capital investments, even if the occupational structure is truncated. And the deep pockets of the parent corporation may sustain a branch plant during economic downturns while the smaller, locally owned firms go under. Nevertheless, the presumption that growth dynamics of this sort increase the vulnerability of individual regions remains a recurrent theme in the literature.

Integration, disintegration and spatial structures

Despite a certain preoccupation with large firms as the central unit of analysis, the persistence of small firms, and a growing appreciation of their complex relationships with their larger counterparts, led a number of researchers to focus on the dynamics underlying the tendency to vertical and horizontal integration (disintegration) and the consequences for linkages, the structure of labour markets, and spatial patterns of growth. Vertically integrated firms, for example, control most or all of the various phases of a production process. They behave quite differently from disintegrated or smaller, functionally specialized firms producing intermediate goods sold on the market to other firms for further processing.

Sabel and Piore, for example, propose that the differentiation of output markets into primary and secondary segments influences the firm's choice of production technique, the demand for labour and, by extension, the location of production in line with geographically differentiated labourmarkets (Piore 1980, Sabel 1982). The primary sector, characterized by reliably stable demand, is the domain of the large, integrated corporation employing standardized mass production techniques based on special-purpose machinery. Workers are typically semi-skilled or unskilled, allowing for the decentralization of production away from traditional industrial agglomerations.

Secondary sector firms, operating in cyclically variable markets, are smaller and functionally disintegrated. Production processes are unstandardized and employment is often skewed towards skilled workers because of the reliance on general purpose machinery that can be employed in a range of tasks. Locational choice is constrained by the need for specialized labour inputs, and firms tend to be clustered in core industrial areas.

The combination of divergent production techniques and geographical differentiation creates quite different patterns of growth. Under certain historical circumstances, secondary sector firms coalesce into mutually reinforcing, flexibly specialized industrial districts characterized by a dense network of interactions among firms that permits rapid adjustment to changing market conditions. The Third Italy is the oft-cited prototype of this model (Sabel 1982, Piore & Sabel 1984, Storper & Scott 1989). Indeed, the focus on firm structure, industry segmentation, and linkages amounts to the rediscovery of agglomeration economies underlying urban and regional growth (Gertler 1987). The reduction of transportation and transaction costs associated with spatial proximity and the efficiency advantages of a more refined division of labour across firms contributes greatly to the competitiveness of these densely articulated production complexes.

Scott, following Coase and Williamson, proposes a different theoretical logic for the analysis of functional integration and disintegration, or organization *scope* in production (Coase 1937, Williamson 1975, Scott 1983a, 1983b, 1984, 1986a, 1986b). The approach centres on the relative efficiencies of internalizing transactions within the firm versus externalizing them on to the open market, given the character of the production process. The kinds of linkages produced under conditions of integration or disintegration shape the spatial allocation of activity.

Intra- and inter-sectoral variations in firm (or plant) organization are in part a function of technology and scale of production. These are influenced by the character of output markets in terms of both quantitative (volume and stability over time) and qualitative (standardized versus unstandardized) aspects. Firms producing unstandardized products at varying levels of output tend to be small, labour-intensive, and functionally disintegrated. Firms producing large volumes of standardized products with standardized, mass production techniques, tend to be large, capital intensive, and vertically integrated because they benefit from both economies of scale and economies of integration. Where several linked but technically distinct manufacturing processes have different optimal scales, the likelihood of vertical disintegration increases.

Economies of integration arise in several ways. They are significant in the presence of technical complementarities where joint production generates savings, as in the case of integrated steel mills or petrochemical complexes.

Economies of scope are generated by reduced costs of management and communications under a unified corporate structure. Or integration may reduce uncertainties and transaction costs associated with obtaining inputs or marketing outputs – for example, by ensuring the availability of raw materials or avoiding spatially dependent transportation and transaction costs.

The linkage patterns generated by different forms of organization are quite distinctive and are likely to produce divergent spatial patterns. Functional disintegration implies irregular, small-scale, and unstable linkages among firms, creating pressures to locate near suppliers and customers in a spatially concentrated pattern. By contrast, factors associated with vertical integration tend to create high-volume, standardized, and stable linkages, permitting geographical dispersion away from industrial concentrations (see Moriarty 1986, for empirical confirmation). Decentralization is further reinforced by the desire to avoid certain diseconomies of agglomeration, including high wage and infrastructure costs and the threat of unionization. The former are outweighed for small units by transaction cost savings and the latter is less likely for small firms in general.

Scott stresses that processes of integration and disintegration are quite likely to be associated with transformations of the labour process. For example, as scale increases with integration (or as integration increases with scale), greater mechanization may lead to the resynthesis of processes that had functioned as discrete operations under different conditions (Scott 1983a, cf. also Walker 1988a). Indeed, diseconomies of agglomeration may encourage a strategy of organizational restructuring and transformation of the production process, including capital deepening, resynthesis and deskilling of labour tasks.

The logic behind the resulting spatial patterning is subtly different from that proposed in the earlier discussion of corporate structure and linkages because the point of departure is different. Scott seeks to understand the origins of industrial organization rather than taking the archetypal large, multi-establishment firm as given. The spatial division of labour is mediated through the relationship between organizational and technical dynamics. Developmental outcomes, however, may be quite similar. Decentralized plants are likely to be largely independent of the territorial economies in which they are implanted, generating only minor spillover effects. By contrast, the agglomerations of small, densely inter-linked plants generate intense activity flows within the local area. These seem the most probable loci for technical innovation (Scott 1986a, Storper & Scott 1988).

Questions remain concerning the relationship between large, integrated firms and their functionally disintegrated subcontractors and suppliers. For example, the shift to the just-in-time (JIT) system, which requires smaller, more frequent shipments from specialized suppliers to their large customers, appears to increase the pressure for proximity between the two (Sheard 1983, Estall 1985, Sayer 1985b, Holmes 1986). In this case, the nature of inter-firm linkages may change independently of organizational characteristics and may give rise to denser linkage networks in the vicinity of the large integrated plant, altering the regional growth patterns associated with this type of facility. Similarly, the nature of technical interdependencies between elements of the production system, the non-linearity of production processes, and the variety of approaches to structuring the division of labour and the organization of firms and industries may be underestimated by the model (Walker 1988b). Further, as Gertler (1987) notes, the universality of the model of flexibly specialized production agglomerations and the degree to which they are spatially bounded in fact remain to be demonstrated. Nevertheless, while the logic of integration/disintegration may not entirely explain the spatial allocation of production, it provides a powerful vantage point from which to analyze spatial dynamics.

Industry concentration and regional development

A focus on the presence of oligopoly offers a different perspective on the spatial behaviour of firms and regional development. Drawing on theories of oligopolistic competition, elements of the product cycle, and a Schumpeterian model of technical change, Markusen (1985) posits the existence of a profit cycle with distinctive spatial correlates. By introducing the possibility of sustained market power, the emphasis on oligopolistic behaviour incorporates strategic behaviour by firms. Freed from the determinant exigencies of pure price competition, at least under certain circumstances, production costs are no longer viewed as the unique location criterion: market power, in so far as it is successfully maintained, can lead to self-reinforcing regional growth based on high wages and high returns to capital.

Crucially, these arguments are set within a technologically dynamic framework that acknowledges the rôle of competition and the drive to accumulate capital in continually transforming product *and* process technology. These features allow the model to be less mechanistic and unidirectional than the product cycle. In particular, much depends on whether firms in an industry can stave off excess competition over the course of the cycle.

The profit stages are associated with changes in other key variables, including output, employment, occupational structure, and spatial patterns. In essence, changing corporate strategies over the profit cycle (ranging from a focus on design and engineering in the initial stage to developing efficient mass production techniques and, eventually, rationalization, cost cutting, and disinvestment) alter factor demand and sensitivity to market proximity. These, in turn, suggest rather distinct spatial patterns. Generally, the pattern of spatial succession is early concentration followed by dispersal. The exact pattern is regulated in the first instance by the changing demand for labour (e.g. from technical and skilled production labour to unskilled), although the timing is heavily influenced by the way the industry proceeds through the cycle. Thus, oligopolization can retard this dispersion, leading in effect to the relative overdevelopment of an industry in a particular region (see also Chinitz 1960). This prepares the ground for an eventual regional crisis as the advent of rationalization and disinvestment resulting from the loss of market power is also highly confined geographically.

Two different kinds of regional growth may be anticipated. Regions hosting the early stages of the cycle are in a favourable position as rapid employment expansion is concentrated in highly qualified, high-income occupational sectors (both blue- and white-collar). Regions that inherit an industry in later stages experience growth, but employment is biased towards low-skill, low-income occupations. Growth may also be short-lived as the industry may already be in decline.

While profit cycle theory is more flexible than product cycle, it shares certain weaknesses of a cyclical approach, as its author acknowledges. Notably, its sectoral specificity, which is in many respects a great strength of the model, means that inter-industry and inter-firm linkages in both their spatial and aspatial aspects are largely left aside. Sector-based models are also relatively insensitive to structural and cyclical changes in the macro-economy. The sectors are inevitably viewed as having relatively autonomous dynamics. Nevertheless, to the extent that oligopolistic structures constitute a strategically important target for analysis given their prominence in the broader economy, the model helps us to come to grips with critical aspects of the connections between corporate strategic behaviour and regional development.

Labour and the geography of production

The advent of an explicitly Marxist geography dramatically altered the basis of regional analysis. At the most general level, the focus shifts to the dynamics of capital accumulation in which geographical processes are deeply implicated (Harvey 1982). The very unevenness of spatial development patterns and the continual restructuring of spatial relationships are part and parcel of this process (Walker 1978, Harvey 1982, Massey 1984, Smith 1984, Scott & Storper 1986, Peet 1987, Storper & Walker 1989). Within this general context, the point of departure for most Marxist regional geography is the problem of production in two critical dimensions – the labour process and the capital–labour relation.

The labour process and the geography of production

The labour process encompasses the tasks workers perform in order to create goods and services, how they are subdivided and organized, and how the relationship of workers to machines is structured. The division of labour occurs in several dimensions, including: the social, referring to different commodities produced by different production units; the technical or detail, encompassing the various tasks that go into the production of a given commodity; and the spatial, or the geographical allocation of sectors, firms, and functional activities (Marx 1967, Massey 1984; Smith 1986, Walker 1988a). In part, the problem facing regional geographers is to understand how these are interrelated and how and why they change over time.

The material basis of production across industries shapes the labour process and, consequently, the demand for labour. A range of skills and of levels of automation are found. These include craft-based batch production requiring a high level of technical skills and general problem-solving ability, as in aircraft production or the construction industry. Continuous flow processes, characteristic of petrochemicals and oil refining, require specific technical skills and the ability to deal with non-routine events. Because they are extremely capital-intensive activities, labour costs constitute a small share of total costs. By contrast, standardized, semi-automated assembly lines, found in the home appliance and automobile industries, rely on low- to semi-skilled workers performing highly repetitive tasks with the intensity of work essentially controlled by the pace of the assembly line. Electronics assembly and garments production traditionally rely on extremely labour intensive, unskilled, manual assembly processes (Storper & Walker 1984). These are just a few examples, but they indicate the persistent diversity of labour processes and forms of investment across sectors (Massey & Meegan 1982, Massey 1984, Walker 1988a, 1988b).

Moreover, changes in the labour process consequent upon the introduction of new technologies or principles of organization can have a variety of effects. Braverman's influential work on the evolution of the labour process under capitalism highlights only one of these – the progressive deskilling of labour to bring it more firmly under the control of capital (Braverman 1974, cf. also Noble 1978, 1984). This thesis has been criticized by a number of writers (cf. Edwards 1979, Burawoy 1979, Harvey 1982). Most pertinent here is the argument that the transformation of labour processes can be accompanied by the resynthesis of tasks and a renewed demand for skilled labour as when, for example, the application of computer controls creates a demand for skilled technicians and programmers. There is no unique evolutionary path (Walker 1988a, 1988b).

As a general principle, the more specialized and skilled the demand for labour, the more constrained are the potential location choices for the firm. Production is likely to remain in regions with a well articulated industrial structure, despite high wages and even high levels of unionization. The lowest skilled activities are the most susceptible to decentralization to peripheral areas. Growth in these regions is based on low wages and a compressed occupational and social structure.

Despite this diversity, attention has tended to focus on the semi-automated, mass production assembly line. This process is associated with the deskilling of labour and the decentralization of production to the periphery in order to escape expensive and highly unionized labour. The relocation of capital in this way means that growth in one region occurs at the expense of employment and investment in another region (Bluestone & Harrison 1982, Peet 1983, 1987).

The labour process argument stresses factors exogenous to a given region in that it tends to look first at the broader dynamics of technical and industrial change in explaining spatial change. The labour relations analysis, by contrast, tends to focus on conditions internal to the region seeking to understand, for example, how problems in one spatially bounded labourmarket lead to shifts in spatial growth patterns.

Labour relations

In this approach, the prospects for accumulation in a given region are seen to hinge principally on the state of class relations (Frobel *et al.* 1980, Peet 1987). Where labour is relatively strong (as reflected in high wages, militant union traditions, labour-market rigidities, and state policies that are favourable to labour), capital has an incentive to shift investment geographically in search of a more compliant, lower cost labourforce to bolster profits. Associated with the analysis of corporate structure and spatially separable functional activities, we arrive at the notion of a new international division of labour (NIDL) with regional growth patterns shaped by the way a particular region is incorporated into this functional hierarchy (Frobel *et al.* 1980, Moulaert & Salinas 1983, Susman 1984).

As suggested, this analysis has been primarily concerned with the relocation of mass production and consequent shifts in the patterns of regional growth and decline. Changes in the labour process are quite important to the analysis as they provide the means for such a shift to be implemented. The thesis of the progressive deskilling of the labourforce is central here. In sum, the underlying dynamic of both technical and spatial change revolves around the problem of labour control.

This analysis is reflected in a number of different cases. For example, Bluestone & Harrison's (1982) influential analysis of deindustrialization in the US stresses the role of capital shift both within the US and abroad in search of more compliant labour. Harrison (1984) also shows, for New England, Mass., how a period of deindustrialization and regional crisis can be succeeded by renewed regional growth precisely because the local labourforce has been severely disciplined by the crisis.

Similarly, others have shown that the growth of employment in the US Sunbelt is largely a function of the availability of low-wage, non-union workers, together with good business climates, i.e. favourable local government attitudes towards capital rather than labour reflected in, for example, so-called 'right-to-work' legislation that inhibits union organizing (Perry & Watkins 1978, Mollenkopf 1981, Clark 1981, Peet 1983).

In Britain, Massey and Meegan have identified various categories of industrial change associated with differing geographical patterns of growth and decline (Massey & Meegan 1982, Massey 1984). Crucially, in sectors characterized by investment in new production techniques, growth is often associated with spatial shifts of investment to new areas (i.e. suburbs or peripheral regions) to capitalize on the availability of non-union, often female, hence low-wage and relatively unorganized labour. Massey emphasizes that neither capital nor labour are undifferentiated categories. Capital can be distinguished according to type of firm, sector, and the like, while the character of local labourforces is shaped by the locality's history and may encompass a number of segments that are drawn into the production process on different terms. A series of British case studies has taken up the theme of local differentiation to explain divergent outcomes (Martin & Rowthorn 1986, Cooke 1988). Similarly, for Italy, Del Monte & Giannola (1978) and Martinelli (1985) have argued that a large part of the investment in the depressed southern region in the early 1970s was motivated by the desire to escape a legacy of acute labour conflict in the industrialized north.

A more formal model of these processes is presented by Clark *et al.* (1986). In contrast to analyses fixing the date of the reversal in regional growth patterns at the end of the 1960s and the beginning of the 1970s, and particularly in opposition to the catastrophe theory model proposed by Casetti (1981), they argue that the shift in regional growth patterns was gradual and already underway during the boom period of the 1950s. Their central thesis is that regional economic differentiation and disequilibrium are the product of the accumulation of short-run adjustments in, for example, output and employment. The adjustments reflect the allocation of the burden of economic uncertainty between capital and labour, emphasizing the importance of contractual relations and the rigidities these impose on the way adjustments can be made. Key issues concern conditions of employment and economic security and the distribution of income shares between capital and labour.

Notably, the approach does not assume a wholesale abandonment of mature regions which, in fact, continued to attract investment over this period. However, due to technical changes in production processes that tended to be increasingly labour-saving with each new vintage of capital equipment, a given amount of capital investment employed fewer and fewer people over time. Thus, continued investment co-exists with rising unemployment, lending a different aspect to our understanding of the nature of the regional crisis (Varaiya & Wiseman 1981, Gertler 1984b, Luger 1986, Clark *et al*. 1986).

Fordism, spatial divisions of labour, and regional growth

The two issues of labour process and labour relations are closely intertwined and their separation is partly for ease of exposition. For example, the objective of exerting greater control over labour can inspire *both* changes in the labour process and geographical shifts of production. And changes in the labour process may be a necessary precondition for the spatial reallocation of production in favour of low-cost, less militant labour markets.

A good example of how the two factors come together is provided by the analysis of the crisis of Fordism associated with the French Regulationist School which identifies historically specific regimes of accumulation (Aglietta 1979, Lipietz 1982, 1986). These can be described as 'ensembles of productive forces and relations' (Scott & Storper 1986) associated with different forms of spatial organization (Swyngedouw 1987). Fordism most narrowly refers to the articulation between the technology and organization of production and the social basis of consumption characteristic of a particular historical epoch. In the postwar period, the ascendency of mass production based on the semi-automatic assembly line gave rise to huge productivity gains with capital accumulation driven primarily by the extraction of relative surplus value. The basis for mass consumption, to absorb the tremendous output of the production system, arises chiefly from a transformation of the conditions of the working class, especially through the institutionalization of the capital–labour relation in collective bargaining, where relative labour peace is traded for wage increases in line with productivity growth. In effect, productivity gains are shared with labour. The situation remains stable – and accumulation proceeds apace – so long as productivity continues to increase and the maintenance of a relatively stable oligopoly permits firms to engage in administered pricing strategies which allow them to pay high wages without threatening profits (Aglietta 1979).

Crisis occurs when these conditions break down. The system hits technical limits that cannot be ameliorated by further refinements of technique or the division of labour (Aglietta 1979). At the same time, the intensification of the labour process eventually leads to strong worker resistance, despite the collective bargaining agreement (Lipietz 1982, 1986). High wages and stagnant productivity undermine the accumulation process, a situation which is exacerbated by the rise of international competition from low-wage countries consequent on the geographic spread of Fordist production methods (Gramsci 1971, Aglietta 1982).

While the system holds, production and accumulation can continue in high-wage regions since capital is protected from labour unrest and profit margins remain high. Indeed, this forms the basis of growth in the industrial core regions. In crisis, capital seeks to free itself from the constraints of its contractual relationship with labour by shifting to peripheral areas in search of lower-cost, unorganized labour that can be brought into an essentially unchanged production system on much harsher terms in order to restore profits (Lipietz 1982, 1986). Industrial decline in the core is paralleled by the growth of investment in peripheral regions (domestically and abroad), but this growth is characterized by much less favourable income and social structures.

The early versions of the NIDL thesis have been extensively criticized for not taking into account the prospects for transforming production processes to increase productivity and ameliorate the problem of high wages in core industrial areas, allowing growth to continue in these regions. As with the product cycle, there is a tendency to treat the evolution of industry and spatial structure in an excessively mechanistic manner (Jenkins 1984, Storper 1985a, Sayer 1985a, Herold & Kozlov 1987, Schoenberger 1988b). The regimes of accumulation approach, however, does provide an explanation for why this route is foreclosed within the boundaries of the Fordist production system, suggesting that a spatial strategy to circumvent the problem of stagnant productivity and profits is the prime alternative.

It should be recalled, however, that there is a wide range of industries that is not organized according to Fordist technical principles and whose labour processes are not susceptible to the pattern of deskilling and decentralization stressed in the NIDL (Walker 1988a, Sayer 1988). Further, the prospect of a reorganization of the technical basis of production that bypasses the limits inherent to Fordism suggests that yet another restructuring of spatial relationships may occur.

This appears to be the case with the advent of flexible technologies and new ways of organizing production such as just-in-time (Sayer 1985b, Holmes 1985, Schoenberger 1987, 1988b). More broadly, the spread of the electronics revolution through the economic system, reconstituting traditional industries and creating new ones, may signal the emergence of a new accumulation regime with its own characteristic patterns of economic, social, and spatial organization (Swyngedouw 1987, Harvey 1987a, 1987b).

Our ability to predict the precise outlines of this reorganization is limited. It has been hypothesized, however, that traditional industrial regions, that have borne the brunt of Fordist restructuring, are hampered by the legacy of social, institutional, political, and physical (the built environment) traditions that were developed under Fordism. The shift to a new regime of accumulation is likely to be accompanied by the incorporation of new territories on the periphery of the old industrial agglomerations that can be structured according to the requirements of the new system, heralding a new pattern of regional growth and decline (Swyngedouw 1987, Storper & Scott 1988).

This has unleashed a new wave of territorial competition for industrial growth because where the new industrial complexes will be established is still open to question. In favour of traditional industrial agglomerations is the fact that their labourforces have been severely chastened by the crisis, thus perhaps tipping the balance in the capital–labour relationship. In many cases, they still afford extensive networks of specialized and flexible suppliers that can be drawn into reorganized production networks. Moreover, cities and regions in the old industrial core have embarked on a process of restructuring their built environments and reshaping political and institutional structures in an effort to lure investment (Harvey 1987b). Whether they can succeed in renewing the conditions for accumulation remains to be seen.

Conclusion

In recent decades regional theory has sought to understand spatial change as an historical and dynamic phenomenon: regions are the historical products of a complex interplay of forces internal and external to the regions themselves that create the conditions for further rounds of profitable investment or, alternatively, create barriers to continued growth. These conditions include the character of technological change, the form and organization of firms and industries, the creation and transformation of labour markets, the development of geographically specific patterns of social, political, and institutional practices, and the enduring nature of the built environment. These conditions are themselves shaped by broader structural changes and macro-economic conditions at the national and international level. The historical development of places and their characteristics are both a product of, and in turn influence, these broader processes (Massey 1984). For these reasons, the continual restructuring of industry and space is understood as a permanent feature of capitalist economies (Walker 1978, Harvey 1982, Smith 1984, Massey 1984).

Within this context, the new approaches have focused on capital and labour, in all their dimensions, as active agents of change. Capital is analyzed in terms of the kind of industry in which it is invested, the kind of firm controlling it, the kind of equipment in which it is embodied, and, therefore, the kind of labour it requires to produce goods and services. Labour as a factor input and labour markets are distinguished according to skills, experience, wage levels, attitudes, and cultural and political practices. The strategies and behaviour of capital and labour are intimately linked with the pattern and form of regional growth.

To analyze these processes, recent theory has cut across the boundaries compartmentalizing traditional analyses. The conceptual divides between industrial location and regional development, the local and the global, the micro and the macro, the spatial and the aspatial, have all been eroded. Instead, the interrelationships among these have been brought to the fore.

If we ask if this enterprise has met with success in the form of the creation of a definitive model of regional growth and change, the answer must be no. If anything, the boundaries of analysis have grown so wide that it is sometimes difficult to keep a firm grasp of the essential dynamics that have been identified. The world seems to have grown messier rather than neater.

Indeed, one of the underlying lessons of this body of work may be that models of regional change of the sort that claim predictive power concerning specific regional outcomes are not what we should be seeking to construct. This does not mean that theory is beyond our reach and that our only recourse is to an endless series of regional case studies in order to trace out the particular complex of circumstances that led to a particular result. Rather, it means that theorizing processes of regional change is very difficult because the processes themselves are not wholly determined *a priori*.

We have made considerable progress in identifying and analyzing the fundamental processes and dynamic relationships underlying regional change. But we are still far from resolving the way these are articulated over time and space. Notably, if the divides between local and global or micro and macro have been eroded, as suggested above, they have not been entirely vanquished. The relationship between local processes and structural changes stubbornly resists satisfactory theorization.

By the same token, historical awareness has sometimes resulted in a certain historical boundedness where the characteristics of a particular historical epoch (monopoly capitalism, late Fordism, etc.) have been interpreted as, in some sense, the ultimate expression of capitalist developmental processes and spatial tendencies. The relationship between historically specific circumstances and the principles underlying capitalist development in general remains a thorny one to investigate.

In the final analysis, definitive theories of regional change elude us because the primary agents of change, capital and labour, are so extraordinarily creative, each in their own way. Firms pursue a seemingly endless variety of strategies to overtake their rivals, to restructure production, and to manage their relationship with labour, all to the end of steadily increasing the profits they are able to accumulate. The use of space is part of this strategic array and it would be surprising indeed to find a unique spatial strategy in the midst of this diversity. By the same token, labour brings to the process of production, to its relationship with capital, to the process of its own reproduction, all the ingenuity of which the human mind is capable. But, despite appearances, this diversity is neither random nor unstructured. At bottom lie the exigencies of production and reproduction of the system. It is the goal of regional theory, as yet unfulfilled, to bring the various strands of analysis together in a coherent analysis of processes of regional change.

Notes

Particular thanks are owed to Flavia Martinelli and Dick Peet for their comments and suggestions on an earlier draft. Responsibility for the contents of this chapter, however, remains with the author.

1 Some care should be taken to distinguish between two different kinds of regional-problem. The first, stressed here, has to do with 'deindustrialization', or the crisis of traditional manufacturing areas, which has manifested itself with increasing severity since the late 1960s. The second concerns traditionally underdeveloped, agricultural regions that failed to make the transition to fully capitalist industrialization in the 19th and 20th centuries. While in the US, the formerly underdeveloped southern region has lately experienced considerable growth, this has been less true for lagging areas in other advanced countries, such as the Italian Mezzogiorno (Martinelli 1985, Holland 1976).

2 Alternatively, adaptations of international trade theory hold capital and labour immobile with the flow of goods across spatial boundaries serving to equalize factor returns in different places.

3 Diverse causes for this tendency are posited. For example, the leading regions become increasingly efficient producers as the scale of output grows, while high returns to both labour and capital induce resource flows from the lagging regions to the core.

4 Mechanization of production, by reducing the firm's demand for labour, may itself attenuate the importance of seeking low-wage labour in the periphery.
5 Vernon (1974) recognizes this possibility for a limited number of what he terms 'senescent oligopolies'.

References

Aglietta, M. 1979. *A theory of capitalist regulation*. London: New Left Books.

Aglietta, M. 1982. World capitalism in the eighties. *New Left Review* **134**, 3–41.

Armington, C. 1986. The changing geography of high technology businesses. In *Technology regions and policy*, J. Rees (ed.). Totowa, NJ: Rowman & Littlefield.

Armington, C., C. Harris & M. Odle 1983. Formation and growth in high technology business: a regional assessment. Washington, DC: The Brookings Institution (mimeo).

Berry, B. J. L. 1972. Hierarchical diffusion: the bases of developmental filtering and spread in a system of growth centers. In *Growth centers in regional economic development*, N. Hansen (ed.). New York: Free Press.

Blackaby, F. (ed.) 1979. *De-industrialization*. London: Heinemann.

Bluestone, B. & B. Harrison 1982. *The deindustrialization of America*. New York: Basic Books.

Borts, G. H. 1960. The equalization of returns and regional economic growth. *American Economic Review* **50**, 319–47.

Borts, G. H. & J. L. Stein 1968. Regional growth and maturity in the United States: a study of regional structural change. In *Regional analysis*, L. Needleman (ed.). London: Penguin.

Braverman, H. 1974. *Labour and monopoly capital: the degradation of work in the twentieth century*, New York: Monthly Review Press.

Burawoy, M. 1979. *Manufacturing consent: changes in the labor process under monopoly capitalism*. Chicago: University of Chicago Press.

Carney, J., R. Hudson & J. Lewis (eds) 1980. *Regions in crisis: new perspectives in European regional theory*. London: Croom Helm.

Casetti, E. 1981. A catastrophe model of regional dynamics. *Annals of the Association of American Geographers* **71**, 572–9.

Castells, M. 1985. High technology, economic restructuring and the urban-regional process in the United States. In *High technology, space and society*, M. Castells (ed.). Beverly Hills: Sage Publications.

Chandler, A. 1962. *Strategy and structure: chapters in the history of American industrial enterprise*. Cambridge, Mass.: MIT Press.

Chinitz, B. 1960. Contrasts in agglomeration: New York and Pittsburgh. *American Economic Review*, papers and proceedings, **50**, 279–89.

Clark, G. L. 1981. The employment relation and the spatial division of labour. *Annals of the Association of American Geographers* **71**, 412–24.

Clark, G. L., M. S. Gertler & J. Whiteman 1986. *Regional dynamics: studies in adjustment theory*. Boston: Allen & Unwin.

Coase, R. H. 1937. The nature of the firm. *Economica* **4**, 386–405.

Cooke, P. 1983. *Theories of planning and spatial development*. London: Hutchinson.

Cooke, P. 1986. The genesis of high technology complexes: theoretical and empirical considerations. Paper presented at the Anglo-American Workshop on the Growth and Location of High Technology Industry, Cambridge, June. Mimeograph.

Cooke, P. 1988. *Localities*. London: Hutchinson.

Darwent, D. 1974. Growth poles and growth centers in regional planning: a review. In *Regional policy: readings in theory and applications*, J. Friedmann & W. Alonso (eds). Cambridge, Mass.: MIT Press.

Del Monte, A. & A. Giannola 1978. *Il Mezzogiorno nell'economia italiana*. Bologna: Il Mulino.

Dicken, P. 1976. The multiplant business enterprise and geographic space: some issues in the study of external control and regional development. *Regional Studies* **10**, 401–12.

Dicken, P. 1986. *Global shift: industrial change in a turbulent world*. London: Harper & Row.

Durand, J.-P., J. Lojkine, C. Mahieu & J. Durand 1984. *Formation et informatisation de la production: le cas de l'automobile*. Vitry: Centre d'Etudes Sociales sur l'Informatisation de la Production.

Edwards, R. 1979. *Contested terrain: the transformation of the workplace in the 20th century*. New York: Basic Books.

Erickson, R. & T. Leinbach 1979. Characteristics of branch plants attracted to non-metropolitan areas. In *Non-metropolitan industrialization*, R. E. Lonsdale & L. Seyler (eds). New York: Halsted Press.

Estall, R. C. 1985. Stock control in manufacturing: the just-in-time system and its locational implications. *Area* **17**, 129–32.

Flynn, P. M. 1986. Technological change, the 'training cycle' and economic development. In *Technology, regions and policy*, J. Rees (ed.). Totowa, NJ: Rowman & Littlefield.

Freeman, C., J. Clark & L. Soete 1983. *Unemployment and technical innovation*. Westport, CT: Greenwood.

Frobel, F., J. Heinrichs & O. Kreye 1980. *The new international division of labor: structural unemployment in industrialized countries and industrialization in developing countries*. Cambridge: Cambridge University Press.

Gertler, M. S. 1984a. the dynamics of regional capital accumulation. *Economic Geography* **60**, 150–74.

Gertler, M. S. 1984b. Regional capital theory. *Progress in Human Geography* **8**, 50–81.

Gertler, M. S. 1987. Capital, technology and industry dynamics in regional development. *Urban Geography* **8**, 251–63.

Glasmeier, A. 1985. Innovative manufacturing industries: spatial incidence in the United States. In *High technology, space and society*, M. Castells (ed.). Beverly Hills: Sage Publications.

Gramsci, A. 1971. Americanism and Fordism. In *Selections from the prison notebooks*. London: Lawrence & Wishart.

Gross, H. T. & B. L. Weinstein 1986. Technology, structural change and the industrial policy debate. In *Technology, regions and policy*, J. Rees (ed.). Totowa, NJ: Rowman & Littlefield.

Hall, P. 1985. The geography of the fifth Kondratieff. In *Silicon landscapes*, P. Hall & A. Markusen (eds). Boston: Allen & Unwin.

Hall, P. and A. Markusen (eds) 1985. *Silicon landscapes*. Boston: Allen & Unwin.

Hamilton, F. E. I. (ed.) 1974. *Spatial perspectives on industrial organization and decision making*. London: Wiley.

Hamilton, F. E. I. & G. J. R. Linge (eds) 1979. *Spatial analysis, industry and the industrial environment*. New York: Wiley.

Hansen, N. 1970. Development pole theory in a regional context. In *Regional economics*, D. McKee, R. Dean & W. Leahy (eds). New York: Free Press.

Hansen, N. 1979. The new international division of labour and manufacturing decentralization in the United States. *Review of Regional Studies* **9**, 1–11.

Harrison, B. 1984. Regional restructuring and good business climates: the economic transformation of New England since World War II. In *Sunbelt/snowbelt*, W. Tabb & L. Sawers (eds). Oxford: Oxford University Press.

Harvey, D. 1982. *The limits to capital*. Oxford: Basil Blackwell.

Harvey, D. 1987a. The geographical and geopolitical consequences of the transition from Fordist to flexible accumulation. Presented at the conference on America's New Economic Geography, Washington, DC, April 29–30.

Harvey, D. 1987b. Flexible accumulation through urbanization: reflections on 'post-modernism' in the American city. *Antipode* **19**, 260–86.

Herold, M. W. & N. Kozlov 1987. A new international division of labor? The Caribbean example. In *The year left 2*, M. Davis (ed.). London: Verso.

Hirschman, A. O. 1958. *The strategy of economic development*. New Haven, Conn.: Yale University Press.

Holland, S. 1976. *Capital vs. the regions*. London: Macmillan.

Holmes, J. 1983. Industrial reorganization, capital restructuring and locational change: an analysis of the Canadian automobile industry in the 1960s. *Economic Geography* **59**, 251–71.

Holmes, J. 1985. Industrial change in the Canadian automotive products industry, 1973–1984: the impact of technical change in the organization and locational structure of automobile production. Paper presented to the IGU Commission on Industrial Change Conference on Technology and Industrial Change, Nijmegen, The Netherlands, August 19–24, and to the IBG/CAG Industrial Geography Symposium on Technical Change in Industry, University College, Swansea, Wales, August 22–6.

Holmes, J. 1986. The organization and locational structure of production subcontracting. In *Production, work, territory: the geographical anatomy of industrial capitalism*, A. Scott & M. Storper (eds). Boston: Allen & Unwin.

Hoover, E. & R. Vernon 1959. *Anatomy of a metropolis*. Cambridge, Mass.: Harvard University Press.

Hymer, S. 1972. The multinational corporation and the law of uneven development. In *Economics and world order*, J. Bhagwati (ed.). London: Macmillan.

Jenkins, R. 1984. Divisions over the international division of labor. *Capital and Class* **22**, 28–57.

Kaldor, N. 1970. The case for regional policies. *Scottish Journal of Political Economy* **17**, 337–48.

Krumme, G. 1969. Toward a geography of enterprise. *Economic Geography* **45**, 30–40.

Lipietz, A. 1982. Towards global Fordism? *New Left Review* **132**, 33–47.

Lipietz, A. 1986. New tendencies in the international division of labor: regimes of accumulation and modes of regulation. In *Production, work, territory: the geographical anatomy of industrial capitalism*, A. Scott & M. Storper (eds). Boston: Allen & Unwin.

Luger, M. 1984. Does Carolina's high tech development program work? *Journal of the American Planning Association* **50**, 280–9.

Luger, M. 1986. Depreciation profiles and depreciation policy in a spatial context. *Journal of Regional Science* **26**, 141–59.

McDermott, P. J. & D. Keeble 1978. Manufacturing organization and regional employment change. *Regional Studies* **12**, 247–66.

Malecki, E. 1979. Locational trends in R&D by large US corporations, 1965–1977. *Economic Geography* **55**, 309–23.

Malecki, E. 1980. Corporate organization of R&D and the location of technical activities. *Regional Studies* **14**, 219–34.

Malecki, E. 1983. Technology and regional development: a survey. *International Regional Science Review* **8**, 89–126.

Malecki, E. 1984. High technology and local economic development. *Journal of the American Planning Association* **50**, 262–9.

Malecki, E. 1986. Technological imperatives and modern corporate strategy. In *Production, work, territory: the geographical anatomy of industrial capitalism*, A. J. Scott & M. Storper (eds). Boston: Allen & Unwin.

Mandel, E. 1978. *Late capitalism*. London: Verso.

Markusen, A. R. 1985. *Profit cycles, oligopoly and regional development*. Cambridge, Mass.: MIT Press.

Markusen, A. R. 1986. Defense spending and the geography of high technology industries. In *Technology, regions and policy*, J. Rees (ed.). Totowa, NJ: Rowman & Littlefield.

Markusen, A. R. & R. Bloch 1985. Defensive cities: military spending, high technology and human settlements. In *High technology, space and society*, M. Castells (ed.). Beverly Hills: Sage Publications.

Markusen, A. R., P. Hall & A. Glasmeier 1986. *High tech America: the what, how, where and why of sunrise industries*. Boston: Allen & Unwin.

Martin, R. & R. Rowthorn (eds) 1986. *The geography of deindustrialization*. London: Macmillan.

Martinelli, F. 1985. Public policy and industrial development in southern Italy: anatomy of a dependent industry. *International Journal of Urban and Regional Research* **9**, 47–81.

Martinelli, F. 1986. Producer services in a dependent economy: their role and potential for regional economic development. Unpublished PhD dissertation, University of California, Berkeley, Department of City and Regional Planning.

Marx, K. 1967. *Capital*, vol. I. New York: International Publishers.

Massey, D. 1984. *Spatial divisions of labor: social structures and the geography of production*. New York: Methuen.

Massey, D. 1985. Which 'new' technology? In *High technology, space and society*, M. Castells (ed.). Beverly Hills: Sage Publications.

Massey, D. & R. Meegan 1982. *The anatomy of job loss: the how, why and where of employment decline*. London: Methuen.

Mensch, G. 1979. *Stalemate in technology*. Cambridge, Mass.: Ballinger.

Mollenkopf, J. 1981. Paths toward the post-industrial city: the northeast and southwest. In *Cities under stress*, R. Burchell & D. Listokin (eds). Piscataway, NJ: Rutgers University, Center for Urban Policy Research.

Moriarty, B. 1986. Productivity, industrial restructuring and the deglomeration of American manufacturing. In *Technology, regions and policy*, J. Rees (ed.). Totowa, NJ: Rowman & Littlefield.

Moulaert, F. & P. W. Salinas (eds) 1983. *Regional analysis and the new international division of labor*. Boston: Kluwer-Nijhoff.

Myrdal, G. 1957. *Economic theory and underdeveloped regions*. London: Duckworth.

Noble, D. 1978. Social choice in machine design. *Politics and Society* **3/4**, 313–47.

Noble, D. 1984. *Forces of production*. New York: Knopf.

North, D. 1974a. Location theory and regional economic growth. In *Regional policy: readings in theory and applications*, J. Friedmann & W. Alonso (eds). Cambridge, Mass.: MIT Press.

North, D. 1974b. A reply. In *Regional policy: readings in theory and applications*, J. Friedmann & W. Alonso (eds). Cambridge, Mass.: MIT Press.

Norton, R. D. and J. Rees 1979. The product cycle and the spatial decentralization of American manufacturing. *Regional Studies* **13**, 141–51.

Oakey, R. 1985a. High technology industry and agglomeration economies. In *Silicon landscapes*, P. Hall & A. Markusen (eds). Boston: Allen & Unwin.

Oakey, R. 1985b. *High technology small firms*. London: Frances Pinter.

Oakey, R., A. Thwaites & P. Nash 1980. The regional distribution of innovative manufacturing establishments in Britain. *Regional Studies* **14**, 235–53.

Office of Technology Assessment (OTA) 1984. Technology, innovation and regional economic development. Washington, DC: US Congress, OTA.

Park, S. O. & J. O. Wheeler 1983. The filtering down process in Georgia: the third stage in the product life cycle. *Professional Geographer* **35**, 18–31.

Peet, R. 1983. Relations of production and the relocation of US manufacturing industry since 1960. *Economic Geography* **59**, 112–43.

Peet, R. 1987. *International capitalism and industrial restructuring*. Boston: Allen & Unwin.

Perloff, H., E. Lampard & R. Muth 1960. *Regions, resources and economic growth*. Baltimore, MD: Johns Hopkins University Press.

Perrons, D. C. 1981. The role of Ireland in the new international division of labor: a proposed framework for regional analysis. *Regional Studies* **15**, 81–100.

Perroux, F. 1950. Economic space: theory and applications. *Quarterly Journal of Economics* **64**, 89–104.

Perroux, F. 1970. Note on the concept of growth poles. In *Regional economics: theory and practice*, D. McKee, R. Dean & W. Leahy (eds). New York: Free Press.

Perry, D. & A. Watkins (eds) 1978. *The rise of the sunbelt cities*. Beverly Hills: Sage Publications.

Piore, M. 1980. The technological foundations of dualism and discontinuity. In *Dualism and discontinuity in industrial societies*, S. Berger & M. Piore (eds). Cambridge; Cambridge University Press.

Piore, M. & C. Sabel 1984. *The second industrial divide*. New York: Basic Books.

Pred, A. 1966. *The spatial dynamics of US urban-industrial growth, 1800–1914*. Cambridge, Mass.: MIT Press.

Pred, A. 1977. *City systems in advanced economies*. London: Hutchinson.

Pred, A. 1980. *Urban growth and city systems in the United States, 1840–60*. Cambridge, Mass.: Harvard University Press.

Rees, J. 1979. Technological change and regional shifts in American manufacturing. *Professional Geographer* **31**, 45–54.

Rees, J. (ed.) 1986. *Technology, regions and policy*. Totowa, NJ: Rowman & Littlefield.

Rees, J. & H. A. Stafford, 1986. Theories of regional growth and industrial location: their relevance for understanding high technology complexes. In *Technology, regions and policy*, J. Rees (ed.). Totowa, NJ: Rowman & Littlefield.

Rees, J., G. F. D. Hewings & H. A. Stafford (eds) 1979. *Industrial location and regional systems*. Cambridge, Mass.: Bergman/Ballinger.

Richardson, H. 1973. *Regional growth theory*. New York: Wiley.

Sabel, C. 1982. *Work and politics: the division of labour in industry*. Cambridge: Cambridge University Press.

Saxenian, A. 1984. The urban contradictions of Silicon Valley: regional growth and restructuring in the semiconductor industry. In *Sunbelt/snowbelt*, W. Tabb & L. Sawers (eds). Oxford: Oxford University Press.

Saxenian, A. 1985. Silicon Valley and Route 128: regional prototypes or historic exceptions. In *High technology, space and society*, M. Castells (ed.). Beverly Hills: Sage Publications.

Sayer, A. 1985a. Industry and space: a sympathetic critique of radical research. *Environment and Planning D, Society and Space* **3**, 3–29.

Sayer, A. 1985b. New Developments in Manufacturing and their Spatial Implications. Working paper 49. University of Sussex, Department of Urban and Regional Studies.

Sayer, A. 1986. Industrial location on a world scale: the case of the semiconductor industry. In *Production, work, territory: the geographical anatomy of industrial capitalism*, A. J. Scott & M. Storper (eds). Boston: Allen & Unwin.

Sayer, A. 1988. Paper presented at the annual meetings of the American Association of Geographers, Phoenix, AZ.

Schoenberger, E. 1985. Foreign manufacturing investment in the United States: competitive strategies and international location. *Economic Geography* **61**, 3.

Schoenberger, E. 1986. Competition, competitive strategy, and industrial change: the case of electronic components. *Economic Geography* **62**, 321–33.

Schoenberger, E. 1987. Technological and organizational change in automobile production: spatial implications. *Regional Studies* **21**, 199–214.

Schoenberger, E. 1988. From Fordism to flexible accumulation: technology, competitive strategies and international locations. *Environment and Planning D, Society and Space* **6**, 245–62.

Schoenberger, E. 1989. Multinational corporations and the new international division of labor: a critical appraisal. *International Regional Science Review* (forthcoming).

Schumpeter, J. 1939. *Business cycles: a theoretical, historical and statistical analysis of the capitalist process.* New York: McGraw-Hill.

Schumpeter, J. 1961. *The theory of economic development.* Cambridge, Mass.: Harvard University Press.

Scott, A. J. 1983a. Industrial organization and the logic of intra-metropolitan location I: theoretical considerations. *Economic Geography* **59**, 223–50.

Scott, A. J. 1983b. Location and linkage systems: a survey and reassessement. *Annals of the Regional Science Association* **17**, 1–39.

Scott, A. J. 1984. Industrial organization and the logic of intra-metropolitan location II: a case study of the printed circuit industry in the greater Los Angeles area. *Economic Geography* **60**, 3–27.

Scott, A. J. 1986a. Industrialization and urbanization: a geographical agenda. *Annals of the Association of American Geographers* **76**, 25–37.

Scott, A. J. 1986b. Industrial organization and location: division of labor, the firm and spatial process. *Economic Geography* **62**, 215–31.

Scott, A. J. & M. Storper (eds) 1986. *Production, work territory: the geographical anatomy of industrial capitalism.* Boston: Allen & Unwin.

Shaiken, H. 1984. *Work transformed: automation and labor in the computer age.* New York: Holt, Rinehart, Winston.

Sheard, P. 1983. Auto production systems in Japan: organizational and locational features. *Australian Geographical Studies* **21**, 49–68.

Smith, A. 1986. *The wealth of nations, Books I–III.* Harmondsworth, Middlesex: Penguin.

Smith, N. 1984. *Uneven development: nature, capital and the production of space.* Oxford: Basil Blackwell.

Storper, M. 1985a. Oligopoly and the product cycle: essentialism in economic geography. *Economic Geography* **61**, 260–82.

Storper, M. 1985b. Technology and spatial productive relations: disequilibrium, interindustry relationships and industrial development. In *High technology, space and society*, M. Castells (ed.). Beverly Hills: Sage Publications.

Storper, M. & A. J. Scott 1989. The geographical foundations and social regulation of flexible production complexes. In *The power of geography*, J. Wolch & M. Dear (eds), 21–40. London and Winchester, Mass.: Unwin Hyman.

Storper, M. & R. Walker 1983. The theory of labor and the theory of location. *International Journal of Urban and Regional Research* **7**, 1–41.

Storper, M. & R. Walker 1984. The spatial division of labor: labor and the location of industries. In *Sunbelt/snowbelt*, W. Tabb & L. Sawers (eds). Oxford: Oxford University Press.

Storper, M. & R. Walker 1989. *The capitalist imperative.* Oxford: Basil Blackwell.

Susman, P. 1984. Capital restructuring and the changing regional environment. In *Regional restructuring under advanced capitalism*, P. O'Keefe (ed.). London: Croom Helm.

Swyngedouw, E. 1987. *High technology development and regional space: an assessment of the role of European telecommunications industries and their implications for regional development patterns.* Belgium: Catholic University of Leuven, Institute for Urban and Regional Planning.

Tabb, W. & L. Sawers (eds) 1984. *Sunbelt/snowbelt.* Oxford: Oxford University Press.

Taylor, M. J. 1975. Organizational growth, spatial interaction and locational decision making. *Regional Studies* **9**, 313–23.

Taylor, M. J. & N. J. Thrift 1982a. Industrial linkage and the segmented economy: 1. Some theoretical proposals. *Environment and Planning A* **14**, 1601–13.

Taylor, M. J. & N. J. Thrift 1982b. Industrial linkage and the segmented economy: 2. An empirical reinterpretation. *Environment and Planning A* **14**, 1615–32.

Taylor, M. J. & N. J. Thrift 1983. Business organization, segmentation and location. *Regional Studies* **17**, 445–65.

Thomas, M. D. 1975. Growth pole theory, technological change and regional economic growth. *Papers of the Regional Science Association* **34**, 3–25.

Thomas, M. D. 1985. Regional economic development and the role of innovation and technological change. In *The regional economic impact of technological change*, A. T. Thwaites & R. P. Oakey (eds). London: Frances Pinter.

Thwaites, A. T. & R. P. Oakey 1985. *The regional economic impact of technological change*. London: Frances Pinter.

Tiebout, C. M. 1974a. Exports and regional economic growth. In *Regional policy: readings in theory and applications*, J. Friedmann & W. Alonso (eds). Cambridge, Mass.: MIT Press.

Tiebout, C. M. 1974b. Rejoinder. In *Regional policy: readings in theory and applications*, J. Friedmann and W. Alonso (eds). Cambridge, Mass.: MIT Press.

US Congress, Joint Economic Committee 1982. *Location of high technology firms and regional economic development*. Washington, DC: Government Printing Office.

Varaiya, P. & M. Wiseman 1981. Investment and employment in manufacturing in US metropolitan areas, 1960–76. *Regional Science and Urban Economics* **11**, 431–69.

Vaughan, R. & R. Pollard 1986. State and federal policies for high technology development. In *Technology, regions and policy*, J. Rees (ed.). Totowa, NJ: Rowman & Littlefield.

Vernon, R. 1966. International trade and international investment in the product cycle. *Quarterly Journal of Economics* **80**, 190–207.

Vernon, R. 1974. The location of economic activity. In *Economic analysis and the multinational enterprise*, J. H. Dunning (ed.). New York: Praeger.

Walker, R. 1978. Two sources of uneven development under advanced capitalism: spatial differentiation and capital mobility. *Review of Radical Political Economics* **10**, 28–38.

Walker, R. 1985. Technological determination and determinism: industrial growth and location. In *High technology, space and society*, M. Castells (ed.). Beverly Hills: Sage Publications.

Walker, R. 1988a. Machinery, labour and location. In *The transformation of work?* S. Wood (ed.). London: Hutchinson.

Walker, R. 1988b. The geographical organization of production systems. *Environment and Planning D, Society and Space* **6**, 377–408.

Walker, R. & M. Storper 1981. Capital and industrial location. *Progress in Human Geography* **5**, 473–509.

Watts, H. D. 1980. *The large industrial enterprise: some spatial perspectives*. London: Croom Helm.

Watts, H. D. 1981. *The branch plant economy*. London: Longman.

Weinstein, B. & R. Firestone 1978. *Regional growth and decline in the United States*. New York: Praeger.

Westaway, J. 1974. The spatial hierarchy of business organizations and its implications for the British urban system. *Regional Studies* **8**, 145–55.

Williamson, O. 1975. *Markets and hierarchies*. New York: Free Press.

6 *Uneven development and location theory: towards a synthesis*

Neil Smith

If a room full of geographers in the 1960s was pressed to identify the part of their discipline which most exemplified their pretensions to practising a science, the majority would have pointed first to location theory. Location, it was felt by many, was the essential vocation of the geographer and the more mathematical the methodology the more evident the truth of the claim of science. It was apt, therefore, that location theory was a primary target for critique in the radical ferment that opened a variety of post-positivist approaches in the early 1970s. The lines of this critique are now well established. In so far as traditional location theory took neoclassical economics as its inspiration, it conveyed a narrow, ahistorical vision of the world, which abstained from questioning deeper societal assumptions. Under the guise of scientific objectivity, it accepted, indeed recapitulated, a social, economic, and political status quo which, in the words of Kenneth Boulding, 'was nothing to quo about'. The critique of location theory attempted to replace an essentially positivist and introverted focus with an understanding of the historical dynamism of locational change as an integral part of broader processes of geographical development.

Since much traditional location theory took as its basic building block the decisions of the individual firm, one of the first imperatives of critique and reconstruction involved an exploration of the broader structural forces and processes that constrained and guided corporate decisions. In geographical terms this implied a search not for the rules of individual investment decisions, but for the larger patterns and processes of geographical development. In short, researchers were looking for the 'big picture', the connection between society and space; instead of trying to identify the optimum site for a new supermarket, many geographers turned to identify the broader processes in which whole landscapes were made and remade, the gamut of location rules for different social functions and land uses, and the resulting geographical patterns. There was a sense that the human landscape was indeed ordered, but by the specific and contingent historical relationships of capitalist society rather than by the universal rules of neoclassical economics. From this perspective, the traditional effort to construct an autonomous location theory – autonomous, that is, from any theory of societal constitution and change – was misconceived. In the last decade and a half much of the innovative research following the alternative, critical tradition has sought to apply the broader societal perspective of theories of uneven development towards rebuilding location theory.

This project continues. It should not be conceived of as an abandonment of locational decision making at the local scale, but as a shift of focus allowing individual decisions to be situated in a broader social and economic context. The synthesis of location and uneven development theories seeks to create a geographical as well as historical continuum between macro and micro conceptions of economic location. It has proven more challenging than first envisaged, but with the dramatic geographical restructuring of cities, regions, and the international economy taking place since the early 1970s, the search for a synthesis of locational questions and theories of uneven development has also proven an exciting research agenda with immediate relevance. From the stitching of shirts to the accumulation of national debts and surpluses, things are not where they used to be, and from Taiwan to Cleveland existing places have changed almost beyond recognition. The power of theory is its ability to suggest the specific connections and relationships between different places and experiences, to supply a necessary explanatory framework, and to guide the process of asking more concrete empirical questions.

This chapter begins with a brief survey of the critique of traditional location theory; it then proceeds to a consideration of theories of uneven development; next some of the attempts at a new foundation for location theory are reviewed; and the final section focuses on problems within the emerging synthesis between locational and uneven development theories.

Critiques of location theory

The earliest influential English-language critique of location theory came in an article by Doreen Massey (1973) in the fledgling radical journal of geography *Antipode* (see also Schmidt-Renner 1966). Massey identified four different threads to location theory. The first, deriving largely from the work of Weber (1929), examined the locational choices and constraints of individual and independent manufacturing firms. The second involved a limited number of firms with interdependent locations (Hotelling 1929). The third represented a more recent behavioural approach to location decisions (Stevens 1961, Wolpert 1964). The fourth approach involved attempts to analyze whole economic landscapes, and in this category we might include a variety of theorists from von Thünen (Hall 1966) to Christaller (1966) to Lösch (1954). This is a useful typology of location theories, but it is important to add that specific contributions are rarely fully contained under any one of the four headings. The works of Isard (1956, 1960), Smith (1966), and Haggett (1965) represent explicit attempts to combine some, or all, of these approaches.

From Massey's critique and others (e.g. Gregory 1981, Webber 1984, Sayer 1985a), it is possible to distill four specific criticisms of location theory. First, in terms of the ideological roots of location theory, Massey contrasted its universal and scientific claims with its actual basis in the needs of the contemporary capitalist economy. Describing location theory as a technology of location, Eliot Hurst (1973) made a similar charge. Firm-centred like its neoclassical progenitor, location theory is more a manual of location policy for corporate executives than a detached science of economic landscape.

Second, positivist location theory is inherently ahistorical. The transformation of locational patterns and structures is conceived as resulting from essentially *ad hoc* alterations in the constraints, conditions, and motivations behind location decisions. Thus the energy crisis or loss of competitiveness are treated as *deus ex machina* factors that explain locational change. Whole landscapes, to the extent they are considered, are seen as the simple arithmetical sum of the individual location decisions and represent space economies of greater or lesser efficiency. But the same basic rules of location are deemed to apply in every society, however imperfectly translated on to the landscape through local culture and political filters. Thus, according to an earlier Japanese critic, location theory attempted to abstract 'a "pure" theory that would be applicable to every society' (Ohara 1950; quoted in Mizuoka 1986, p. 47).

Third, the neoclassical assumption that the economy tends inherently toward equilibrium is reflected in location theory. Individual firm decisions are presumed to result in the most efficient locations in geo-economic equilibrium with the larger economy. In the context of whole landscapes, Lösch's theory of location is an explicit attempt to derive a spatial equilibrium landscape. The possibility of crisis or extended disequilibrium – of sharp ruptures in the economic system and the logic of location – is not entertained; yet such discontinuities in the growth of the space economy are a periodic norm, and certainly typify the post-1973 experience. Here the ahistorical and equilibrium-centred character of location theory are mutually reinforcing.

Fourth, the crude conceptualization of geographical space in positivist geography also afflicts location theory (Harvey 1973, pp. 22–49). Positivist geography, according to Anderson (1973), indulged in a certain 'fetishism of space' that obscured the social relationships between people by treating them as spatial relationships between places. This is a practical issue as much as an academic one. To take just one example, the problem of rural poverty was and still is routinely conceived as the rural problem with the clear implication that it is spatial rather than social relationships that account for poverty. Acting on precisely this assumption, President Kennedy sought to alleviate Appalachian poverty by building roads into the region, thereby enhancing its geographical access and, he assumed, its economic wellbeing. The roads were well trafficked indeed, but in the opposite direction to that envisaged, as thousands of Appalachians migrated to escape poverty.

In location theory geographical space is reduced to an inert field of activity, due in large part to the fact that the theory's inspirational source – neoclassical economics – is deliberately aspatial. Locational analysis treats space as an adjunct to economic relationships, as merely the surface on which certain economic relationships are worked out. There is little real symbiosis between space and economy; from Weber to Lösch distance is the only active component of geographical space. As Massey (1973, p. 34) puts it, 'location theory deals essentially with some form of "abstract" space.' In fact, of course, 'the space of industrial location is the product of a complex historical process.'

The gap between abstract theories of location and the dynamically changing patterns of geographical development became increasingly evident in the 1970s. The resort to abstract modelling was matched by an opposite retreat into a descriptive empiricism (Keeble 1976). For those who attempted to retain some link between theory and empirical process, the relaxing of sometimes severe assumptions proved an empirical nightmare. The actual diversity of possible location decisions by interdependent firms very quickly led, as Massey suggested, into a 'cul-de-sac of complexity' (1973, p. 34). Political and intellectual frustration with location theory, above all with its inability to account for real geographical processes, patterns, and events, led to a search for new models of geographical location. The 1970s critique provoked a clear intellectual break with earlier questions and concerns, and the ensuing radical research involved an intense theoretical enquiry aimed largely at the broader geographical dimensions of economic and social development. The focus shifted decisively from firm-centred spatial abstractions and ahistorical formulations toward more general theoretical visions of development and, perhaps more significantly, from the regional scale on which location theory and a now-rejected regional geography had focused, towards the higher and lower scales of global and urban processes. At both scales there ensued a range of theoretical investigations accompanied by a more general intellectual 'breakthrough to Marxism' (Peet 1977, p. 16). By the 1980s the resulting theoretical ferment would reorient toward locational questions at the regional scale. In the meantime, however, a broad ranging enquiry into the causes and anatomy of geographically uneven development under capitalism (although not always called such) was especially fertile in generating theoretical concepts and frameworks that would be adapted for later use. Without wishing to diminish the importance of emerging urban theory, we shall focus here on theoretical developments at the global scale.

Theories of uneven development

In addition to rapid change in real world geographical processes, relationships, and landscapes, a dual theoretical inspiration underpinned the fashioning of theories of uneven development. In the first place, the classic texts of Marxism contained a few spartan phrases about the potency of the law of uneven development. These turned out to be less useful than was first assumed, and at times were even harmful (Smith 1984a, pp. xi–xii). More important by far was the vigorous debate over development and underdevelopment that impinged on most of the social sciences in the 1970s. This debate led to a variety of conceptions of uneven development, and geographers found a literature sympathetic to many of their critiques of positivism in general and location theory in particular. Underdevelopment theory was inherently historical, laid emphasis on questions of context and interrelationship, and aimed to comprehend not the locational rationale of individual capitals but the way in which capital in general, and specific capitals operating collectively, inspired a highly differentiated geography of development. What this literature did not provide was a sophisticated and well developed conception of geographic space.

Chronologically if not always intellectually, the development debate began with radical critiques of mainstream modernization theory. These critiques emanated largely from Latin America, which had been made to bear the brunt of modernization theory in practice with the Monroe Doctrine and the postwar Pax Americana. Dependency theory succeeded in connecting apparently opposite experiences; development and underdevelopment were deemed two sides of the same coin (Frank 1967, 1969, 1972). A powerful core of the world economy exploited a weak periphery, using its economic power to foster an enforced dependency. Underdevelopment resulted not from neglect but from active peripheral engagement in the world economy.

There emerged a critique of dependency theory (see especially Laclau 1971, Brenner 1977, Browett 1981, and for a good overview Blomström & Hettne 1984), which effectively confronted Marxist and radical theorists with the need for a rigorous understanding of underdevelopment. Directly or indirectly, this spawned a further array of different approaches to uneven development which we shall examine under the following headings: (a) unequal exchange; (b) the Regulation School; (c) theories of global production; (d) uneven geographical development. The link with location theory may at first seem obscure. This is no illusion, but rather testimony to the different origins and inspirations of Marxist theories of development and academic geography. The essentially geographic interest of these earlier debates should, however, be evident.

Unequal exchange

In his initial formulation of the theory of unequal exchange, Emmanuel (1972) argues that inequality between core and peripheral countries in the world economy should be understood in terms of unequal terms of trade. He rebuts the Ricardian theory of comparative advantage, on which conventional trade theory is based, and which argues that the exchange of commodities is by definition an exchange of equal values to the mutual benefit of both parties. In a highly technical argument, he also criticizes Marx's analysis of commodity exchange. When products are exchanged between low- and high-waged nations, there is a hidden transfer of value from the periphery to the core. Because of proportionately higher labour costs, commodities from the core generally sell at a higher price than those from peripheral countries. At a specific price level, the high-waged nation receives a greater quantity of 'labour value' from the low-waged economy than the periphery can buy from the core at the same price. The so-called 'exchange of equal labour values' on the market is therefore structurally unequal in favour of high-waged core countries.

Emmanuel's theory of unequal exchange sought to explain the exploitation and underdevelopment of peripheral regions in terms of differential wage structures in an otherwise global market for capital. But as many critics pointed out, this explanation attributes to Third World countries very little responsibility for their own development. Like dependency theory, unequal exchange theories assign Third World societies a subordinate and ultimately submissive role. Amin (1974a, 1976, 1977), a trenchant critic of Emmanuel, acknowledges the fact of wage differentials between developed and underdeveloped economies, but rejects the assumption that wages can be treated as an independent variable. Rather, there are clear historical reasons for international wage and productivity differentials integral to the processes of unequal development, and Emmanuel at best elaborates one mechanism perpetuating this inequality. It is necessary, according to Amin, to look beneath the asymmetries of trade to the fundamental relations of production that engender not just unequal exchange but a more pervasive unequal development. Thus Amin retains the notion of a structural asymmetry in the world economy but locates it in the sphere of production rather than the sphere of commodity exchange. He identifies two basic models of capitalist development – central and peripheral. In the central

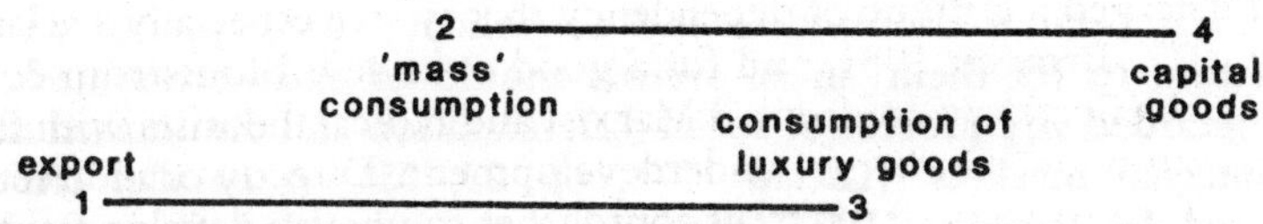

Figure 6.1 Models of capitalist development according to Amin.

model, economic development is propelled by the relationship between mass consumption and the production of capital goods; in the peripheral model the consumption of luxury goods (obviously limited to a narrow band of the population) together with exports propel development. This relationship is summarized in Figure 6.1 (Amin 1974b).

Other critics of unequal exchange also stress that a theory located in the exchange sphere captures only the epiphenomenal results of unevel development rather than the root causes or mechanisms. Thus Shaikh (1980, p. 50) takes Amin's argument one step further, arguing that 'it is perfectly possible for all of the structural patterns of international uneven development . . . to exist, while at the same time there is a zero or even positive net transfer of value for the U[nderdeveloped] C[apitalist] R[egion] export sector as a whole.' There is no such unidirectional relationship between wage differentials, value transfer, and uneven development as Emmanuel suggests.

Foot & Webber (1983, p. 292) more explicitly attempt to 'place unequal exchange within a theory of uneven development', while simultaneously stressing the implicit geographical fabric of development. The division of national economies into core and periphery is not as rigid as Emmanuel, or even Amin, concede, but is historically fluid, and at best semi-permanent. Following a technical critique of Emmanuel's conceptions of value, price, and price of production, Foot & Webber (1983) argue that unequal exchange between different sectors of production, which may or may not be geographically separate, functions to keep the profit rate high in the more developed sectors, but this unequal exchange is by no means a *sine qua non* of uneven development. The underdevelopment of specific economies can occur even under conditions of net positive terms of trade and an inflow of value to the underdeveloped economy.

A further critique emphasizes the actual historical experience of Third World societies in relation to the world market, and seeks to disavow the binary vision whereby underdeveloped societies are classified as either fully capitalist or still feudal. Only by appreciating specific transformations in indigenous class systems and social relations of production, in the context of the internalization of capital, can one truly begin to account for underdevelopment. Laclau (1971) initiated this line of criticism arguing that in Latin America the penetration of capital neither swept existing social relations of production away nor established a separate sphere of production (a formal as opposed to an informal sector). Capital penetrated sufficiently that 'even the most backward peasant regions are bound by fine threads . . . to the "dynamic" sector of the national economy and, through it, to the world market.' Yet the effect of the external market has been to 'accentuate and consolidate' pre-capitalist relations of production, to fix them 'in an archaic mould of extra-economic coercion, which retarded any process of social differentiation and diminished the size of their internal markets' (Laclau 1971, pp. 23, 35). A parallel critique was elaborated by Brenner (1977) concerning the transition from feudalism to capitalism in Europe.

The Regulation School

The so-called Regulation School presents a theory of the different historical regime of capitalism which, while not specifically intended to elaborate a theory of uneven development, is pregnant with insights into geographical unevenness (Lipietz 1977, 1987). A regime of accumulation is a relatively stable arrangement in the relationship between production and consumption, and between production and the reproduction of labour, in a given economy. The Regulation School distinguishes between extensive and intensive regimes. Under the extensive regime, which dominated capitalism up to the 20th century, development was driven primarily through the extension of the scale and sphere of capitalist production in search of absolute surplus value. Under the more recent intensive regime, there has been a closer integration of production and consumption orchestrated by a more rapid reorganization of work relations and the capital–labour relation. Relative surplus value, facilitated by more or less continual technological innovation, became the driving force of the intensive regime, also known as Fordism (Gramsci 1971) because of its characteristic combination of mass production technologies and mass consumption habits. Fordism is also marked by an intensification of concern over the control of the working class in the sphere of consumption as well as production. As regimes of accumulation are not spontaneously stable, the state must establish, maintain, and reproduce the requisite mechanisms of social regulation (Aglietta 1979).

Implied in this formulation is the notion that with shifts in the regime of accumulation come distinct geographical shifts. Dunford & Perrons (1983) use a regulationist framework to investigate the stages of uneven geographical development in the British space-economy from 1780 to 1945. At the international scale, Lipietz (1982, 1984, 1986) argues that since the early 1970s, the rapid industrialization of some previously Third World nations has extended Fordism worldwide (from its core in the developed world) and the emergence of 'peripheral fordism'. Yet in the developed capitalist nations themselves, the Fordist regime increasingly disintegrates. The relative stability of the various compromises between labour and capital has been superseded in many of the developed capitalist countries (especially Britain, France, and the US) by wholesale employer and governmental assaults on labour amid calls for a new social contract. The arrangement of relations between production and consumption, production and reproduction, is restructured. The marginalization of some sectors of the industrial working class, the employment in larger absolute (if not relative) numbers of previously marginal groups such as women and minorities (usually still in service and low-waged production jobs), the partial shift to flexible rather than assembly line production, the commodification of numerous aspects of social reproduction, the rude integration of working-class housing into national and international markets (through gentrification and associated processes), and above all the failure of Keynesian state policies to accomplish the overall regulation of the regime of accumulation amid crisis – all wreak havoc on postwar Fordism. We can associate these shifts with an anticipation of the 'post-Keynesian city', and at the regional scale with processes of flexible accumulation that are responsible for the outlines of a new 'space economy' (Harvey 1985, pp. 211–26; 1987). None the less, the regulationist perspective suggests correctly that the current transformation is more deeply rooted than is generally captured by the notion of post-industrialism.

The attraction of the Regulation School is that it provides a comprehension of the different historical periods of capitalism while simultaneously offering a way of integrating production, consumption and reproduction, and the state in a single framework. Further, it is a vision of social change that incorporates the geographical unevenness of development, although its understanding of the present period of intense uneven geographical development remains largely suggestive. To its detriment, the regulationist theory of uneven development remains tied to a conception in which the different experiences of the nation states far outweigh in importance the internationalization of the economy. In an astonishing reversal of his earlier insights on the differentiated coherence and transformation of the internal economic system as a whole, Lipietz (1986, p. 27) makes explicit this priority of the nation–state, claiming that general theory can have no possible rationale in the face of 'the dead-end of the specificity' of national regimes of accumulation. Thereby the world economy is rendered virtually unauthorizable, indicating that it may be the nationalist rather than internationalist perspective that leads to a dead-end.

Theories of global production

In the preceding approaches, the analysis has focused on the world market rather than production systems. A further set of theories emphasizes the central role of the internationalization of production in the contemporary world economy. Hymer (1972) emphasizes the importance of multinational corporations in establishing global patterns of uneven development. More recently a number of theorists stress a new international division of labour, by which is meant the pattern of 'changes in the international location of industrial production' (Jenkins 1984, p. 53) as well as other economic functions. Three specific transformations have inspired this approach (Fröbel *et al.* 1980, pp. 34–6). The development of the division of labour in the traditional manufacturing sectors has resulted in the deskilling of workers and fragmentation of the work process (Braverman 1974); the rapid development of the means of transportation and communication has substantially cheapened the movement of goods and information, extending the locational field of many industries; and a large reservoir of potential wage labourers has been created (largely but not exclusively) in the periphery of the world system. The new international division of labour is therefore characterized by two specific geographical shifts. First, a more thoroughly international system of production is clearly emerging: nationally determined divisions of labour within a larger international system are giving way to a fundamentally international division of labour incorporating identifiable national patches – that is, the priority of geographical scales is reversed. Second, as decline in the old centres of production is matched by expansion disproportionately located at the edges, there is a relative shift of traditionally core activities – most obviously a range of manufacturing processes – towards some of the hitherto underdeveloped economies.

The new international division of labour can also be understood in the context of the tendency towards a falling rate of profit (Marx 1967, Chs 13–15). Falling rates of profit are experienced unevenly according to sector, location, and the profile and actions of specific capitals. The new international division of labour is in part harbinger, in part response, to this falling rate of profit and ensuing economic crisis, and traditional locational patterns of capital investment are dramatically altered in the search for 'superprofits'. The present crisis of capital involves a marked centralization of capital (Andreff 1984), and a parallel decentralization as certain peripheral economies are partly integrated into the world economy. There is both an internationalization of capital and an internationalization of individual national economies sufficiently dramatic to challenge the conceptual rigidity of the new international division of labour. That this restructuring is partly rooted in a crisis of capital should not, however, blind us to the importance of the changing social relations of production in many Third World countries since the 1960s and the contributions of local capitals and Third World states to local development patterns. There is 'no single emerging pattern which characterizes the integration of Third World countries into the international division of labour' (Jenkins 1984, p. 46).

A somewhat different approach is taken by theorists of the globalization of capital. Gibson & Horvath (1983) argue that a profound change is underway: what is identified by some as a new international division of labour can be understood as part of the transformation toward a new epoch in the development of capitalism, a new 'submode of production'. They argue that 'the root of the current crisis is to be traced to changes going on in the form of capitalism created by the prior "monopoly submode of production" and the geographically specific system it created.' The current crises of monopoly capitalism are 'creating a new variant of capitalism, the global submode of production, in the process creating also a new international, national, regional and urban spatial order' (Gibson & Horvath 1983, p. 179; see also Peet 1983a, Ross 1983, Susman & Schutz 1983). Capital now is fundamentally global, not merely international, and the sharp historical transition to global capitalism brings an equally sharp geographical transformation at all spatial scales. While globalization theorists share with the Regulation School in an attempt to understand the central historical shifts occurring within capitalism, the former maintain a sharper focus on questions of value and production and adopt a more completely international perspective.

Uneven geographical development

The fourth approach to uneven development attempts to make explicit the geographical dimensions and implications of underdevelopment and uneven capital accumulation. This can be done in various ways. Slater (1975) attempted to connect patterns of geographical expansion to the underlying processes of underdevelopment, relating internal spatial inequalities in Tanzania to the country's position in global capitalism. At the same time, David Harvey (1975, 1982, Chs 12 & 13) sought a more theoretical road toward a similar destination. He argued that the complex and contradictory relations engendering the accumulation and circulation of capital are expressed in the specific spatial configurations of social and economic development and that, conversely, the inherently plastic geography of capitalism contributes to the central dynamic of capitalist growth and crisis. Specifically, Harvey (1981, 1982, pp. 431–8, 442–5) argues that one available resolution to crisis-provoking falling rates of profit is for capital to shift geographically in search of places of production where profit rates are higher – there is a search for a spatial fix to the inherent contradictions of capital. Put differently, the accumulation of capital both facilitates and engenders the development of specific spatial configurations – landscapes for production and consumption – which in turn hinder later phases of expansion which have different geographical requirements and impulses. The fundamental contradiction between the physical fixity and mobility of capital is played out in geographical terms.

Harvey's work has been amplified, extended and criticized (Walker 1978, Soja 1980, Webber 1982) with an attempt made at deriving a more explicit theory of uneven geographical development from the dynamics of capitalist production (Smith 1984a). In the first place, it is important to situate an understanding of uneven development in relation to the more traditional geographical concern with human effects on the environment; uneven development is not an accidental alteration of the environment, but a highly systematic (not to be confused with determinate) process of the production of nature. The production of nature (and of space) is accomplished in practice by the continual, if never permanent, resolution of opposing tendencies toward the ***geographcial equalization and differentiation*** of the conditions and levels of production. The search for a spatial fix is continually frustrated, never realized, creating distinct patterns of geographical unevenness through the continual seesaw of capital. At the most abstract level, one can derive a tendency for capital to develop some spaces at the expense of others yet, in the process, to diminish the very conditions that made initial development attractive, viz., the absorption of cheap labour, exhaustion of resources, congestion of land uses, unionization of the labourforce, etc. Previously underdeveloped areas thereby became relatively more attractive for investment; in one place development is superseded by relative underdevelopment, while in another underdevelopment engenders development. The potential is established for a continual geographical seesaw or oscillation in the location of intensive capital investment which, while never fully resolving problems of profit rate and crisis, continually structures and restructures the economic landscape (Smith 1984a, 1986a).

This integration of the theory of capital accumulation with uneven geographical development has been challenged, especially by Browett (1984; see also Smith 1986b), who argues that far from being systematic and integral to capitalist development geographical unevenness may be an accidental byproduct resulting from contingent forces. Others have sought to explore further the theoretical basis of a geography of capitalism. Mizuoka (1986) gives many of these ideas a technical foundation with the help of Japanese mathematical economics, identifying a progressively expanding and deepening real subsumption of space to capital. In a fascinating series of papers, Webber and others

(Webber 1987a, 1987b, 1988, Webber & Rigby 1986, Foot & Webber 1983) provide a mathematical underpinning to several key concepts in uneven development theory and a sympathetic critique of common assumptions. A number of researchers confront theoretical analyses with actual historical patterns of change. Dunford & Perrons (1983), for example, sketch the continually transforming geographical arena of domestic British capital. Yet the most evocative historical accounts of uneven geographical development come from researchers less concerned with the abstract theoretical task of integrating space and society, and more concerned with a political evaluation of contemporary capitalist development. Lipietz (1982, 1984) and Harris (1983, 1987) both analyze the rise of the newly industrialized countries (NICs) in vividly geographical terms that point toward the decline of the traditional geographical definition of the difference between developed and underdeveloped nations. As Harris (1987) puts it, the industrialization of significant sectors of the underdeveloped world, however incomplete and unbalanced, has invalidated the ideology of the Third World geographically defined (see also Worsley 1979, Brett 1985).

No neat lines strictly divide the different approaches to uneven development surveyed in this brief exposition which has omitted much that deserves inclusion. None the less, the discussion gives a sense of the range of theoretical analyses contributing directly, or less directly, toward a reformulation of location theory.

A new location theory?

Theories of uneven development emerged not merely as a product of intellectual debates but in the context of sustained economic and geographical restructuring. Regional decline in the old industrial areas of many of the advanced capitalist nations was hardly new, but by the early 1970s was sharper and more widespread, affecting a broader array of industrial spaces. If chronic unemployment and depression in Clydeside and industrial New England had seemed comparatively invisible amid a more general postwar expansion, the sharpness of decline throughout the English north and the US northeast and midwest in the 1970s provoked the trenchant perception of an emerging regional problem (Holland 1976a, 1976b, Community Development Project 1977) and the recognition of a division between Sunbelt and Frostbelt (Sale 1975). Traditional location theory was revealed as incapable of explaining such a dynamic and richly diverse world, while the importance of uneven development theory was recognized as a link to the larger structure of capital accumulation and crisis. But even the latter was limited, in so far as it spoke to the most general rudiments of the geography of capitalism. Increasingly, a search was made for a middle-level industrial geography, theoretically informed, yet empirically specific as regards contemporary changes. Much of the theoretical work in this new location theory revolved around the concept of spatial divisions of labour.

Spatial divisions of labour

The recent interest in spatial divisions of labour adopts and adapts a concept proposed, but insufficiently elaborated, by classical Marxism. In his voluminous 1899 study of the development of capitalism in Russia, Lenin noted the emergence of a 'territorial division of labour' (paralleling the division of labour in society), which he defined as 'the specialisation of certain districts in the production of some one product, of one sort of product and even of a certain part of a product'. The territorial division of labour was weakly developed in Russia prior to the development of large-scale manufacturing: 'the small industries did not produce such extensive districts; [but] the factory broke down their seclusion and facilitated the transfer of establishments and masses of workers to other places. Manufacture not only creates compact areas, but introduces specialisation within these areas' (Lenin 1956, pp. 436–7). This idea of a territorial, or spatial, division of labour was later reintroduced in the geographical literature by a variety of theorists (Schmidt-Renner 1966, Buch-Hansen & Nielson 1977), but has been pursued most consistently by Massey (1979, 1984) and dominates current efforts to develop new theories of geographical location.

In her original critique of traditional location theory, Massey (1973) confronted its simplifying assumptions about space, history and society. The first part of that critique is re-directed today at Marxist theory which, she argues, is susceptible to the 'same dichotomisation between formal models on the one hand and empirical description on the other which plagues traditional location theory' (Massey 1984, p. 6). Massey's employment of 'spatial divisions of labour, is therefore intended to provide a middle ground between formal theory and empirical description.

> The term [spatial divisions of labour] is introduced in order to make a point. The normal assumption is that any economic activity will respond to geographical inequality in the conditions of production, in such a way as to maximise profits. While this is correct it is also trivial. What it ignores is the variation in the way in which different forms of economic activity incorporate or use the fact of spatial inequality *in order* to maximise profits. This manner of response to geographical unevenness will vary both between sectors and, for any given sector, with changing conditions of production. It may also vary with, for example, the structure of ownership of capital . . . . Moreover, if it is the case that different industries will use spatial variation in different ways, it is also true that these different modes of use will subsequently produce/contribute to different forms of geographical inequality (Massey 1979, p. 234).

Neoclassical location theory emphasized the importance of labour, raw materials, and transport costs as theoretically co-equal factors of production. With an expansion in the range of materials produced by hand, the increased complexity of production and institutional organization, the burgeoning of intermediate processes, and the rapid development of the means of transportation and communication in the postwar period, raw material location and transport costs ceased to be central determinants of industrial location. The new generation of location theories came to emphasize three alternative sets of relations behind the continual structuring and restructuring of the spatial division of labour: industrial organization and corporate strategy; pre-existing characteristics of specific places; and the uniqueness of the labour factor (Storper & Walker 1983).

Massey focuses on the first two of these. Her empirical research on the British electrical engineering and electronics industry demonstrates that different activities within this industry have different locational requirements and patterns, and specific responses to crisis. While much of the research and development activity remained in the southeast of England, where the greatest concentration of specialized labour and technological innovation were found, economic crisis provided an opportunity to reorganize labour-intensive production processes in the light of geographical shifts in wage rates, labour discipline, and state incentives for relocation (Massey 1978).

Massey (1984, p. 76) identifies a three-part typology of multiplant corporate spatial structures. First, in the 'locationally concentrated' organizational form, each of the plants in a corporation is relatively self-contained; the total labour process is performed *in situ* and corporate control is decentralized to the individual plants. Second, there is a 'cloning branch-plant' spatial structure in which ownership and overall corporate control is concentrated at a single headquarters while separate divisions, responsible for product production, have administrative control only over the branch itself. Third, some branch plants produce exclusively for assembly elsewhere, and this represents a 'part-processing' structure. The point of this typology is that different organizational structures both seek and create different attributes of labour and thereby contribute to a highly differentiated spatial division of labour.

Of equal importance for Massey (1984) is the pre-existing nature of places; different regions absorb, translate, and reproduce common global and national impulses toward restructuring in different ways. Places differ according to cultures of work and recreation, relations between work and home, quantities and qualities of reserve labourforce, patterns, habits, and institutions for the reproduction of labour power, and so on, such that different responses to industrial change are elicited. Further, localities are proactive; larger patterns of restructuring represent in part the agglomeration of local experiences. Through tradition as well as struggle, localities strongly influence broader restructuring processes.

While it is necessary, in Massey's words, to retain 'both the general movement and the particularity of circumstance' (1984, p. 8), much recent work has involved a decisive focus on particularity rather than generality. *Spatial divisions of labour* did not, in the end, transcend its illustrative case studies and failed to paint a picture of the general movement involved in contemporary geographical restructuring. The entirely necessary shift to a greater concern with locational difference has proceeded in practice by jettisoning many of the theoretical insights and frameworks that would allow the general movement to be comprehended. Rooted in a realist conception of space as inherently contingent (Sayer 1985b), much of this work has led to the detachment, rather than bridging, of theoretical and empirical investigation (Massey & Meagan 1985). Uneven development, according to this conception, is less a systematic process rooted in the contradictions and structure of capitalism, more an *ad hoc* lack of evenness. As Massey observes approvingly, 'the unique is back on the agenda' (1985, p. 19).

Labour theory of location

While Massey focuses on organizational structure and the peculiarities of place, Walker & Storper (1981, Storper & Walker 1983) propose a labour theory of location. The global spread of production, expansion of raw material sources, reorganization of production, and decrease in costs of transportation imply 'a decline in the importance of. . . non-labour "factors" in the locational calculus' (Walker & Storper 1983, p. 2). The 'labour factor' becomes commensurately more important. And labour, they argue, is no ordinary commodity. Expanding, on Marx's distinction between labour and labour power, they note that the conditions of purchase, performance, and reproduction of labour power differ from those of other commodities making the complexity of this factor increasingly important. Geographical differences in wages, productivity, unionization rates, level of class struggle, labour reserves and skills, internal divisions in the working class, social expenses, patterns of social reproduction – all have a growing influence on location decisions. As the basis for a labour theory of location, they propose a six-part description typology of different labour processes intended to provide a middle ground between more general theoretical frameworks and narrow case studies (Storper & Walker 1984, pp. 34–6). According to this vision, the economic landscape is a complex mosaic of differentiated places.

Mizuoka (1986, pp. 255–62) argues that this 'labour theory of location leads away from a concern with theory, providing less a middle range theory than a sequence of 'empirical generalizations' (Chouinard *et al.* 1984). As with Massey's analysis of spatial divisions of labour, the effect is to enhance our ability to examine specific industries and places and reveal the complexity and richness of detail inherent in individual case studies, but to curtail our ability to recount these specific experiences through the further development of theory. Not only has there been a move back to locational analysis after a decade of efforts aimed at more general theory; the return to location has been accompanied by a re-emphasis on empirical research often distanced from theoretical endeavour.

Empirical and locality studies

The specifics of restructuring lend ample inducement to detailed empirical research. An array of innovations continually alters the locational calculus: flexible specialization and job demarcation, automation with robots and limited worker participation in management, minimum inventory systems, multiple sourcing of parts production, plant downscaling, vertical disintegration. A literature too large and eclectic to allow adequate summary provides case studies of specific sectors (Daniels 1979, Nelson 1984) and industries (Barnett & Schorsch 1983, Rainnie 1984, Jenkins 1985, Sayer 1986, Storper & Christopherson 1987), with a particularly intense focus on locational trends in high-tech industries (Malecki 1979, 1985, Saxenien 1983, Castells 1985, Glasmeier 1985, Markusen 1986a, Hall & Markusen 1985, Scott & Storper 1987). In addition to these industry studies, a series of locality studies has also emerged devoted to equally detailed analyses of individual places. Conceived in the belief that much of the earlier theorizing involved little knowledge of grassroots changes in local communities, these studies largely eschew a theoretical focus, attempting instead to generate a place-specific empirical matrix of social, economic, cultural, and political change.

Critique

Early efforts at a new body of location theory sought a mutual translation between theory and empirical analysis. The dangers of empiricism in continuing locality studies (Cochrane 1987, Gregson 1987, Savage *et al* 1987, Smith 1987, Cox & Mair 1989) and the scattered eclecticism of the empirical project suggest that the dichotomy between theoretical and empirical approaches may be reaffirmed rather than dissolved. Even the more sophisticated work of Massey, Walker, and Storper, while yielding various specific insights, has provided empirical typologies rather than middle-level theory. They alert us to the complexity of location but make few breakthroughs in synthesizing the theoretical vision of global geographical change with detailed restructurings. Massey argues that 'neither theorising nor elaboration of general frameworks can in themselves answer questions about what is happening at any particular time or in any particular place' (1984, p. 9). In the context of a 'retreat from Marxist theory' (O'Keefe 1985, p. 7) this and similar claims have been taken as a renunciation of theory and an endorsement of the new empiricism. In truth, of course, 'what is happening at any particular time and in any particular place' can only be comprehended if one understands the larger societal context and its relationship with the particular. And unless one is to indulge in an infinite circle of ever-widening empirical investigations, theory has a crucial rôle in this process. Theory is the missing ingredient in a gathering empirical eclecticism.

The shift back to an emphasis on location has led in practice to a considerable fragmentation of research. A host of unintended consequences is at work, to be sure, but the result is none the less a research frontier where the most prized studies are first and foremost detailed case studies of individual industries, sectors, or corporations. The linkage between these different experiences is rendered of secondary importance. As Foster-Carter (1978, p. 75) observes in a broader context, the 'recrudescence of marxist analysis is tending, like a tide going out, to create little rock pools increasingly unconnected to one another, in which narrowly circumscribed issues [and localities, we might now add] are discussed separately and without thought of their mutual implication.'

As I wish to be clear on this point, I shall repeat more explicitly what was said earlier. The shift toward empirical research was not only highly desirable but necessary if our knowledge of locational patterns, processes, and transformations was to be advanced. Quite unnecessary and already debilitating is the fact that re-emphasis on the empirical was achieved by burying theory. The real danger, already apparent, is that the new locational analysis may grow to be as autonomous from broader theoretical and contextual concerns as its predecessor.

Conclusion: towards synthesis

If a fragmented focus and minimalist scope are today fashionable, they do not go unchallenged. Keinath (1985) has investigated the spatial dimensions of post-industrial society while Peet (1983, 1984, 1987) interprets locational changes in the United States in terms of a geography of class struggle. Differences in the social relations and conditions of production and employment are indeed important, but there is no intrinsic need to interpret these only at the local scale. The twin processes of deindustrialization/reindustrialization are responsible for a whole new regional geography of North America (Markusen 1986b, Smith & Ward 1987), and can only be understood in relation to broader experiences of uneven development at the global and national scales. Yet this work also does not provide the desired middle-level theory, since it is often difficult to integrate with detailed industry and locality studies; the black box between theory and broad empirical pattern is not necessarily illuminated. But this work does demonstrate in starkly empirical terms that the theory is far too suggestive to be quietly buried.

We can safely asume that, in the short term, industry and locality studies will proliferate. But if a synthesis of uneven development theory and locational analysis is to be achieved, it will be necessary to identify a number of research themes around which the empirical information can be interpreted. By way of an anticipatory conclusion, we can focus on three such themes: the production of space; a theory of geographical scale; and a new regional geography.

The notion of the production of space originated in the work of Lefebvre (1974). For Lefebvre, geographical space has become the central arena of social and economic reproduction, and the production of space the primary means by which contemporary capitalist society reproduces itself. This concept has the advantage of highlighting geographical difference due to human activity, and linking specific geographical patterns with social processes. In the context of industrial location one can also conceptualize the production of space in terms similar to Schmidt-Renner's (1966) *Standortkomplexe der Produktion (Locational production complexes)* and *Territorialstruktur (Territorial structure)*. This should not be seen as denying the importance of consumption patterns in the production of space, the constitutive role of community, culture, systems of reproduction, or place-bound uniqueness. Rather, as Scott & Storper (1986, p. 302) suggest, 'production and work constitute the fundamental reference points of the entire human landscape . . . Geographical unevenness is socially and historically produced out of the basic dynamics of commodity production as such.'

A second promising focus of research concerns the question of scale. The current critique holds that the world is too complex to be captured by general theory, but this would seem to misconstrue the power and purpose of theory. Theory is not a tool for dimissing complexity but for making sense out of it. To be sure, a general theory of uneven development cannot explain the complexity of events involved in the closure of a steel mill. The problem, however, lies not so much in the theory as in the *expectation* that it *could* account for a single local event. Its limited scale of applicability does not render theory irrelevant; rather it highlights the need for translation rules between different scales of reality and levels of enquiry (Alford & Friedland 1985). The theory may not provide a custom-packed explanation for a mill closure, but an explanation that takes no cognizance of the theory will hardly be adequate.

Theory distinguishes complexity from incoherence. The point is that with a more sophisticated sense of scale much of the complexity perceived in narrow case studies becomes manageable. Differences of geographical scale are not universal givens but, like geographical spaces, are products of creative human activity (Smith 1984a, pp. 133–48). It has recently been suggested, for example, that the combination of deindustrialization and the global spread of production is responsible for an expansion of the geographical scale at which regions in the northeastern US are constituted (Smith 1984b, Webber 1986, Smith & Ward 1987). Further, there is a close relationship between the geographical scales engendered in the landscape by human activity and conceptual scales of analysis; generalizations made at the regional scale may not be appropriate at the urban scale. If we are to integrate successfully the pursuit of theory and the performance of empirical research, a more sophisticated theory of scale will alert us against exploring a number of wasteful cul-de-sacs.

A third promising focus of research concerns the possibility of a theoretically founded new regional geography. The question of scale is central to this effort because a theoretically founded regional geography is hardly possible without a more rigorous comprehension of regions. The most useful of the new generation of empirical studies successfully advance our understanding of the development of specific places, and this is the key to the development of a new regional geography (Nelson 1984, Harrison 1984, Martin & Rowthorn 1986, Agnew 1987). As Pudup (1988, p. 3) argues, 'theoretical abstraction and empirical description are mutually dependent constituents of a regional research strategy.' Such a reconstructed regional geography would capture the penchant for concrete analyses, but would also require 'a rigorous definition of its object and methods of study', a more open-ended approach, and the transcendence of 'the spurious division between empirical and theoretical research' (Pudup 1988, p. 3).

Through most of human history, landscape has been a virtually unchanging presence; Macbeth was struck through with the horror of the preternatural when he witnessed Birnam Wood moving steadily up to Dunsinane. Today, however, we accept much more profound humanly induced changes in the landscape with ne'er a comment. Every day a new condominium, a new crane on the skyline, a steel plant suddenly closed, a new office complex in last year's orange grove. Moving landscapes are exactly what we are trying to explain. With such rapid change, the need for a historical/social theory of geographical location has never been greater.

References

Aglietta, M. 1979. *A theory of capitalist regulation: the US experience*. London: New Left Books.

Agnew, J. 1987. *The United States in the world economy. A regional geography*. Cambridge: Cambridge University Press.

Alford, R. R. & R. Friedland 1985. *Powers of theory. Capitalism, the state, and democracy*. Cambridge: Cambridge University Press.

Amin, S. 1974a. *Accumulation on a world scale: a critique of the theory of underdevelopment*. New York: Monthly Review Press.

Amin, S. 1974b. Accumulation and Development: A theoretical model. *Review of African Political Economy* **1**, 9–26.

Amin, S. 1976. *Unequal development: an essay on the social formations of peripheral capitalism*. New York: Monthly Review Press.

Amin, S. 1977. *Imperialism and unequal development*. New York: Monthly Review Press.

Anderson, J. 1973. Ideology in geography. *Antipode* **5**, 1–6.

Andreff, W. 1984. The international centralization of capital and the re-ordering of world capitalism, *Capital and Class* **22**, 59–80.

Barnett, D. F. & L. Schorsch 1983. *Steel: upheaval in a basic industry*. Cambridge, Mass.: Ballinger.
Blomström, M. & B. Hettne 1984. *Development theory in transition. The dependency debate and beyond: Third World responses*. London: Zed Books.
Braverman, H. 1974. *Labor and monopoly capital*. New York: Monthly Review Press.
Brenner R. 1977. The origins of capitalist development: a critique of neo-Smithian Marxism. *New Left Review* **104**, 25–92.
Brett, E. A. 1985. *The world economy since the war: the politics of uneven development*. Basingstoke: Macmillan.
Browett, J. 1981. Into the cul-de-sac of the dependency paradigm with A. G. Frank. *Australia and New Zealand Journal of Sociology* **4**, 57–79.
Browett, J. 1984. On the necessity and inevitability of uneven spatial development under capitalism. *International Journal of Urban and Regional Research* **8**, 155–76.
Buch-Hansen, M. & B. Nielson 1977. Marxist geography and the concept of territorial structure. *Antipode* **9**, 1–12.
Castells, M. (ed.) 1985. *High technology, space, and society*. Beverly Hills: Sage Urban Affairs Annual Reviews.
Chouinard, V., R. Fincher & M. Webber 1984. Empirical research in scientific human geography. *Progress in Human Geography* **8**, 347–80.
Christaller, W. 1966. *Central places in southern Germany*. Englewood Cliffs, NJ: Prentice-Hall.
Cochrane, A. 1987. What a difference the place makes: the new structuralism of locality. *Antipode* **19** 354–63.
Community Development Project 1977. *The costs of industrial change*. London: Community Development Project.
Cox, K. & A. Mair 1989. Levels of abstraction in locality studies. *Antipode* **21**.
Daniels, P. W. (ed.) 1979. *Spatial patterns of office growth and location*. New York: Wiley.
Dunford, M. & D. Perrons 1983. *The arena of capital*. Basingstoke: Macmillan.
Eliot Hurst, M. 1973. Establishment geography. *Antipode* **5**, 40–59.
Emmanuel, A. 1972. *Uneven exchange: a study of the imperialism of trade*. New York: Monthly Review Press.
Foot, S. P. H. & M. J. Weber 1983. Unequal exchange and uneven development. *Environment and Planning D, Society and Space* **1**, 281–304.
Foster-Carter, A. 1978. The modes of production controversy. *New Left Review* **107**, 47–77.
Frank, A. G. 1967. *Capitalism and underdevelopment in Latin America*. New York: Monthly Review Press.
Frank, A. G. 1969. Latin America: underdevelopment or revolution. New York: Monthly Review Press.
Frank, A. G. 1972. *Lumpenbourgeoisie – lumpendevelopment*. New York: Monthly Review Press.
Fröbel, F. J. Heinrichs & O. Kreye 1980. *The new international division of labour*. Cambridge: Cambridge University Press.
Gibson, K. & R. Horvath 1983. Global capital and the restructuring crisis in Australian manufacturing. *Economic Geography* **59**, 178–94.
Glasmeier, A. K. 1985. Innovative manufacturing industries: spatial incidence in the United States. In *High technology, space and society*, M. Castells (ed.) 55–79. Beverly Hills: Sage Urban Affairs Annual Reviews.
Gramsci, A. 1971. *Prison notebooks*. New York: International Publishers.
Gregory, D. 1981. Alfred Weber and location theory. In *Geography, ideology and social concern*, D. R. Stoddart (ed.) 165–85. Oxford: Basil Blackwell.
Gregson, N. 1987. The CURS initiative: some further comments. *Antipode* **19**, 364–70.
Hagget, P. 1965. *Locational analysis in human geography*. New York: St Martin's Press.
Hall, P. 1966. *Thünen's isolated state*. Oxford: Pergamon Press.

Hall, P. & A. Markusen (eds) 1985. *Silicon landscapes*. Boston: Allen & Unwin.
Harris, N. 1983. *Of bread and guns: the world economy in crisis*. Harmondsworth: Penguin.
Harris, N. 1987. *The end of the Third World: newly industrializing countries and the decline of an ideology*. Harmondsworth: Penguin.
Harrison, B. 1984. Regional restructuring and 'good business climates': the economic transformation of New England since World War II. In *Sunbelt/snowbelt: urban development and regional restructuring*, L. Sawyers & W. Tabb (eds), 48–96.
Harvey, D. 1973. *Social justice and the city*. London: Edward Arnold.
Harvey, D. 1975. The geography of capitalist accumulation: a reconstruction of the Marxian theory. *Antipode* **7**, 9–21. (Reprinted 1977 in *Radical geography: alternative viewpoints on contemporary social issues*, R. Peet (ed.), 263–92. Chicago: Maaroufa Press.)
Harvey, D. 1981. The spatial fix – Hegel, Von Thünen and Marx. *Antipode* **13**, 1–12.
Harvey, D. 1982. *The limits to capital*. Oxford: Basil Blackwell.
Harvey, D. 1985. *The urbanization of capital*. Baltimore, MD: Johns Hopkins University Press.
Harvey, D. 1987. Flexible accumulation through urbanisation: reflections on 'postmodernism' in the American city, *Antipode* **19**, 260–86.
Hotelling, H. 1929. Stability in competition. *Economic Journal* **39**, 41–57.
Holland, S. 1976a. *The regional problem*. London: Macmillan.
Holland, S. 1976b. *Capital versus the regions*. London: Macmillan.
Hymer, S. 1972. The multinational corporation and the law of uneven development. In *Economics and world order from the 1970s to the 1990s*, J. Bhagwati, (ed.). New York: Columbia University Press.
Isard, W. 1956. *Location and space economy*. Cambridge, Mass.: MIT Press.
Isard, W. 1960. *Methods of regional analysis*. New York: Wiley.
Jenkins, R. 1984. Divisions over the international division of labour. *Capital and Class* **22**, 28–57.
Jenkins, R. 1985. Internationalization of capital and the semi-industrialized countries: the case of the motor industry. *Review of Radical Politcial Economies* **17**, 59–81.
Keeble, D. 1976. *Industrial location and planning in the United Kingdom*. London: Methuen.
Keinath, W. F. 1985. The spatial component of the post-industrial society. *Economic Geography* **61**, 223–40.
Laclau, E. 1971. Feudalism and capitalism in Latin America. *New Left Review* **67**, 19–38. (Reprinted in his 1977 *Politics and ideology in Marxist theory*, 15–40 London: Verso.)
Lefebvre, H. 1974. *La Production de l'espace*. Paris: Anthropos.
Lenin, V. I. (1956) *The development of capitalism in Russia*. Moscow: Progress Publishers.
Lipietz, A. 1977. *Le Capital et son espace*. Paris: Maspero.
Lipietz, A. 1982. Towards global Fordism? *New Left Review* **132**, 33–47.
Lipietz, A. 1984. How monetarism choked Third World industrialization. *New Left Review* **145**, 71–87.
Lipietz, A. 1986. New tendencies in the international division of labour: regimes of accumulation and modes of regulation. In *Production, work, territory: the geographical anatomy of industrial capitalism*, J. Scott & M. Storper, (eds), 16–40. Boston: Allen & Unwin.
Lipietz, A. 1987. Reflexions autour d'une fable: pour un statut marxiste des concepts de regulation et d'accumulation. CEPREMAP Paper 8530, 142, Rue du Chevaleret, 75013 Paris.
Lösch, A. 1954. *The economics of location*. New Haven, Conn.: Yale University Press.
Malecki, E. J. 1979. Locational trends in R. & D. by large U.S. corporations, 1965–1977. *Economic Geography* **55**, 309–23.

Malecki, E. J. 1985. Industrial location and corporate organization in high technology industries. *Economic Geography* **61**, 345–69.

Markusen, A. 1986a. Defense spending and the geography of high tech industries. In *Technology, regions and policy*, J. Rees (ed.). New York: Praeger.

Markusen, A. 1986b. Industrial heartland: rust bowl or tool kit. Evanston, IL: Northwestern University Center for Urban Affairs and Policy Research. Mimeograph.

Martin, R. & B. Rowthorn (eds). 1986. *The geography of deindustrialization*. Basingstoke: Macmillan.

Marx, K. 1967. *Capital*, vol. 3. New York: International Publishers.

Massey, D. 1973. Towards a critique of industrial location theory. *Antipode* **5**, 33–9 (Reprinted 1977 in *Radical geography: alternative viewpoints on contemporary social issues*, R. Peet (ed.), 181–97. Chicago: Maaroufa Press.)

Massey, D. 1978. The U.K. electrical engineering and electronics industry. *Review of Radical Political Economics* **10**, 39–54.

Massey, D. 1979. In what sense a regional problem? *Regional Studies* **13**, 233–43.

Massey, D. 1984. *Spatial divisions of labour: social structures and the geography of production*. New York: Methuen.

Massey, D. 1985. New directions in space. In *Social relations and spatial structures*, D. Gregory & J. Urry (eds), 9–19. Basingstoke: Macmillan.

Massey, D. & R. Meegan 1985. *Politics and method. Contrasting studies in industrial geography*. London: Methuen.

Mizuoka, F. 1986. Annihilation of space: a theory of Marxist economic geography. Unpublished PhD dissertation. Clark University, Worcester, Mass.

Nelson, K. 1984. Back offices and female labour markets: office suburbanization in the San Francisco–Oakland SMSA. Unpublished PhD dissertation, University of California, Berkeley.

Ohara, K. 1950. *Shakai Chirigaku no Kiso Riron* [*Fundamentals of social geography*]. Tokyo: Kokon Shein.

O'Keefe, P. 1985. Recent developments in British radical geography. *Antipode* **17**, 7–8.

Peet, R. (ed.) 1977. *Radical geography: alternative viewpoints on contemporary social issues*. Chicago: Maaroufa Press.

Peet, R., 1983a. The global geography of contemporary capitalism. *Economic Geography* **59**, 105–11.

Peet, R., 1983b. Relations of production and the relocation of United States manufacturing. *Economic Geography* **59**, 112–43.

Peet, R., 1984. Class struggle, the relocation of employment, and economic crisis. *Science and Society* **48**, 38–51.

Peet, R. 1987. *International capitalism and industrial restructuring*. Boston: Allen & Unwin.

Pudup, M. B. 1988. Arguments within regional geography. *Progress in Human Geography* **12**, 369–83.

Rainnie, A. F. 1984. Combined and uneven development in the clothing industry: the effects of competition on accumulation. *Capital and Class* **22**, 141–56.

Ross, R. J. S. 1983. Facing Leviathan: public policy and global capitalism. *Economic Geography* **59**, 144–60.

Sale, K. 1975. *Power shift: the rise of the southern rim and its challenge to the eastern establishment*. New York: Random House.

Savage, M., J. Barlow, S. Duncan & P. Saunders 1987. 'Locality research': the Sussex programme on economic restructuring, social change and the locality. *Quarterly Journal of Social Affairs* **3**, 27–51.

Saxenien, A. L. 1983. The urban contradictions of Silicon Valley: regional growth and the restructuring of the semi-conductor industry. *International Journal of Urban and Regional Research* **7**, 237–62.

Sayer, A. 1985a. Industry and space. A sympathetic critique of radical research. *Environment and Planning D. Society and Space* **3**, 3–29.

Sayer, A. 1985b. The difference that space makes. In *Social relations and spatial structures*, D. Gregory & J. Urry, (eds), 49–66. Basingstoke: Macmillan.

Sayer, A. 1986. Industrial location on a world scale: the case of the semi-conductor industry. In *Production, work, territory: the geographical anatonomy of industral capitalism*, A. J. Scott & M. Storper (eds). 107–23. Boston: Allen & Unwin.

Schmnidt-Renner, G. 1966. *Elementare Theorie der Ökonomischen Geographie*. Gotha; VEB Herman Haack.

Scott, A. J. & M. Storper (eds) 1986. *Production, work, territory: the geographical anatomy of industrial capitalism*. Boston: Allen & Unwin.

Scott, A. J. & M. Storper 1987. High technology industry and regional development: a theoretical critique and reconstruction. *International Social Science Journal* **22**, 215–32.

Shaikh, A. 1980. Foreign trade and the Law of Value: Part 2. *Science and Society* **44**, 27–57.

Slater, D. 1975. Underdevelopment and spatial inequality. *Progress in Planning* **4**, 2.

Smith, D. 1966. A theoretical framework for geographical studies of industrial location. *Economic Geography* **42**, 95–113.

Smith, N. 1984a. *Uneven development: nature, capital and the production of space*. Oxford: Basil Blackwell.

Smith, N. 1984b. Deindustrialization and regionalization: class alliance and class struggle. *Papers of the Regional Science Association* **54**, 113–28.

Smith, N. 1986a. Uneven development and the geography of modernity. *Social Concept* **3**, 67–90.

Smith, N. 1986b. On the necessity of uneven development. *International Journal of Urban and Regional Research* **10**, 87–104.

Smith, N. 1987. Dangers of the empirical turn. *Antipode* **19**, 59–68.

Smith, N. & D. Ward 1987. The restructuring of geographical scale: coalescence and fragmentation of the northern core region. *Economic Geography* **63**, 160–82.

Soja, E. J. 1980. The socio-spatial dialectic. *Annals of the Association of American Geographers* **70**, 207–25.

Stevens, B. H. 1961. An application of game theory to a problem in location strategy. *Papers and Proceedings of the Regional Science Association* **7**, 143–58.

Storper, M. & S. Christopherson 1987. Flexible specialization and regional industrial agglomeration: the case of the U.S. motion picture industry. *Annals of the Association of American Geographers* **77**, 104–17.

Storper, M. & R. Walker 1983. The theory of labor and the theory of location. *International Journal of Urban and Regional Research* **7**, 1–43.

Storper, M. & R. Walker, 1984. The spatial division of labour: labour and the location of industries. In *Sunbelt/snowbelt: urban development and regional restructuring*, L. Sawyers & W. Tabb (eds), 19–47. New York: Oxford University Press.

Susman, P. & E. Schutz 1983. Monopoly and competitive firm relations and regional development in global capitalism. *Economic Geography* **59**, 161–77.

Walker, R. 1978. Two sources of uneven development under advanced capitalism: spatial differentiation and capital mobility. *Review of Radical Political Economy* **10**, 120–35.

Walker, R. & M. Storper 1981. Capital and industrial location. *Progress in Human Geography* **5**, 473–509.

Webber, M. J. 1982. Agglomeration and the regional question. *Antipode* **14**, 1–11.

Webber, M. J. 1984. *Industrial location*. Beverly Hills: Sage Publications.

Webber, M. J. 1986. Regional production and the production of regions: the case of Steeltown. In *Production, work, territory: the geographical anatomy of industrial capitalism*, A. J. Scott & M. Storper (eds), 197–224. Boston: Allen & Unwin.

Webber, M. J. 1987a. Rates of profit and interregional flows of capital. *Annals of the Association of American Geographers* **77**, 63–75.

Webber, M. J. 1987b. Profits, crises and industrial change 1: theoretical considerations. *Antipode* **19**, 307–28.

Webber, M. J. 1988. Profits, crises and industrial change 2: the experience of Canada 1950–1981. *Antipode* **20**, 1–18.

Webber, M. J. & D. L. Rigby 1986. The rate of profit in Canadian manufacturing, 1950–1981. *Review of Radical Political Economics* **18**, 33–55.

Weber, A. 1929. *Theory of the location of industries*. Chicago: Chicago University Press.

Wolpert, J. 1964. The decision process in spatial context. *Annals of the Association of American Geographers* **54**, 537–58.

Worsley, P. 1979. How many worlds? *Third World Quarterly* **1**, 100–8.

7 *Rural geography and political economy*

Paul Cloke

'Let's have a look', said Eeyore, and he turned slowly round to the place where his tail had been a little while ago, and then, finding that he couldn't catch it up, he turned round the other way, until he came back to where he was at first and looked between his front legs, and at last he said, with a long sad sigh,, 'I believe you're right'.

'Of course I'm right,' said Pooh.

'That Accounts for a Good Deal', said Eeyore gloomily. 'It Explains Everything. No Wonder'.

A. A. Milne (1928, p. 44)
Winnie-the-Pooh

At the risk of using an unwarranted metaphor, my initial contention is that for many rural geographers the search for, and acknowledgement of, theory in their work has been rather similar to the hunt for Eeyore's tail. Seeking after theory has been perfunctory, usually circular, and conducted in a gloomy and often unwilling manner. When faced with the need to underpin research with a conceptual framework many rural geographers have preferred to concentrate on their principal interest, that is empirical investigation of rural issues. Accordingly, such theory that has developed has stemmed principally either from a reluctant and sometimes unknowing acquiescence in the theoretical thoughts of others or, more often, from the theory-laden constraints of an ingrained positivistic methodology.

This contention is both overstated, in that it belittles the longheld theoretical interests of a minority of rural geographers, and understandable given the historical development of the rural subdiscipline and the foremost interests and background of many of its component researchers. Nevertheless, in comparison with urban and regional studies, rural geography's ponderous lurch towards political-economy approaches has been both slothful and grudging. This chapter seeks first to account for the recent development of rural geography as a subject, suggesting that even contemporary acceptance of the need for applied research may represent new hay in old barns rather than theoretical advance. Then, detailed discussion follows of four interrelated themes which are fundamental to any incorporation of rural research into the broader political-economy model: the specificity (or otherwise) of rurality and rural space; the centrality of capital accumulation as the driving force of rural change; power relations in the structuration of the extensive land market, with particular reference to agriculture; and the limits to state planning in rural areas. Discussion of these themes reflects the important initial output of rural geographers and others, but the as yet limited extent of rural geographical expeditions into the political-economy territory inevitably channels the chapter towards an agenda for future research rather than making it a display-case for what has already been done.

The state of rural geography

In order to understand why rural geography has been reluctant to adopt the characteristics of critical social theory, thereby opting for traditional orthodoxy rather than what has almost become a new orthodoxy, we must review the fluctuating fortunes of the rural subdiscipline within postwar geography. Studies of the countryside used to be of central importance to the subject as a whole. Clout (1972) reports that rural studies were at the core of human geography prior to World War II, and Proudfoot (1984) suggests that the core of both historical geography and contemporary human geography up to the mid-1950s was unquestionably concerned with rural areas. This pre-eminent position may partly be ascribed to the economic dominance of agriculture at this time, and partly to the fact that, with a few exceptions, regional geography was based 'either on rural areas, or on regions defined in terms of physical features and the agricultural responses to these, or on the characteristics of rural life in the broadest sense' (Proudfoot 1984, p. 11).

From this dominant institutional position rural geography slid into a wilderness phase lasting 15 years or so. As regionalism met its demise during the 1960s and was replaced first by systematic study and then by applied and relevant brands of systematic study, human geographers were increasingly attracted by the scale and sheer visibility of problems in the *city* environment. During this period rural geography lay fallow, entrenched in its agricultural roots and seemingly oblivious to the theoretical strides being achieved within urban studies (Cloke 1985a).

Some novelty and consequent focus was restored to rural geography during the 1970s by which time rural areas had sunk sufficiently low on the horizon of geographic priority that impetus for rural geography could legitimately be achieved by pointing out how little was actually known about the rural environment and rural people compared to their urban counterparts. This task was performed by Hugh Clout (1972) whose book *Rural geography: an introductory survey* represented a perceptive and timely synthesis of potential areas of study for rural geographers, that importantly steered away from the previously dominating influences of agricultural economics (see, for example, Weller 1967). Clout's book served three important purposes:

(a) It demonstrated that what had gone before was not totally wilderness. The work of Best & Coppock (1962), Bracey (1952, 1970), Coppock (1964), Pahl (1965), and Wibberley (1959, 1967), for example, looms large and is an acknowledged influence.
(b) It brought together a range of disparate material in the name of rural geography, thereby attempting to demarcate a distinct subject area.
(c) It combined various social, economic, and land use aspects of the countryside into a single category of study (Fig. 7.1), although in a later

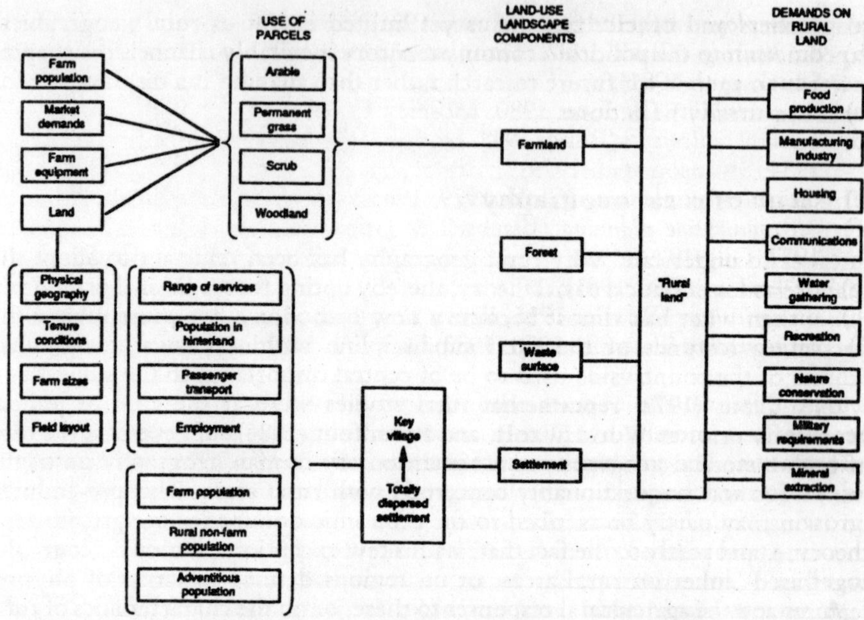

Figure 7.1 Clout's components of rural geography (source: Clout 1972).

> publication Clout admitted that the internal cohesion of rural studies was limited to areal characteristics, 'No specific methodology binds them together and they are united only by the fact that they are concerned with the less densely occupied sections of the earth's surface' (Clout 1977, p. 475).

What followed during the 1970s was a phase of resurgence, particularly in Britain, Europe, Canada, and Australasia but accompanied by a similar trend elsewhere in the developed and developing worlds. Surprisingly, rural *geography* did not assume great importance in the USA, although multidisciplinary rural studies did become more popular during this period. These priorities are marked by a relative paucity of geographical literature on the rural environment of the USA. In Britain an authentic claim was made that fundamental socio-economic changes were occurring in the countryside which had largely been ignored by geographers. In fact, this claim mirrored earlier statements (for example by Bonham-Carter 1951, and Saville 1957) that rural planners (and by implication other interested rural parties) were 'groping in a fog of ignorance', but by the 1970s the need for more information on rural environments became inescapable and rural geography experienced a bandwagon effect whereby it became attractive both to new generations of researchers and to that footloose academic element who are prepared to switch their attentions to fashionable areas of study.

The rural geography literature burgeoned during this phase, with seemingly vigorous coverage of every different aspect of rural life. Examples of important publications which were instrumental in this quenching of the thirst for information include:

(a) on *employment*; (Hodge & Whitby 1981);
(b) on *housing* (Clark 1982, Dunn *et al*. 1981, Phillips & Williams 1982, Shucksmith 1981);
(c) on *accessibility* (Banister 1980, Moseley 1977);
(d) on *rural settlements* (Bunce 1982, Hodge & Qadeer 1983, Johansen & Fuguitt 1984, Swanson *et al*. 1979);
(e) on *settlement planning* (Cloke 1979, 1983, and Woodruffe 1976);
(f) on *countryside planning* (Blacksell & Gilg 1981, Davidson & Wibberley 1977, Gilg 1979);
(g) on *land use* (Best 1981);
(h) on *agriculture* (Bowler 1985, Ilbery 1985);
(i) on *recreation* (Patmore 1970, 1983).

By the early 1980s it became clear that progress in rural geography could not be sustained merely by further attempts to satiate the appetite for rural information. The combined impact of these and other publications had more or less performed that role successfully. In doing so, however, they represented a: 'growing *multidisciplinary* perspective through which a jumble of method, theory, approach and other distinctive disciplinary attributes has been thrown together' (Cloke 1985, p. 2).

A review of potential new emphases for rural geography (Cloke 1980) isolated three common deficiencies, and emphasized the importance of these factors for the future prosperity of rural geography:

(a) If rural geography was to progress by adopting an applied or relevance or welfare approach to the study of rural problems, then some form of conceptual framework would be required on which to found the various strands of investigation, analysis, and problem solving.
(b) Rural geographers had been tardy and uninspired in the introduction of rigorous analytical methods for the study of countryside systems and processes.
(c) There was a need to establish a formal applied emphasis to rural geographical research which would allow a greater degree of integration with rural planners and planning processes.

Certainly, research in rural geography has become more applied and policy-oriented over the last decade, but a vigorous push towards conceptual and theoretical advances (and the analytical matters which hang on the coat-tails of theory) has not occurred. This has been partly due to a strong instinct for institutional survival. While it is reasonable to expect rural geographers to share information and, to a lesser extent, concepts related to positivism with researchers from other disciplines who are also interested in things *rural* (agricultural economists, rural sociologists, rural planners, and the like), it would be far more threatening to adopt structuralist theoretical stances which harbour such institutionally destructive axioms for rural geography as social relations being more important than spatial differentiation, the category rural being unimportant as a classificatory device, and so on. Self-preservation has led many rural geographers, having just undergone a period of justifying their subject's value, to batten down the hatches against such heresies. Another factor militating against any automatic desire to join in the broader movement of critical social theory involves the fundamental reasons why rural geographers are interested in their subject matter. Many have strong family connections with the agricultural sector; others' interests have been kindled through intrinsic appreciation of the attractions of the countryside. These foci on the extensive land use characteristics of 'rural' exert an anti-gravitational force away from any concepts which seek to minimize 'rural' as a differentiating factor. These trends are self-fulfilling. To an extent, young researchers who have been trained in critical social theory will tend to be attracted to work in urban and regional localities where such theoretical constructs are more readily accepted, leaving rural geography with less new blood with which to break down its prejudices.

These attitudes towards the adoption of new theoretical frameworks are ensconced in the recent crop of rural geography textbooks which have appeared as syntheses of work during the post-Clout era. For example, Pacione (1984), while acknowledging the place of economic, social and political processes within rural geography, and devoting a chapter to 'power and decision-making' asserts that: 'it is important to appreciate that rural-based investigations are not simply regional applications of some wider perspective; the rural environment poses new conceptual and methodological questions, and presents unique problems for investigation' (Pacione 1984, p. 1). Equally, Gilg (1985), who confesses that he was prompted to write his book partly because he recognized an omission in other work of an overview of 'the approach of rural geographers to their subject, particularly with regard to . . . the development of theory' (Gilg 1985, p. xi), reaches the conclusion that: 'there is little evidence, therefore, that rural geography has become any more coherent, or that it has produced a widely accepted body of theory, and indeed the subject remains broadly *theory free*. (Gilg 1985, p. 172; emphasis in original). This last suggestion is seemingly contradicted by his observation that 'rural geographers have employed all types of approach, but the most common and widely accepted is still the logical positivist approach' (Gilg 1985, p. 4). The idea of positivist approaches being theory-free is anathema, and may explain why Gilg considers the need for a conceptual framework as a 'passing academic fashion' – a view condemned by Munton (1986) amongst others.

A kind of applied positivism may be discerned in these accounts, and this approach has found favour with some because it obviates the need to despoil technical geography with politics and ideology. One of the most striking critiques of this position has come from Hoggart & Buller (1987) who forcefully suggest that applied positivism in rural geography is neither technically neutral nor theory-free. They claim that the emphasis on applied research:

> implies that researchers should primarily restrict their attentions to identifying the most appropriate means of 'tinkering' with existing socio-economic conditions in order to weaken the impress of malevolent trends

and that the approach thereby:

> fails to recognise that the processes which brought about current maldistributions or malpractices (that 'applied' researchers want to change) are inher-

ent in policy procedures which have to be relied on to alleviate the problems researchers have analysed (Hoggart & Buller 1987, p. 267).

Of recent texts only Phillips & Williams (1984) and Hoggart & Buller (1987) recommend any kind of political-economy approach to their readers. Phillips and Williams suggest that 'an analysis of political and economic changes should provide the starting point for interpreting changes in the countryside' (1984, p. vii). Even though their text focuses on the *implications* of macro-economic and political change on areas and groups in the countryside, rather than the *mechanisms* of such change, Phillips and Williams have identified a direction for theoretical understanding which finds echoes of support from a growing number of rural geographers. The only surprise is that it took this long for rural researchers to grasp the importance of political-economy approaches which have been adopted widely in urban and regional studies for more than a decade. One important catalyst for the increasing adoption of political-economy perspectives has been the establishment of the Rural Economy and Society Study Group, a multidisciplinary forum for the study of the social formation of rural areas in advanced societies. Conference-based publications such as that by Bradley & Lowe (1984) have at least started the task of theoretical cross-fertilization and provided a basis for a deeper understanding of political-economic theory and its implications for the subject matter researched by rural geographers.

It is a central contention of this chapter that compared with the positivism and humanism which have preceded it, the use of a political-economy approach offers rural geographers substantive expectations of progress in a number of important areas: especially the analysis of economic progress; understanding restructuring within society; and understanding both the role played by the state and the connections between economy, civil society, and the state. These areas of progress represent a logical (if radical in rural terms) extension of the existing penchant for applied rural studies. Moreover, a political-economy approach throws up manageable issues for rural researchers – issues which though theoretically separate are in fact contextually cognate with their previous interests. Thus, for example, Healey (1986), in a land use planning context, suggests that political-economy concepts will aid the understanding of:

(a) the way the organization of the economy produces particular forms of investment (and disinvestment) in rural areas;
(b) the variety of social groups; their interests in land, property, and the environment; and the interaction between social groups and economic processes;
(c) the way in which and the reasons why the state operates as it does in response to or as an initiatior of economic reorganization.

Before proceeding to an analysis of the implications of a political-economic approach for rural geographical study, therefore, it is important to appreciate the current status of applied geographical studies in rural areas so as to highlight (if only for followers of the 'theory-freedom' theory) the dangers of simply placing old investigations into new theoretical categories.

Old hay in new barns?

One danger in the potential incorporation of political-economy themes into rural geographical research is that there will be a strong temptation to cobble together some form of theoretical package which paradoxically will both demonstrate an awareness of the theoretical strides issuing forth from critical social theory and permit researchers to carry on much as before. Rural geographers (as others) have been adept at achieving such compromises in the past, and Munton (1986), in a prediction of what future rural research will be like, suggests that:

> Empirical enquiry will almost certainly continue to dominate the work of British rural geographers and although much of that research may continue to depend on a logical positivist methodology, more modest claims will be made for its explanatory powers and greater attention will be paid to the contributions to be derived from critical social theory (Munton 1986, p. 5).

This image of a legitimating and opportunistic theoretical pluralism is founded on evidence from research and synthesis over the last five years or so, much of which has been willing to acknowledge structural explanations, but rarely have research methods been specifically constructed to inform on structural mechanisms.

A useful illustration of this tendency to put old hay into new barns is the issue of rural deprivation. Here, it is all too easy to confirm that 'urban and rural deprivation are manifestations of the same forces emanating from the dynamic of late-industrial capitalism' (Pacione 1984, p. 199, referring to Moseley 1980) whilst at the same time stressing that 'the geographers' major contribution to quality of life research, to date, has been the introduction of a spatial dimension in their work on *territorial* social indicators' (Pacione 1984, p. 214, referring amongst others to Pacione 1980, 1982). Neither should this tendency to acknowledge critical social theory but pursue positivistic investigation be viewed merely as a critique of individual researchers. Until recently there has been widespread slothfulness in rural geographical research to provide new investigative substance with which to fill the more readily available new barns. The recent history of investigations of rural deprivation place this trend into clear perspective.

Although facets of what are now described as deprivation have been the subject of study by rural researchers for 30 years or more (see Bracey 1952), the recognition of a combination of such facets in such a way that the whole is greater than the sum of the parts has only occurred over the last decade. Influence by the welfare and inequality themes which represented the liberal wing of post-positivist thought in human geography (Smith 1974, 1977, Coates *et al.* 1977) researchers began to piece together a jigsaw of low opportunity levels for rural residents. Two collections from conference papers (Walker 1978, Shaw 1979) emerged as the first major collective illustrations of rural deprivation and these were important in helping to break down the popular image of rural areas as being healthy, hearty, and poverty-free.

In theoretical terms, however, these initial studies of deprivation were dependent on the empirical investigations of the *components* which when aggregated became the problem. Thus Shaw (1979) categorized three types of deprivation which when linked together could be viewed as a cyclical and self-sustaining process:

(a) *household deprivation* – including matters of *income* and *housing* which dictate the ability of individuals and families to make use of opportunities available in rural areas;
(b) *opportunity deprivation* – summarizing those elements of rural life which are disappearing, for example employment and service opportunities;
(c) *mobility deprivation* – referring to particular non-mobile groups in rural society.

The combined effects of household, opportunity, and mobility deprivation are to isolate particular groups within rural communities, and to present them with complex and sometimes insurmountable difficulties in obtaining the basic needs for survival in their established place of residence. Here, then, was an ideal opportunity to follow empirical investigations of deprivation components with a pursuit of class-based analyses (with appropriate recognition of racial issues – see Carlson *et al*. (1981), and gender issues – see Little (1985)) of deprivation in rural areas as the first step to an analysis within a political-economy framework. Instead, investigations of rural deprivation became bogged down by the politically inspired notion that levels of service provision were paramount, and that these inadequate levels were being caused by a biased allocation of central government expenditure in favour of urban areas (McLaughlin 1984). Publications by the Association of County Councils (1979) and the Association of District Councils (1980) were widely used by academic researchers as evidence of deprivation, regardless of their main propaganda purpose which was to expose the paucity of funds experienced by their constituent councils. As McLaughlin (1986) has pointed out:

> By focussing the problem analyses and subsequent policy prescriptions on the issue of rural *areas* as poor places and on questions of service decline *per se* the policy debate on rural deprivation has largely ignored crucial questions about the particular groups and individuals *within* rural areas who gain or lose as a result of policies. In contrast to the urban deprivation debate, the key issues of differential standards of living and quality of life in rural areas and the resource distribution processes affecting them have also been ruled off the agenda (p. 292).

It could be argued that many rural geographers were content with an area-based approach, as this suited not only their protective instincts towards 'rural' as an analytical category within geography, but also their predilection for techniques involving model-building and territorial social indicators. These tendencies in turn suited the legitimation tactics of government which permitted the rural service explanation of deprivation to become an issue (and even sponsored research on this aspect) so as to steer attention away from alternative explanations involving the uneven distribution of economic and political power within society. To this extent, then, the new barn of rural deprivation has quite clearly been filled with old hay by rural geographers.

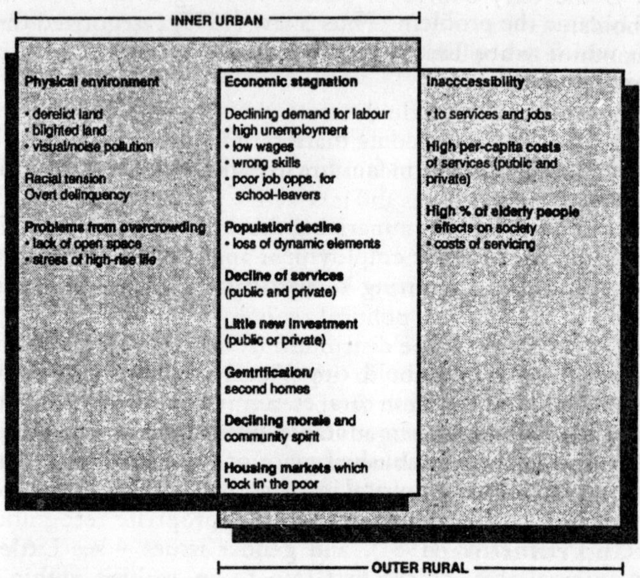

Figure 7.2 Urban and rural Britain: overlapping sets of problems (source: Moseley 1980).

A way forward for this situation has been marked by Moseley (1980a, 1980b) who has suggested that although rural areas do display particular characteristics which make them different from urban areas (a pleasant environment, a spaced-out geographical structure, and a distinctive local political ideology), such aspects of rurality are not the cause of deprivation, which is fundamentally similar whether experienced in an urban or a rural location (Fig. 7.2). Two implications emerge from this simple, but with hindsight important, observation: first, that research into the causes of rural deprivation must view social, environmental, and economic problems in the countryside as localized manifestations of aspatial processes (Cloke & Park 1985); and second, that attention could more profitably be paid to the *producers* rather than *consumers* of deprivation. Of themselves, these simple concepts of emphasizing production in aspatial terms do not in any way constitute a conversion to political-economy frameworks, but the very fact that such statements had to be made in the early 1980s demonstrates the ingrained barriers to structural explanation and investigation which were present at that time. Over the ensuing half-decade there has been a widespread willingness to accept rural deprivation as a product of forces emerging from the dynamics of late industrial capitalism, but few attempts to convert such statements into research programmes. Ironically, the most recent major research on rural deprivation has been McLaughlin's contract research for the Department of the Environment and the Development Commission. Initiated within the context of the legitimation of deprivation as an issue of service provision, this research has begun to publicize results of a markedly different nature. Indeed McLaughlin's (1986) preliminary conclusions suggest that the problems of the rural deprived are unlikely to be solved

(a) if our housing policies continue to support the sale of council housing which is the only source of rented housing available to low-income households;
(b) if we continue to abolish the Wages Councils which attempt to protect the interests of the low paid;
(c) if we continue to dismantle the state welfare system upon which the less able depend;
(d) if we fail to recognize the unequal distributional effects of our rural land use and environmental policies and take action accordingly (McLaughlin 1986, p. 307).

Many of these themes are now being taken up by rural geographers, but clearly a research agenda with political economy to the fore will have to define the problematic in terms of the distributional impacts of economic restructuring and the role of the state in its allocation of resources within such restructuring. Only when these factors are explicitly adopted as the major framework of study will investigative programmes be constructed which will begin to fill the new barns with new hay.

Rurality and rural space

If the political-economy model is to be important in the work of rural geographers, a number of key conceptual themes will have to be explored and developed. Four such themes are discussed in the remainder of this chapter, although these are artificial divisions and the interrelationships between the themes are crucial. The first, and perhaps most difficult, conceptual hurdle for rural geographers is that of overcoming their perpetual identity crisis. Their professional *raison d'être* depends on an ability to differentiate commonly accepted *rural* characteristics. They have attempted to define rurality (Cloke 1977, Cloke & Edwards 1986), promote rural courses within geography, and fill libraries with rural texts (see above) and journals (such as *Journal of Rural Studies* and *Countryside Planning Yearbook*). Such institutional momentum is difficult to decelerate, even if important theoretical advances are to be spurned because they do not fit the pattern of rural separatism.

There has, therefore, developed a culture of rurality. Rural *space* has been defined and preserved in the archives and annals of rural studies as being that area which

(a) is dominated (either currently or recently) by extensive land uses, notably agriculture and forestry;
(b) contains small, lower order settlements which demonstrate a strong relationship between buildings and extensive landscape, and which are thought of as rural by most of their residents;
(c) engenders a way of life which is characterized by a cohesive identity based on respect for the environmental and behavioural qualities of living as part of an extensive landscape (Cloke & Park 1985).

Points (a) and (b) have remained relatively unshaken in the minds of rural geographers but point (c) has been under attack for more than a decade from those (see, for example, Bailey 1975, Cloke & Griffiths 1980) who have argued that the defining parameters of social problems (such as poverty, powerlessness, and so on) should be represented by common variables of social production, thus rendering the rural–urban difference defunct.

The conflict between land use/settlement characteristics and socio-economic characteristics within the culture of rurality has steadily widened, and indeed the progressive debate over socio-spatial relations amongst prominent political-economic theorists has removed the theoretical debate far beyond the mere recognition of the fact that Britain and many other Western countries are by now culturally if not physically urbanized. The degree to which spatial considerations have been prioritized in political-economic theory has ebbed and flowed over recent years. Using Dunleavy's (1980) *Urban political analysis* as an arbitrary marker point of the anti-spatial ebb (it is also notable for the coincidence in time with Moseley's by comparison very simple yet important waymarker in the study of rural deprivation – see above) there is here a clear theoretical argument that both economic and socio-cultural activities in advanced capitalist Britain are organized on an aspatial basis. He admits that spatial variations between inner urban, suburban, and quasi-rural areas may be legitimate objects of enquiry in that the development of collective consumption processes may differ in these areas. Nevertheless, the key issue is that such spatial variations are not due to autonomous spatial factors, be these physical, environmental, or cultural. Rather they represent elements of the differentiation of functions and activities at a national level. Therefore to study rural anything – or urban anything – is a false representation of prevailing socio-economic and political structures (Dunleavy 1982).

Just as human geographers have patently gone overboard in stressing the significance of spatial differentiation in the past, there is now a strong movement by some political-economic theorists to pull back from the equally overboard position that space is merely a social construct. Dunford & Perrons (1983), for example, have stressed that an emphasis on functional and spatial differentiation of social reproduction is of utmost importance and Massey (1984) argues that space may be a social construct but that social relations are also constructed over space. Therefore to ignore space is to ignore the territory of social construction.

If taken to a logical extreme this flow of theoretical argument leads to the conclusion that if space is important to social theory, then the investigation of varying socio-spatial arrangements (including villages and cities) could also be important in social analysis. Giddens (1981) appears to reflect this emphasis on spatial variation when he suggests that the city is at the core of, rather than incidental to, social theory, a point picked up by Saunders (1985) who suggests that

> not unnaturally, many 'urban' sociologists, 'urban' geographers and other academics with a specialist 'urban' interest have begun to seize upon this straw offered by Giddens . . . they are now informed by one of our generation's leading theorists that urban sociology and its object of study – the city – lie at the heart of what the social sciences are all about (p. 68).

For 'rural' researchers, confusion reigns. Does the emphasis on *the city* legitimize a focus on social arrangements in space, in which case they are back on familiar ground fighting the battle of the underdog to raise the profile of the countryside with reference to the city? Or, does placing cities at the core leave no room for considerations of non-cities, reflecting rather that the city incorporates the countryside in the operation of single-space social relations? In either case the future for a specific rural specialism looks gloomy.

Saunders himself interprets Giddens's work as reinforcing a socially rather than spatially constituted mode of study (*non-spatial* sociology). None the less, he counsels alertness over the possibility that spatial organization might act as a secondary device for initiating or facilitating the development of certain societal processes (aspatial sociology). He thus tends to follow the now classic theoretical guidelines laid down by Urry (1981a, 1981b, 1985) that space has no general effects; rather certain potential social actions will be triggered by particular time–space contexts. The study of such contexts, in the now widely recognizable *locality studies*, reintroduces a spatial dimension albeit as an adjunct to the principal themes of social reproduction.

Here again, the rural researcher in search of a future for his or her culture of rurality might wish to stretch a theoretical point and suggest that some localities are more rural than others and are therefore worthy of study on that basis. To do so would be to imbue the locality concept with an unwarranted and anachronistic framework of differentiation. The uniqueness of localities is derived from the recomposition of local society by capitalist restructuring. Having reviewed the evidence relating to the potential existence of specifically rural localities, Urry (1984) concludes that:

> various critical notions – of different, overlapping spatial divisions of labour, of all localities as sites for the reproduction of labour-power, of variations in local social structures etc. – render problematic the notion that there are distinct 'rural' localities (p. 59).

Such a conclusion to the ebb and flow of analysis encapsulates a conundrum for rural geographers which threatens to create far-reaching divisions within the subject. Accept the arguments of most political-economic theorists and the legitimation for *rural* categories of study largely disappears, and if continued entails an over-emphasis on secondary spatial factors which in turn will be detrimental to a proper understanding of the primarily important structural changes which underlie them. Reject these arguments, and the potential explanatory power of the political-economy model is removed.

There would seem to be little room for compromise here, although Phillips & Williams (1984) have summarized four main reasons for retaining rural as an investigative unit:

(a) the need for rural studies to counterbalance the predominance of urban studies;
(b) the pragmatic requirement for analytically convenient categories such as rural and urban;
(c) the need to expose many of the romantic rural myths which have been fostered by an historically anti-urban social science;

(d) the basic belief that rural areas have distinctive characteristics distinguishable from those in urban areas.

It may well be that analytical convenience is the most honest amongst these reasons although there is, or has been, some legitimacy in each of them. The time may well have come, therefore, for rural geographers to forego their rurality culture and be content to work alongside other social scientists in the investigation of the mechanisms and impacts of restructuring and of the rôle of the state both as a catalyst of social reconstruction and as a resource allocation agency. In some respects, the fact that the concept of *rural* localities has been shown to be undermined by the political-economy approach is of little real consequence except for the impact on the institutions of rural studies. After all, there is a clear need for comparative studies of all types of localities including those which were previously constituted as rural. There is equally a clear need for urgent study of the differences between some rural areas, and of the way in which these differences add to the range of potential responses to capital and state restructuring.

The choice for 'rural' researchers interested in political economy is either to accept the relative redundancy of their parochial rurality culture, or to find some specific element of economic restructuring or state involvement which overcomes the loss of the locality as a spatial unit of study. It has been suggested, for example, that different configurations of the local state might occur in which variations in local history have led to different political cultures and in turn to the adoption of different political programmes (Saunders 1985). These differences might be seen to give rural researchers a specific point of entry into political-economic investigations, although there are clearly as many problems in differentiating a rural local state as there are with rural localities. It is therefore to *agriculture* as a particular fraction of capital accumulation, which has been identifiably rural because of its land usage, that rural researchers have turned for specificity with political-economy frameworks. Consequently, it is necessary to review the impact that critical rural sociology has had on the thinking of rural geographers, before widening the perspective to other fractions of capital, and the impact of accumulation, restructuring, and the recomposition of civil society in previously constituted rural areas.

The contribution of critical rural sociology

If rural geography has somewhat belatedly experienced the trauma of potentially radical conceptual change, rural sociologists have also gone through what Bradley (1981) has termed 'a profound sense of inner disquiet' over conceptual issues. Faced with the increasingly unsuitable nature of the rural-urban continuum and diffusionist tools of analysis, sociologists have undoubtedly faced up to the need for critical perspectives earlier than their geographical counterparts. One problem for rural sociologists, which also underlies the geographical predicament, is that the historical roots of the Marxian tradition which underlie critical social theory give little emphasis to rural areas. As Bradley notes:

> Marx's dictum about the peasantry being insignificant in historical terms has been applied to rural society, *tout court*, so that the social fabric of the countryside has been interpreted, in a static manner, as a residual category. To this extent much contemporary marxism has swallowed its own ideology, in accepting 'the idiocy of rural life' (1981, p. 586).

Nevertheless during the late 1970s and early 1980s there emerged a new research agenda in rural sociology which was heavily influenced by neo-Marxist and kindred perspectives (Buttel & Newby 1980). Five main themes were highlighted as crucial to this new critical rural sociology:

(a) the structure of agriculture within advanced capitalism;
(b) state agricultural policy;
(c) labour and agriculture;
(d) regional inequalities and agriculture;
(e) agricultural ecology.

The new approach was implemented, notably by Howard Newby's Essex School, so as to focus on agrarian class relations in general, and more particularly on the organization of property relationships (rather than divisions of labour) as the principal shaping factor of these class relations. As Newby (1986) has noted:

> The exodus of the rural working population from the countryside had led to a marked expansion of ex-urban middle-class commuting population within rural areas, which produced new cleavages and conflicts within the rural population. It was therefore possible to follow a chain of events which led from the continual reorganisation of property relations in agriculture (sponsored, it should be added, in large part by the state) through to changes in social composition of rural areas and on into an analysis of emergent social conflicts in the countryside, of which issues relating to environmental conservation, employment growth, and housing may be regarded as emblematic (p. 212).

Such an approach tends to treat rural sociology and agricultural sociology as synonymous, and in being influenced by such themes, some rural geographers have also tended to retrench into the agricultural heartland in order to retain a rural (= agricultural) specificity. In this way the locality concept is also imbued with agricultural characteristics in order to produce 'rural' localities. As Rees (1984) has pointed out:

> the emerging emphasis appears to lie in understanding rural society in terms of the analysis of *agricultural* production and the social relations thereby generated. The specificity of rural localities is seen to derive from the peculiarities of the capitalist production of agricultural commodities, with issues of landownership consequently occupying a central analytical position (p. 32).

The appeal of critical rural sociology to rural geographers has been due in no small part to the strides made by Howard Newby and his colleagues. In *The deferential worker*, Newby (1977) explains the apparatus which farmers have manipulated during the postwar period in order to ensure their hegemony over agrarian labour; in *Property, paternalism and power*, Newby *et al.* (1978) extend this analysis to an account of the property relations within capitalist agriculture, with special reference to the maintenance of political power; and in *Green and pleasant land*? Newby (1980) widens these themes in relation to environmentalism and countryside planning.

Inevitably this work has its detractors. Bradley (1981), for example, regrets the acceptance of the ideology which constructs the farmer and his poverty as pre-existent ahistoric subjects. This tendency, he claims, leads to an overemphasis on the activities of farmers in the market place to the detriment of a focus on the processes by which particular property relations and relations of appropriation and distribution are created and reproduced. A similar critical theme is pursued by Barlow (1986) who suggests that the argument that property should be the defining principle of rural society is limited in its explanatory potential:

> Private property, then, is necessary for the existence of class struggle in capitalist society, but its actual *form* in a given society is both a cause and an outcome of class and social struggles. Therefore landownership *can* have an effect on social 'stratification', but this is likely to evolve as the role of land in the accumulation of capital changes. Property relations are *created* and *reproduced* and are not historically constant (Barlow 1986, p. 311).

Despite these theoretical hiccups and a self-confessed immaturity of the political-economy approach, there is no doubt that both the Essex School and its North American counterpart (Buttel 1982) have been significant influences in the recent direction of rural geographical work aimed at the agricultural sector. In their excellent review, Marsden *et al.* (1986) isolate four major areas of theoretical and empirical enquiry:

(a) the process of capital penetration of agriculture and the reasons for unsuccessful penetration;
(b) agrarian class structure;
(c) the transformation of the family labour farm in a capitalist context;
(d) the relations between agriculture and the state.

They suggest that two major difficulties have been encountered in applying the political-economy approach to agriculture. First, they argue that the status of the family farm has not yet been successfully explained within capitalist development – a failure of crucial significance given the importance of this specific form of agricultural production. Second, they assert that the radical switch in emphasis towards explanation of social change according to structural processes and class formation has diverted attention from the variety of forms and processes in productive relations which are reflected in the unevenness and specificity of agricultural production in time–space realms. As a result they call for a more careful integration of structural themes with those historically and locationally specific actions of farm production businesses.

In many ways these difficulties will be readily acceptable to those rural geographers who cling to more traditional behaviouralist methods such as Ilbery's (1985) approach to agricultural geography which is founded on sub-aggregate analyses of production and consumption. It must therefore be argued that while the *integration* of structural and human agency themes is much to be desired, many rural geographers need no second bidding to focus their work on sub-structural phenomena and that it is the political economy paradigm which remains in need of promotion. There are, however, an increasingly important number of studies which do seek such integration, notable amongst these being the studies of capitalist agriculture, agrarian class structures, and the family farm by Marsden (1984) and Winter (1984).

The potential dangers of pursuing the directions taken by critical rural sociology, therefore, are that an overemphasis on agriculture can lead not only to an unwarranted synonymity between 'rural' and 'agricultural', but also that an assertion of the need for sub-structural study with which to explain the unevenness of agricultural production can lead to undue focus on agricultural-localities to the detriment of other elements of social reproduction and capital restructuring. So to do would be to concentrate on the outcome of *past* rounds of investment (Massey 1979). As Rees (1984) has stressed:

> by focussing upon one particular aspect of capitalist production (agriculture) because this may have been the historically determining influence in given (rural) areas, attention may be diverted from the totality of contemporary processes operating in areas thus defined (p. 34).

The potential benefit of pursuing the critical rural sociologist's path into political-economy approaches is that agricultural production can be viewed as but one fraction of capital accumulation, which may be placed most compatibly alongside other such fractions operating in the same localities. This subjugation of the agricultural as one of many elements of capital restructuring the social recomposition rather than *the* rural element opens up many other integrative themes which as yet have been explored in only a pioneering manner. For example, the role of the state in agriculture has been much studied (see, for example, Bowler 1985, Hill 1984, Lowe *et al*. 1986, Marsh & Swanney 1980), but more in an analytical framework which merely recognizes that 'agriculture and forestry operate in an economic climate heavily determined by political considerations and decisions' (Lowe *et al*. 1986, p. 50) than within a more demanding critique of the state as constrained primarily by its relations with society as a whole. Similarly, the important research on environmental pressure groups (see, for example, Lowe & Goyder 1983) has (perhaps understandably) tended to be framed within pluralist models of power thereby assuming a rôle for the state which takes less than full account of more structuralist views of the state as representative of the current balance of class influence and furnisher of the needs of capital interests. The major strengths of the political-economy approach lie in the primacy attached to capital restructuring within all fractions of capital. The interests of rural geographers therefore appear to be best served by seeking to investigate non-agricultural fractions with similar energy to that expended on the agricultural theme. It is to these non-agricultural fractions that this chapter now turns.

The centrality of capital restructuring

Over the last decade dramatic and far-reaching changes have occurred in what have previously been constituted as rural areas. At the very simplest level, two major non-agricultural trends have dominated the interests of rural geographers and other rural researchers. First, the phenomenon of *counterurbanization* was marked by an upheaval in population trends whereby remote rural areas, having previously been characterized as population donors during a century of depopulation, were suddenly shown by the 1981 Census in Britain to have gained population over the previous decade (Champion 1981, Robert & Randolph 1983). These trends mirrored those already occurring in North America (Berry 1976, Brown & Wordwell 1981, Morrill 1980) and were also noticed in Western Europe (Fielding 1982), Australasia (Smailes & Hugo 1985), and in other international situations (Vining & Kontuly 1978, Vining & Pallone 1982).

Second, *urban-to-rural industrial shifts* were highlighted (Fothergill & Gudgin 1982, 1983, Keeble 1980, Keeble *et al.* 1983, Massey & Meegan 1982), again in line with the North American experience. Rural areas, variously and sometimes carelessly defined, were shown to have been the major recipients of new manufacturing growth during the 1970s, in stark contrast to the declining manufacturing sectors of the inner cities and metropolitan areas more generally.

Inevitably, counterurbanization trends came to be explained by trends in manufacturing growth. Why were rural populations growing? Because of the growth in manufacturing industry. Why was the manufacturing sector in rural areas growing? Well, much of the growth was seen to be represented by small, seemingly independent local firms, and so rather than answers to this question being rooted in concepts of capital restructuring, neoclassical economic explanations were advanced. For example, Fothergill & Gudgin (1982) suggested a *constrained location theory* whereby the constraints on factory floorspace in cities were instrumental in driving those firms that wished to expand away from inner cities and out into areas of less constraint. With the steady displacement of labour by machinery, increasing amounts of factory floorspace per worker employed were required even without expansion of production, and so rural locations were seen to benefit from the migration of firms wishing to expand their premises so as either to maintain or increase their workforce. Such an explanation is the advance factory builders' dream. Presumably all you have to do is to build more premises, and unemployment will disappear! Other neoclassical explanations, for example the influence of lower operating costs in rural areas as an attraction to manufacturing industry (see Keeble 1984), were also offered as reasons why the urban–rural shift had occurred.

Combinations of these explanations, based as they were on the usual pluralist theoretical mix so often accepted within rural geography, were patently unsatisfactory in at least two major respects:

(a) The urban–rural shift of manufacturing could not be explained away by one or other factor of managerial decision making. Evidently these neoclassical symptoms were underlain by more fundamental processes relating to the continuation of capital accumulation under changing conditions, and of the new structures of capital which were emerging in response to those conditions. Neither was the shift specifically an urban-to-rural one. Leaving aside the rather arbitrary population thresholds used as definitions of rural by some of the researchers involved, the patterns of growth and decline were far more complex than the categories rural and urban suggest. Certain localities were obviously better able to respond to the broader structural imperatives than were other, seemingly similar localities.

(b) The urban–rural shift of population was similarly not caused in all locations by shifting manufacturing labourmarkets. To use broad examples, population growth in marginal locations has been influenced in various combinations by: new service sector employment, often state-sponsored; higher levels of stay-at-home or in-migrant unemployed; higher levels of retirement – often fed by greater incidences of early retirement; a search by increasing number of households for an alternative non-city lifestyle, ranging from agricultural self-sufficiency to more sophisticated self-employment in rural areas; and so on (Cloke 1985b).

Partly because the manifestations of these changes were so startling, and partly because neoclassical and pluralist explanations of change were so impoverished, the 1980s has proved to be a significant watershed in the acceptance of political-economic frameworks for understanding phemonena of change. With excellent external guidance from such authorities as Fielding (1982) on counterurbanization and Massey & Meegan (1982) on capital–labour relations, rural geographers have for the first time exhibited a widespread recognition of the need to explore political-economic fundamentals as the route to understanding socio-spatial phenomena.

There is nothing to be gained in the context of this chapter from a lengthy exposition of now familiar concepts of political-economic processes, restructuring, and state-society relations. These have been excellently reviewed elsewhere (see, for example, Cooke 1983, Dunleavy 1980, Harvey 1982). There does, however, seem to be some merit in isolating key themes from the political-economy paradigm which are most applicable to the interests of rural geography groupings and thereby functional in the development of a research agenda for the appropriate development of that paradigm.

First, it is important to stress the centrality of *capital accumulation* as the driving force of social formation. The continuous process of reinvestment undertaken by individual capital units in search of surplus value is capable of generating unbalanced and unregulated trends of growth and decline. At various stages of this process deterrents to accumulation are experienced which require a restructuring of production to ensure a continuation of acceptable profit. Mechanization, labour control, and market manipulation can be brought about without a spatial shift of production, but *capital restructuring* can also involve a relocation of production to a more favourable accumulation environment. Bradley & Lowe (1984) have stressed the mobility of Western capitalism which in its current monopoly phase tends to be relatively unconstrained in spatial terms. Some areas previously constituted as rural will currently often represent favourable accumulative environments, due to compliant and relatively inexpensive labour pools and, in a more localized fashion, to state subsidy of various factors of production in particular policy areas such as that currently operating in mid Wales (Pettigrew 1987). Relocation is often twinned with restructuring of capital:

> capital restructuring alters and reconstitutes various technical divisions of labour, particularly whereby higher and lower level functions of specific firms and enterprises are spatially split, for example through the location of branch plants in peripheral regions (Bradley & Lowe 1984, p. 11).

Despite governmental exhortions to pedal-power, labour by contrast has been shown to be relatively immobile leading to *recomposition* impacts within local civil society as restructuring leads to societal conflict.

Urry (1984) lists six important elements of capital restructuring analysis:

(a) different patterns of economic restructuring occur in relation to different spatial divisions of labour;
(b) restructuring follows changing patterns of accumulation;
(c) locational changes should be explained according to restructuring complexities rather than traditional economic or political criteria;
(d) restructuring both follows class struggles and transforms the conditions for reproduction of class relations;
(e) the relevance of an area is a product of its position *vis-à-vis* the overlap of spatial divisions of labour;
(f) uneven development results which again should be explained according to restructuring rather than to traditional regional criteria.

Two further themes emerge as key issues from this analysis. The first (point (f)) relates to the uneven nature of capital relocation, restructuring, and recomposition, and refers to the idea of favourable accumulative environments introduced above. Clearly, localities which have to date been constituted as rural and the social relations therein which have been similarly labelled, are being reproduced because of their current favourable status for relocation due to the wider discontinuities of capital activity. This realization has further implications for the assumptions laid down by critical rural sociology that land and property ownership (particularly in connection with agricultural fractions of capital) constitute the major power-pack for social change in these localities. Any such viewpoint is contradicted by analysis of restructuring, as Rees (1984) stresses:

> Changes in rural employment structures are central to any understanding of the reality of rural social life. On the one hand they reflect profound shifts in the nature and organisation of capitalist production and, more specifically, the widely differing impacts of these shifts on different types of locality. On the other, employment changes themselves have resulted in radical developments in terms of rural class structures, gender divisions, the forms of political conflict occurring in rural areas and, indeed, of the complex processes by which 'rural cultures' are produced and reproduced (p. 27).

The second key and closely related issue to arise from Urry's analysis of restructuring refers to class relationships and social reproduction (point (d)).

Clearly, restructuring does not take place in a societal vacuum. Different localities will have different histories of political and class conflict, will have experienced varying forms of social reproduction, and will exhibit particular contemporary class compositions. As such, class relations are not only the end-product of foregoing rounds of capital accumulation and restructuring, but also serve to mould the characteristics of ensuing iterations of these processes; hence Rees's conviction (quoted above) that political-economic analysis should deal with class structures, political change, and state dynamics as well as with the more obvious shifts in the nature of production.

There is one further and important implication of accepting a materialist analysis of political economy. In doing so, we accept a research framework which dictates that society is structured by the imperatives of capitalism and that classes representing capital and labour are locked into an irresolvable conflict over the production and distribution of the surplus value created by labour. A major outcome of such premises is that the typical social subdivisions adopted by rural geographers as the basis for analysis of rural people and rural planning are thrown into disarray. As Healey (1986) has pointed out in the context of land use planning: 'The notion of irresolvable conflict thus undermines many concepts dear to planners, such as the "general interest" or a "balanced strategy" except as temporary phenomena' (p. 187). Equally:

> From these general premises, then, we are offered principles for grouping society into segments (structuring 'interests') which look quite different to those typical in community power studies or even land use planners' lists of who to consult in public participation exercises (Healey 1986, p. 187).

Rural researchers will therefore be required not only to adopt new conceptions of the driving force behind changes in their research constituency but also to reclassify society in order to incorporate the end-product and fore-shaping roles played by class composition in the processes of capital restructuring.

If the dramatic changes associated with counterurbanization and manufacturing shifts have promoted both an abstract realization of neo-classical theoretical poverty and a widespread acknowledgement of the need to invoke the political-economy paradigm, the actual progress made by rural geographers towards rescheduling their research to these ends has (not unexpectedly) been painfully slow. At the end of their important book proposing a political-economy orientation for research in rural social geography, Phillips & Williams (1984) reflect that

> the theoretical understanding of political economy as it affects rural areas is inadequate to develop far such a research methodology, and data are lacking with which to test empirically many propositions. . . . Therefore, many of the substantive chapters are based upon managerialist and behavioural studies of social processes, with the implicit (rather than explicit) notion that these are located within the larger social formation (p. 237).

Rural research in the political-economy paradigm thus has infant status according to this review, and despite the likelihood of a significant growth in output in this area over the next decade, the current base level for such growth is small within the rural geographical fraternity.

The development of the political-economic approach with reference to what have previously been constituted rural areas has therefore so far been sporadic and multidisciplinary. Aside from research dealing with agricultural fractions of capital as the core of rural analysis (see above), there are a small but growing number of contributions of direct relevance to the key themes outlined in this chapter. For example, Bolton & Grafton (1986) have studied the counterurbanization phenomenon in rural Devon with reference to the restructuring of manufacturing capital in the area; Markusen's (1980) research in the United States has revealed elements of restructuring and recomposition in western boom towns; Lovering (1978, 1982) and Bradley & Lowe (1984) have made important contributions to the dynamics and interrelationships of local labour-markets; and Pratt (1986) has produced some interesting evidence of political-economic processes underlying the development of industrial estates as sites for potential restructured capital accumulation. Aside from the manufacturing sector, there are now a number of contributions which are also beginning to tackle the distributional impact of service capital and state-derived service development (see Cooke 1983, Daniels & Thrift 1985, Owen & Green 1985) and which offer class-based analysis to fill the vacuum existing during a long phase of post-Pahl hangover. On the latter theme, Little's (1987) microscale examination of the interrelationships between labour and housing markets and class-distributed impacts, Cloke & Thrift's (1987) exploration of the issues of social reproduction and reformulation attendant on the emergence of a multi-faceted new middle class in rural areas; and Little's (1985) exposure of new gender relations, all represent interesting starting points for the development of research programmes within a political-economy paradigm. As Healey (1986) suggests, however, 'we have barely begun to explore what the political economy paradigm offers' (p. 117) and there is a clear need to pursue with vigour what Barnes (1982) has called *paradigm development*.

The role of the state

If structuration of labourmarkets and to a large extent housing markets can be explained with recourse to the restructuring of capital accumulation by different capitalist fractions, any concerted use of political-economy approaches must in addition account for the rôle played by the state in influencing production and consumption. One of the principal changes to have occurred in rural geographical study over the last decade is an emphasis on applied research with special reference to planning and policy making. Indeed, much discussion has focused on the ability of planning to perform particular regulatory or interventionist strategies, and thereby on the degree to which planners and policy makers should be blamed for perceived shortcomings and problems in rural areas. Although such issues lead directly to political questions of the accountability of, and constraints on, the planning system within a wider societal context, rural geographers (unlike their urban and regional counterparts) have again been shy to delve into such overtly ideological matters as power relations within the state (see Cloke 1988).

Nevertheless, an analysis of the state is a fundamental facet of the adoption of political-economy models for rural research (Wolfe 1977) and it is therefore necessary to view planning and policy making as aspects of state activity, being equally subject to constraints experienced by other such activities. The forum is not, therefore, central–local government relations (although these can be important) but rather the relationship between the entire capitalist state and its host society (Hanrahan & Cloke 1983). If the state is viewed as an arena for the independent arbitration of competing interests in society, then planning and policy making will reflect such neutrality. If, however, as political-economic theory suggests, the state performs other functions, notably the preservation of the societal status quo and the creation and continuation of favourable conditions for capital accumulation (see, for example, Poulantzas 1973), then it should be expected that planning and policy making will be constrained by these functions.

The issues of power relations and the state have been excellently reviewed elsewhere (see, for example, Cooke 1983, Clark & Dear 1984, Dunleavy 1980, Ham & Hill 1984, Saunders 1979). Different conceptualizations of power (pluralist, élitist, managerialist, structuralist, and so on) are crucial to the detailed theoretical framework to be adopted within political-economic models, but discussion here will be restricted to the specific questions of: first, the apparent failure of state planning to achieve progressive socio-economic impacts in what have been traditionally viewed as rural areas; and second, the nature and function of the local state functioning in these areas.

It has been argued that there are at least two levels of constraint experienced by policy makers with responsibility for rural areas (Cloke & Hanrahan 1984). First, the acceptance of an art of the possible which is conditioned by the overall state–society relationship presents decision makers with an artificially narrow range of policy options. If social production and investment are designed to aid capital accumulation by the minority, and social consumption is confined by the need not to disturb the societal status quo, then it is the state–society constraint which underpins the continuing uneven distribution of power, wealth, and opportunity. A secondary set of constraints arises from the complex interagency relations within and between the public and private sectors. Three such arenas of conflict can be highlighted, although such categorizations inevitably present an artificially delimited picture of the processes involved.

Central-local relationships

Recent evidence from studies of intergovernmental relations (see, for example, Goldsmith 1986) has clearly highlighted the trends of increasing government involvement with, and penetration into society, resulting in a proliferation and increasing complexity of the institutions of government. Such complexity involves not only the traditional central and local elements of the state but has been extended to a 'regional state' often acting in a mediating capacity between these two elements (Saunders 1985b).

In attempting to understand the constraints on policy making which arise from central–regional–local turbulence, many of the political assumptions inherent in urban political analysis have been found to be ill-suited to localities which include rural areas as traditionally constituted. The typical illustration of a monetarist conservative central government resolving to reduce the respective powers of socialist local government, trade unions, and consumer agencies (Saunders 1982) is not immediately applicable in the local political environment of many shire counties. These are often dominated by an innate conservatism which can serve to neutralize the expected ideological divisions between local and central government. Nevertheless, friction does occur between the conservatism expressed at central and local levels, most notably over the allocation of financial resources (Association of County Councils 1979). The key question, therefore, is that of the degree of discretion permitted to the local state by the centre, and the vigour with which any available discretion is exercised by those holding political power in 'rural' government agencies.

What is clear from the work of rural researchers and others is that local government has consistently and increasingly found itself in a position of subordination to definitions and priorities which are externally imposed by the central state, rather than experiencing any particular freedom in the generation of its own policy agenda. For example, Wright (1982) drawing from a study of decision making in all sectors of the rural environment concludes that investment programmes are developed 'with at least one eye on the central department's policies and priorities, and are implemented with the knowledge that the central department monitors the implementation of programmes closely' (pp. 6–7). In many cases this central control over local policy activity is highly formalized. Hambleton (1981), for instance, has stressed that the apparent devolution of planning responsibilities to lower tier authorities in Britain in fact represents a concern for greater regulation on the part of central government rather than any wish to permit new initiatives. Regulatory controls operate alongside legal and financial controls which have become increasingly important over recent years. The result is that the central–local relationship has a significant constraining effect on the range of policy decisions to be made by politicians and professionals at the local level.

Inter-agency relationships

With the complex network of agencies involved in policy making for rural areas, difficulties of co-ordination and co-operation between agencies have become an important feature of the critique of postwar rural planning. Inter-agency relations are thus viewed as another significant arena in which the effectiveness of state activity can be constrained. Organizational theory suggests that each agency will tend jealously to guard its own decision-making domain, and will, therefore, be reluctant to permit undue external interference with their decision-making responsibilities once these have been handed down to them from the centre. Friend *et al.* (1974) and Clegg & Dunkerley (1980) have highlighted the naïvity inherent in any expectation that agencies which are constantly seeking to legitimize both their own existence and the independence of their resource base will wish to agree to anything more rigid than short-term and *ad hoc* liaisons in the name of corporate planning. As Leach (1980) has pointed out, 'authorities will only co-operate when it suits them, or when they have to, and then very much on their own terms, and in line with their own interests' (p. 293).

Evidence of inter-organizational conflict in rural areas has been readily available, particularly in the sphere of the required planning liaisons between different levels of local government in Britain, and specifically concerned with the integration of strategic and local plans (Bruton 1983, Healey 1979, Leach & Moore 1979). Outside of the specific planning function, there is further evidence of a failure on the part of all agencies concerned to tackle positive action against clearly defined negative trends such as the loss of rural services (Packman & Wallace 1982). It would seem, therefore, that where local scale regulation is required in order to reform the patterns of production and consumption dictated by the market, inter-agency conflicts often serve to stifle initiatives and reinforce the limitations imposed by the centre on the scope of, and available finance for, progressive policy making. By contrast, in their studies of rural areas under greater urban pressure, Healey *et al.* (1982) were able to recognize few conflicts between organization in the public sector, and Bell (1987) found good working relations between water and planning agencies in matters of infrastructural provision. Inter-agency relations should, therefore, be recognized as being different in various growth/no-growth contexts, and also as varying with particular configurations of political and professional power within agencies (Laffin 1986). Such relations do, however, serve to constrain the achievement of progressive corporate action in many rural areas, a factor clearly illustrated by the fact that in general terms, the adoption of corporate planning mechanisms has not achieved the co-ordination of policies for rural areas that are radically dissimilar from those which have preceded them.

Public–private sector relationships

The passivity of the state in terms of positive policy measures aimed at regulating market-led change cannot be appreciated fully unless the dependency of policy makers on the private sector is understood. In short, public policy has become increasingly reliant on the availability of private capital for its realization. As a consequence, policy making is constrained by the necessity to prescribe planning outcomes which pragmatically are likely to fit in with the objectives of private capital. Without the availability of such capital many apparently positive policy intentions turn out to be rather tame rhetoric. Clearly, then, industrialists, shopkeepers, retail capital, bus operators, housing development capital, and the financial backing that underpins all of these areas of activity, are crucial to the objectives of policy makers.

Some commentators have suggested that the way forward in this situation is via partnership between the public and private sectors *within* policy-making mechanisms (Mawson & Miller 1983). To some extent, the economic proclivities of the public sector which have tended to result in policies favouring centralized locations for service delivery have mirrored the search for economies of scale through rationalization that has been pursued by the private sector. This, however, constitutes a *partnership of decline*, and where policy makers have sought to channel private sector resources into specific locations where need has been recognized, the equivalent *partnership of development* has been far less readily forthcoming. Indeed, Murie's (1980) research into planning controls over housing suggests the mythical nature of much of the partnership debate:

> Much housing planning and policy-making gives little attention to the private sector and perpetuates a myth of competence and control. One con-

> sequence is often a very wide gap between plans and practice and between the intention and consequence of policy (p. 310).

Blacksell & Gilg's (1981) study of local authority resistance to residential development in supposedly protected areas exposes just such a gap within the context of rural Devon.

The ability of private sector interests to dictate policy options far exceeds these obvious channels of direct involvement. Austin (1983), for example, has shown that in order to reduce their own tax payments, corporate interests will use every available opportunity to influence decision makers in the direction of minimum collective expenditure on service delivery. With private interests so dominant, the potential for constructive social and economic planning in what have traditionally been constituted as rural areas is severely curtailed. Decision makers should thus be viewed as bound in on all sides by the state–society relationship discussed above.

Having outlined, albeit briefly, the primary and secondary constraints operating on state activity in capitalist society, it is important now to review the questions of whether the local state in rural localities is in any way different from that elsewhere, and if so, whether the nature of local state take-up of available policy discretion is thereby altered. One immediate problem here is that many local state theorists have assumed political opposition to the centre from the periphery. Dunleavy (1982), for example, stresses the potential utility of local government as a defence of working-class interests against central government attacks, either through attempts to reintroduce market disciplines or by disruptive tactics in local politics intended to influence the central state agenda on behalf of issues such as social insurance, pension provision, and hospital care. In both cases, these perspectives are predicated on a significant level of working-class involvement in local government and within the wider local state.

In fact, local authorities embracing rural areas, being politically conservative in nature, have certainly not exhibited these expected trends of working-class political involvement. Furthermore, many analysts of central–local relations (see Saunders 1984) have suggested that the local state as a whole has not been particularly successful either as a defence against central state attacks or in attaining reforms in the central state policy agenda. Rather, the institutional nature of local government has been reorganized by the centre so as to ease the implementation of centrally derived uniform policies; financial support from the centre to the periphery has been reconstituted so as to reduce expenditure on social consumption, strategies of local corporatism have outflanked local democracy; and an anti-state culture along with private enterprise solutions has been mobilized to reduce popular expectations of welfare provision.

There are many conceptual frameworks on which to found these trends. Dear & Clark (1981), for example, sum up this unholy marriage between central control and local democracy thus:

> It is only via the local state system that social and ideological control of a spatially extensive and heterogeneous jurisdiction becomes possible. In this manner, local needs are anticipated and answered, and state legitimacy ensured. Although the existence of the local state is functional for capitalism,

it is also in keeping with the principles of local self-determination in democracy (p. 1280).

This viewpoint offers little hint of significant discretion or autonomy for the local state. By contrast, Cawson & Saunders (1983) have sought to unravel the same knot by suggesting a dualistic model of the state which identifies:

(a) the corporate sector of government, located centrally and producing social investment policies;
(b) the competitive sector operating principally at the local level and producing social consumption policies.

Inherent in this model is the existence of relative autonomy available to different localities reflecting pressures from local élites, pressure groups, and individuals. A third standpoint is offered by Hirsch (1981) who suggests that the local state exhibits occasional discretionary powers, but that these are heavily constrained by the art of the possible externally defined by the state–society relationships.

According to which of these different concepts of available discretion is adopted, the local state in rural areas can be attributed different levels of importance. Although it might have been expected that the conservative rural local authorities would have attained a sympathetic political partnership with the conservative central state, they have in fact suffered a number of important handicaps. First, they have been passive agents in the central state's direct control over rural resources such as agriculture and forestry (Lowe *et al.* 1986), and in its indirect control over the promotion of industrial development through non-democratic organizations such as the Development Commission (Healey & Ilbery 1985). Second, they have been equally subject to the financial penalties imposed by central government on their more radical urban counterparts. Third, they have been circumspect in the exercise of any available discretion because of the political conservatism inherent in a majority of their constituents (Gilg 1984).

Studies of the policy-making procedures present within local governments in rural localities have demonstrated that the environment of political conservatism has proved to be a comfortable breeding ground for élitist power within the local state. Glover's (1985) work on transport policy in Bedfordshire illustrates this point in relation to the powers of key officers, 'the officers clearly demonstrate an awareness of how to use council rules and procedures to exert influence on the political process and how to further their aims with persuasive and sound technical arguments' (p. 145). Other studies, including that by Buchanan (1982) in Suffolk and by Cloke & Little (1987a, 1987b) in Gloucestershire, have revealed that the joint activities of these senior professional actors in tandem with gatekeeping political figures hold the key to an understanding of the exercise of available discretion by the local state in rural localities. As Blowers (1980) succinctly summarizes:

Within the limitations prescribed by the necessity to ensure the maintenance of the prevailing pattern of social relationships, planners exert considerable influence and power. And it is a power that is unequally distributed, being

> concentrated among a few officials and politicians . . . the distinction between administration and politics becomes irrelevant; what matters is the relationship between those leading politicians and officials who are responsible for the development of policy and its implementation, and their interaction with the powerful interests in society at large (p. 37).

Use of the political-economy paradigm will lead to an ever greater attention to these issues, especially to the nature of political conservatism within the local state, the availability and manipulation of discretion and autonomy, the changing constitution of state organizations in seeking the legitimation of central state aspirations, the role of the professions, and so on. Political conservatism and a small scale of activity should not be categorized in any way as rural phenomena, but they do contribute significantly towards the variety of politico-cultural localities within which the process of capital accumulation is being worked out.

Conclusion

There are two important dangers to be recognized in the adoption of political-economy approaches in rural geography. Continuing the metaphor used in the introduction to this chapter, there are first those who will argue that Eeyore's tail has been there all the time – not lost but unobtrusively in the background. Thus the search for theoretical concepts on which to base the research effort will be viewed by some as an unnecessary distraction, as these concepts have been implicitly recognized by researchers for many years and have already found their way into current research. This attitude might be summarized as 'political economy is common sense'. It is often linked with the idea that positivistic methods and political-economy concepts are compatible, and a retrospective analysis of research in rural geography permits political-economy themes to be read into existing work. Such attitudes can stem only from a lack of understanding of the demands of a political-economy approach, both in conceptual and methodological terms.

The second danger relates to an over-confidence in the initial explanatory powers of political-economy approaches – Eeyore's 'it explains everything' attitude. It will be obvious from the limited examples used in this chapter that the use of the political-economy paradigm by rural researchers is as yet immature and undeveloped. A considerable task of total conceptual immersion and careful subsequent programming of research techniques will have to be achieved before significant explanatory strides can be made. The greatest barrier to such achievements will be a dissipation of energy in the continuing fight to legitimize and protect 'rural' as a significant analytical category. Those whose overriding interest is in land use *per se* will not be persuaded that extensive rural activities should not be treated differently from intensive urban uses. Those, however, whose interest lies more in explaining why land use changes occur and what interactions exist between economy and society in these areas of extensive land use are far more likely to succumb to the social construction of space and to concentrate their research on wider structural phenomena within familiar localities.

Given that the mire of rurality legitimation can be avoided, a skeletal agenda may be suggested for future research in the political-economy mode. Perhaps four main areas require initial attention:

(a) The historical treatment of localities of interest to rural geographers has tended to be archival, if painstaking. If a political-economy approach is to be pursued at all, it will have to be grounded in a materialist view of the past. In particular, histories of social reproduction and reformulation will be required in order to recognize the iterative relations between economic restructuring and host society.
(b) The processes of economic restructuring require specific attention, with particular emphasis on the unevenness of capital expansion. Relocation, restructuring, and recomposition should be analysed in terms of particular fractions of capital, and the interrelations between agriculture and other capital fractions should be fully explored. The *mobility* of capital and labour should figure prominently here.
(c) The impacts of restructuring are also a crucial research target. Recognizing that class composition and political, economic, and cultural configurations in a particular locality can *shape* restructuring as well as be affected by it, there are important changes to be investigated resulting from particular iterations of the restructuring process. Changing class structures, particularly the infiltration of different fractions of the middle class (with marked impacts in local economies, political representation, and so on); changing gender divisions; and changing cultural characteristics represent just some of the major issues requiring serious attention here.
(d) The recognition of the role of the state will necessitate a reworked analysis of planning, policy making and management in these localities. Specific recognition of local power relations and mechanisms of conflict resolution should be linked with the overriding national and international contexts of state involvement. Moreover, the central and local aspects of policy and power should be interrelated conceptually by means of a focus on *policy networks* and *policy communities* which transcend traditional interorganizational hierarchies.

To the experienced political-economy theoretician, such an agenda will appear simple and rather basic. For most rural geographers, however, it is a question of not running before we can walk.

References

Association of County Councils 1979. *Rural deprivation*. London: Association of County Councils.

Association of District Councils 1980. *Rural recovery: strategy for survival*. London: Association of District Councils.

Austin, D. M. 1983. The political economy of human services, *Policy and Politics* **11**, 343–59.

Bailey, J. 1975. *Social theory for planning*. London: Routledge & Kegan Paul.

Banister, D. J. 1980. *Transport mobility and deprivation in inter-urban areas*. Farnborough: Saxon House.

Barlow, J. 1986. Landowners, property ownership and the rural locality. *International Journal of Urban and Regional Research* **10**, 309–29.

Barnes, B. 1982. *T. S. Kuhn and social science*. London: Macmillan.

Bell, P. J. 1987. Implementation and the role of the water industry. In *Rural planning: policy into action?* P. J. Cloke (ed.). London: Harper & Row.

Berry, B. J. L. (ed.) 1976. *Urbanisation and counterurbanisation*. Beverly Hills: Sage Publications.

Best, R. H. 1981. *Land use and living space*. London: Methuen.

Best, R. H. & J. T. Coppock 1962. *The changing use of land in Britain*. London: Faber.

Blacksell, M. & A. Gilg 1981. *The countryside: planning and change*. London: Allen & Unwin.

Blowers, A. 1980. *The limits to power: the politics of local planning policy*. Oxford: Pergamon.

Bolton, N. & D. Grafton 1986. Counterurbanisation and economic restructuring in the rural periphery: some evidence from North Devon. Paper presented to the 2nd British–Dutch Symposium on Rural Geography, Amsterdam.

Bonham-Carter, V. 1951. *The English village*. Harmondsworth: Penguin.

Bowler, I. 1985. *Agriculture under the common agricultural policy*. Manchester: Manchester University Press.

Bracey, H. E. 1952. *Social provision in rural Wiltshire*. London: Methuen.

Bracey, H. E. 1970. *People in the countryside*. London: Routledge & Kegan Paul.

Bradley, T. 1981. Capitalism and countryside: rural sociology as political economy. *International Journal of Urban and Regional Research* **5**, 581–7.

Bradley, T. & P. Lowe (eds) 1984. *Locality and rurality*. Norwich: GeoBooks.

Brown, D. L. & J. M. Wardwell 1981. *New directions in urban–rural migration*. New York: Academic Press.

Bruton, M. J. 1983. Local plans, local planning and development plan schemes in England 1974–1982. *Town Planning Review* **54**, 4–23.

Buchanan, S. 1982. Power and planning in rural areas. In *Power, planning and people in rural East Anglia*, M. J. Moseley (ed.). Norwich: University of East Anglia Press.

Bunce, M. 1982. *Rural settlement in an urban world*. London: Croom Helm.

Buttel, F. 1982. The political economy of agriculture in advanced industrial societies. *Current Perspectives in Social Theory* **3**, 27–55.

Buttel, F. & H. Newby (eds) 1980. *The rural sociology of the advanced societies: critical perspectives*. London: Croom Helm.

Carlson, J. E., M. L. Lassey & W. R. Lassey 1981. *Rural society and environment in America*. New York: McGraw-Hill.

Cawson, A. & P. Saunders 1983. Corporatism, competitive politics and class struggle. In *Capital and politics*, R. King (ed.). London: Routledge & Kegan Paul.

Champion, A. G. 1981. Population trends in rural Britain. *Population Trends* **26**, 20–3.

Clark, G. 1982. *Housing and planning in the countryside*. Chichester: Wiley.

Clark, G. & M. Dear 1984. *State apparatus: structures and language of legitimacy*. Boston: Allen & Unwin.

Clegg, S. & D. Dunkerley 1980. *Organisation and society*. London: Routledge & Kegan Paul.

Cloke, P. J. 1977. An index of rurality for England and Wales. *Regional Studies* **11**, 31–46.

Cloke, P. J. 1979. *Key settlements in rural areas*. London: Methuen.

Cloke, P. J. 1980. New emphases for applied rural geography. *Progress in Human Geography* **4**, 181–217.

Cloke, P. J. 1983. *An introduction to rural settlement planning*. London: Methuen.

Cloke, P. J. 1985a. Whither rural studies? *Journal of Rural Studies* **1**, 1–10.

Cloke, P. J. 1985b. Counterurbanisation: a rural perspective. *Geography* **70**, 13–23.

Cloke, P. J. (ed.) 1987. *Rural planning: policies into action?* London: Harper & Row.

Cloke, P. J. 1988. Planning, policy-making and state intervention in rural areas. In *Policies and plans for rural people: an international perspective*, P. J. Cloke (ed.). London: Allen & Unwin.

Cloke, P. J. & G. W. Edwards 1986. Rurality in England and Wales, 1981. *Regional Studies* **20**, 289–306.

Cloke, P. J. & M. J. Griffiths 1980. Planning and the rural–urban relationship – the case of south-west Wales. *Tijdschrift voor Economische en Sociale Geografie* **71**, 255–63.

Cloke, P. J. & P. J. Hanrahan 1984. Policy and implementation in rural planning. *Geoforum* **15**, 261–9.

Cloke, P. J. & J. K. Little 1987a. Rural policies in the Gloucestershire Structure Plan I: a study of motives and mechanisms. *Environment and Planning A* **19**, 959–81.

Cloke, P. J. & J. K. Little 1987b. Rural policies in the Gloucestershire Structure Plan II: implementation and the county–district relationship. *Environment and Planning A* **19**, 1027–50.

Cloke, P. J. & C. C. Park 1985. *Rural resource management*. London: Croom Helm.

Cloke P. J. & N. J. Thrift 1987. Intra-class conflict in rural areas. *Journal of Rural Studies* **3**, 321–34.

Clout, H. D. 1972. *Rural geography: an introductory survey*. Oxford: Pergamon.

Clout, H. D. 1977. Rural settlements. *Progress in Human Geography* **1**, 475–80.

Coates, B. E., R. J. Johnston & P. L. Knox 1977. *Geography and inequality*. Oxford: Oxford University Press.

Cooke, P. 1983. *Theories of planning and spatial development*. London: Hutchinson.

Coppock, J. T. 1964. *An agricultural atlas of England and Wales*. London: Faber.

Daniels, P. & N. J. Thrift. *The geographies of the service sector*. Economic and Social Research Council, Changing Urban and Regional System Working Paper 1.

Davidson, J. & G. P. Wibberley 1977. *Planning and the rural environment*. Oxford: Pergamon.

Dear, M. & G. Clark 1981. Dimensions of local state autonomy. *Environment and Planning A* **13**, 1277–94.

Dunford, M. & D. Perrons 1983. *The arena of capital*, London: Macmillan.

Dunleavy, P. 1980. *Urban political analysis: the politics of collective consumption*. London: Macmillan.

Dunleavy, P. 1982. Perspectives on urban studies. In *Urban change and conflict. An interdisciplinary reader*, A. Blowers, C. Brook, P. Dunleavy & L. McDowell (eds). London: Harper & Row.

Dunn, M. C., M. Rawson & A. Rogers 1981. *Rural housing: competition and choice*. London: Allen & Unwin.

Fielding, A. J. 1982. Counterurbanisation in Western Europe. *Progress in Planning* **17**, 1–52.

Fothergill, S. & G. Gudgin 1982. *Unequal growth: urban and regional employment change in the U.K.* London: Heinemann.

Fothergill, S. & G. Gudgin 1983. Trends in regional manufacturing employment: the main influences. In *The urban and regional transformation of Britain*, J. B. Goddard & A. G. Champion (eds). London: Methuen.

Friend, J. K., J. M. Power & C. J. L. Yewlett 1974. *Public planning: the inter-corporate dimension*. London: Tavistock.

Giddens, A. 1981. *A contemporary critique of historical materialism*. Vol. 1: *Power, property and the state*. London: Macmillan.

Gilg, A. W. 1979. *Countryside planning: the first three decades 1945–76*. London: Methuen.

Gilg, A. W. 1984. Politics and the countryside: the British example. In *The changing countryside*, G. Clark, J. Groenendijk & J. Thissen (eds). Norwich: GeoBooks.

Gilg, A. W. 1985. *An introduction to rural geography*. London: Edward Arnold.

Glover, R. 1985. Local decision-making and rural public transport. In *Rural accessibility and mobility*, P. J. Cloke (ed.). Lampeter, Wales: Centre for Rural Transport.

Goldsmith, M. (ed.) 1986. *New research in central–local relations*. Aldershot: Gower.

Ham, C. & M. Hill 1984. *The policy process in the modern capitalist state*. Brighton: Wheatsheaf Press.

Hambleton, R. 1981. Policy planning systems and implementation: some implications for planning 'theory'. Paper presented to the Planning Theory in the 1980s Conference, Department of Town Planning, Oxford Polytechnic.

Hanrahan, P. & P. J. Cloke 1983. Towards a critical appraisal of rural settlement planning in England and Wales. *Sociologia Ruralis* **23**, 109–29.

Harvey, D. 1982. *The limits to capital*. Oxford: Basil Blackwell.

Healey, D. & B. Ilbery (eds) 1985. *The industrialisation of the countryside*. Norwich: GeoBooks.

Healey, P. 1979. On implementation. Some thoughts on the issues raised by planners' current interest in implementation. In *Implementation – views from an ivory tower*, C. Minay (ed.). Oxford Polytechnic: Department of Town Planning.

Healey, P. 1986. Emerging directions for research on local land-use planning. *Planning and Design* **13**, 103–20.

Healey, P., J. Davis, M. Wood & M. Elson 1982. *The implementation of development plans*. Oxford Polytechnic: Department of Town Planning.

Hill, B. 1984. *The common agricultural policy: past, present and future*. London: Methuen.

Hirsch, J. 1981. The apparatus of the state, the reproduction of capital, and urban conflicts. In *Urbanization and urban planning in capitalist society*, M. Dear & A. Scott (eds). London: Methuen.

Hodge, G. & M. Qadeer 1983. *Towns and villages in Canada*. Toronto: Butterworth.

Hodge, I. & M. Whitby 1981. *Rural employment: trends, options, choices*. London: Methuen.

Hoggart, K. & H. Buller 1987. *Rural development: a geographical perspective*. London: Croom Helm.

Ilbery, B. 1985. *Agricultural geography: a social and economic analysis*. Oxford: Oxford University Press.

Johansen, H. E. & G. V. Fuguitt 1984. *The changing rural village in America*. Cambridge, Mass.: Ballinger.

Keeble, D. E. 1980. Industrial decline, regional policy and the urban–rural manufacturing shift in the United Kingdom. *Environment and Planning A* **12**, 945–62.

Keeble, D. E. 1984. The urban–rural manufacturing shift, *Geography* **69**, 163–6.

Keeble, D. E., P. L. Owens & C. Thompson 1983. The urban–rural manufacturing shift in the European Community. *Urban Studies* **20**, 405–18.

Laffin, M. 1986. *Professionalism and policy: the role of the professions in the central–local government relationship*. Aldershot: Gower,

Leach, S. 1980. Organisational interests and inter-organisational behaviour in town planning. *Town Planning Review* **51**, 286–99.

Leach, S. & N. Moore 1979. County/district relations in shire and metropolitan counties in the field of town and country planning: a comparison. *Policy and Politics* **7**, 165–79.

Little, J. K. 1984. *Social change in rural areas: a planning perspective*. Unpublished PhD thesis, University of Reading.

Little, J. K. 1985. Feminist perspectives in rural geography: an introduction. *Journal of Rural Studies* **2**, 1–8.

Little, J. K. 1987. Rural gentrification and the influence of local-level planning. In *Rural Planning: Policy into Action?* P. J. Cloke (ed.). London: Harper & Row.

Lovering, J. 1978. The theory of the internal colony and the political economy of Wales, *Review of Radical Political Economics* **10**, 55–67.

Lovering, J. 1982. Gwynedd in British capitalism. Paper presented to Plaid Cymru Summer School, Carmarthen, Wales.

Lowe, P. & J. Goyder 1983. *Environmental Groups in Politics*. London: Allen & Unwin.

Lowe, P., G. Cox, M. MacEwen, T. O'Riordan & M. Winter 1986. *Countryside conflicts: the politics of farming, forestry and conservation*. London: Temple Smith/Gower.
McLaughlin, B. P. 1984. Rural deprivation: from administrative convenience to policy respectability. In *Rural Deprivation and Planning Research Report No. 6*. Gloucestershire College of Arts and Technology.
McLaughlin, B. P. 1986. The rhetoric and the reality of rural deprivation, *Journal of Rural Studies* **2**, 291–308.
Markusen, A. 1980. The political economy of rural development: the case of the western U.S. boomtowns. In *The rural sociology of the advanced societies: critical perspectives*, F. Buttel & H. Newby (eds). London: Croom Helm.
Marsden, T. 1984. Capitalist farming and the farm family: a case study. *Sociology* **18**, 205–24.
Marsden, T., R. Munton, S. Whatmore & J. Little 1986. Towards a political economy of capitalist agriculture: a British perspective. *International Journal of Urban and Regional Research* **10**, 498–521.
Marsh, J. S. & P. J. Swanney 1980. *Agriculture and the European Community*. London: Allen & Unwin.
Massey, D. 1979. In what sense a regional problem? *Regional Studies* **13**, 233–44.
Massey, D. 1984. *Spatial divisions of labour*. London: Macmillan.
Massey, D. & R. Meegan 1982. *The anatomy of job loss*. London: Methuen.
Mawson, J. & D. Miller 1983. *Agencies in regional and local government*. University of Birmingham: Centre for Urban and Regional Studies.
Milne, A. A. 1928. *Winnie the Pooh*, 6th edn. London: Methuen.
Morrill, R. L. 1980. The spread of change in metropolitan and non-metropolitan growth in the United States. *Urban Geography* **1**, 118–29.
Moseley, M. J. 1977. *Accessibility: the rural challenge*. London: Methuen.
Moseley, M. J. 1980a. Is rural deprivation really rural? *Planner* **66**, 97.
Moseley, M. J. 1980b. *Rural development and its relevance to the inner city debate*. SSRC Inner Cities Working Paper No. 9.
Munton, R. J. C. 1986. Research in rural geography in Britain: some reflections on future directions. Paper presented to the 2nd British–Dutch Symposium on Rural Geography, Amsterdam.
Murie, A. 1980. The housing service. *Town Planning Review* **51**, 309–15.
Newby, H. 1977. *The deferential worker*. London: Allen Lane.
Newby, H. 1980. *Green and pleasant land?* Harmondsworth: Penguin.
Newby, H. 1986. Locality and rurality: the restructuring of rural social relations. *Regional Studies* **20**, 209–15.
Newby, H., C. Bell, D. Rose & P. Saunders 1978. *Property, paternalism and power*. London: Hutchinson.
Owen, D. & A. Green 1985. *A comparison of the changing spatial distribution of socio-economic groups employed in manufacturing and services*. Centre for Urban and Regional Development Studies Discussion Paper 70.
Pacione, M. 1980. Quality of life in a metropolitan village, *Transactions of the Institute of British Geographers* **5**, 185–206.
Pacione, M. 1982. The use of objective and subjective measures of life quality in human geography. *Progress in Human Geography* **6**, 495–514.
Pacione, M. 1984. *Rural geography*. London: Harper & Row.
Packman, J. & D. Wallace 1982. Rural services in Norfolk and Suffolk: the management of change. In *Power, planning and people in rural East Anglia*, M. J. Moseley (ed.). Norwich: Centre of East Anglian Studies.
Pahl, R. E. 1965. *Urbs in rure*. Department of Geography, London School of Economics, Geographical Paper No. 2.
Patmore, I. A. 1970. *Land and leisure in England and Wales*. Newton Abbot: David Charles.

Patmore, J. A. 1983. *Recreation and resources*. Oxford: Basil Blackwell.

Pettigrew, P. 1987. A bias for action: industrial development in Mid Wales. In *Rural planning: policy into action?* P. J. Cloke (ed.). London: Harper & Row.

Phillips, D. & A. Williams 1982. *Rural housing and the public sector*. Farnborough: Gower.

Phillips, D. & A. Williams 1984. *Rural Britain: a social geography*. Oxford; Basil Blackwell.

Poulantzas, N. 1973. The problems of the capitalist state, In *Power in Britain*, J. Urry & J. Wakeford (eds). London: Heinemann.

Pratt, A. 1986. Realism and development: the reproduction of the industrial built environment. Paper presented to the Annual Conference of the Rural Economy and Society Study Group, Oxford.

Proudfoot, B. 1984. Rural geography in Britain: some aspects of rural problems and policies. In *The changing countryside*, G. Clark, J. Groenendijk & F. Thissen (eds). Norwich: GeoBooks.

Rees, G. 1984. Rural regions in national and international economies. In *Locality and rurality*, T. Bradley & P. Lowe (eds). Norwich: GeoBooks.

Robert, S. & W. G. Randolph 1983. Beyond decentralisation: the evolution of population distribution in England and Wales, 1961–1981. *Geoforum* **14**, 75–102.

Saunders, P. 1979. *Urban politics: a sociological interpretation*. London: Hutchinson.

Saunders, P. 1982. Why study central–local relations? *Local Government Studies* March/April, 55–66.

Saunders, P. 1984. Rethinking local politics. In *Local socialism?* M. Boddy & C. Fudge (eds). London: Macmillan.

Saunders,P. 1985a. Space, the city and urban sociology. In *Social relations and spatial structures*, D. Gregory & J. Urry (eds). London: Macmillan.

Saunders, P. 1985b. The forgotten dimension of central–local relations: theorising the 'regional state'. *Government and Policy* **3**, 149–62.

Saville, J. 1957. *Depopulation in England and Wales 1851–1951*. London: Routledge & Kegan Paul.

Shaw, J. M. (ed.) 1979. *Rural deprivation and planning*. Norwich: GeoBooks.

Shucksmith, M. 1981. *No homes for locals?* Farnborough: Gower.

Smailes, P. J. & G. J. Hugo 1985. A process view of the population turnaround: an Australian rural case study. *Journal of Rural Studies* **1**, 31–44.

Smith, D. M. 1974. Who gets what where and how: a welfare focus for human geography. *Geography* **59**, 289–97.

Smith, D. M. 1977. *Human geography: a welfare approach*, London: Edward Arnold.

Stacey, M. 1969. The myth of community studies. *British Journal of Sociology* **20**, 34–47.

Swanson, B. E., R. A. Cohen & E. P. Swanson 1979. *Small towns and small towners*. Beverly Hills: Sage Publications.

Urry, J. 1981a. Localities, regions and social class. *International Journal of Urban and Regional Research* **5**, 455–73.

Urry, J. 1981b. *The anatomy of capitalist societies*. London: Macmillan.

Urry, J. 1984. Capitalist restructuring, recomposition and the regions. In *Locality and rurality*, Norwich: GeoBooks.

Urry, J. 1985. Social relations, space and time. In *Social relations and spatial structures*, D. Gregory & J. Urry (eds). London: Macmillan.

Vining, D. R. & T. Kontuly 1978. Population dispersal from major metropolitan regions: an international comparison. *International Regional Science Review* **3**, 49–73.

Vining, D. R. & R. Pallone 1982. Migration between core and peripheral regions: a description and tentative explanation of the patterns in twenty-two countries. *Geoforum* **13**, 339–410.

Walker, A. (ed.) 1978. *Rural poverty: poverty, deprivation and planning in rural areas*. London: Child Poverty Action Group.

Weller, J. 1967. *Modern agriculture and rural planning*. London: Architectural Press.

Wibberley, G. P. 1959. *Agriculture and urban growth*. London: Michael Joseph.
Wibberley, G. P. 1967. The pressures on Britain's rural land. In *Economic change and agriculture*, J. Ashton & S. T. Roberts (eds). Edinburgh: Oliver & Boyd.
Winter, M. 1984. Agrarian class structure and family farming. In *Locality and rurality: economy and society in rural regions*, T. Bradley & P. Lowe (eds). Norwich: GeoBooks.
Wolfe, A. 1977. *The limits of legitimacy*. New York: Free Press.
Woodruffe, B. J. 1976. *Rural policies and plans*. Oxford: Oxford University Press.
Wright, S. 1982. Parish to Whitehall: administrative structures and perceptions of community in rural areas. Gloucestershire Papers in Local and Rural Planning No. 16.

8 *The restructuring debate*

John Lovering

Introduction

Societies do not usually present themselves to us with their workings exposed. We have to make an effort to make sense of the world, and we do this by adopting a particular approach. An approach imposes some order on our investigations, determining how we carve up reality into bits, and what kind of theories are built to explain what we find. A good approach will direct attention systematically to relevant features, and call forth theories which can justly claim to be scientific.[1] The restructuring debate in human geography is concerned with a particular approach, increasingly influential over the last decade.

The word restructuring used to be largely confined to academic texts and the financial pages, but in recent years has become almost commonplace. It suggests a qualitative change from one state, or pattern of organization, to another. In the discourse of radical social science, restructuring refers to qualitative changes in the relations between the constituent parts of a capitalist economy. These changes arise from conscious decisions.

> in response to changed conditions of accumulation induced by class struggle in the workplace or transmitted through the competitive conditions endemic to capitalism. They may involve sectoral switches of capital . . . geographical change . . . or scale economies as a consequence of the centralisation of capital. Restructuring has implications for, or may be undertaken through, changes in the labour process or the division of labour. (Lee 1987, p. 411).

The very fact that there is such a thing as the restructuring approach – sometimes alternatively known as the structural approach (Johnston 1986, Boddy 1986) – reflects the fact that many people believe profound changes are afoot in the way society is organized. It is argued that the relationship between the capitalist organization of the economy and the spatial patterning of activities is changing (for overviews, see Dunford & Perrons 1986, Gregory & Urry 1985, Lash & Urry 1987, Massey 1978a, Massey 1984, Massey & Meegan 1982, Massey & Allen 1984, Massey & Meegan 1985, Peet, 1987, Scott & Storper 1986). The roots of the restructuring approach lie in Marxist analysis, which highlights the pervasive implications of the fact that economies are organized on a capitalist basis. Under capitalism, production is governed by profit-seeking, and profit arises out of the domination of labour by capital. A capitalist economy is organized through agents (firms) whose *raison d'être* is to generate profits. The fact that firms respond to demand only when it is backed by purchasing power, means that production is an expression of 'accumulation for accumulation's sake' (Harvey 1978, p. 102), rather than human needs. Many needs may not be capable of expression in terms of demand for commodities, and in any case, given the unequal distribution of income under capitalism, some groups will exert much more influence over output than others. Social (and geographical) change in capitalist society is profoundly conditioned by this. Capitalism is also characterized by pervasive social conflicts, some of which erupt visibly and shape the course of economic development. The restructuring approach attempts to keep these realities at the forefront of analysis.

The most familiar concept from the restructuring approach is the notion of a spatial division of labour. This refers to the way different tasks in production are allocated to particular groups of people in particular locations; 'our primary interest is in employment, and occupational and social structures – in spatial divisions of labour' (Massey 1984, p. 9). At any point in time the spatial division of labour is both a legacy of past investment and an influence of future investment. This suggests that, 'not only does production shape geography, the historically evolved geographical configuration has its influence on the course taken by accumulation' (Massey 1978a, p. 119).

We can break down the concept of restructuring as it is used in this literature, into three distinct dimensions. First, it refers to the way capitalist enterprises respond to changing competition by altering their products or services and the way production and distribution are organized. Periodically, these changes will significantly alter the number and kinds of jobs these firms provide. Second, it refers to the way that these changes result in consequential changes in the way economic activity is organized across geographical space, through the creation and destruction of spatial divisions of labour (Peet 1983, Massey & Meegan 1982, Massey 1984, Cooke 1987b). However, the tasks performed by workers in an area, and the rewards they earn as a result, are inevitably associated with other aspects of social life in those areas. Work is more than just a matter of pay packets. The character of employment in a town or a region tends to be related in various ways to the pattern of relationships within households; between the sexes; between classes; and between populations and political authorities. So economic restructuring is bound up with socio-spatial change. Third, the restructuring approach is concerned to explicate some of the links between the spatial division of labour and the geographical pattern of social relations.

This focus leads to a particular view on the appropriate spatial unit of analysis to be used in research. A concern with restructuring in this sense implies that it is not sensible to focus on a town or a region in isolation, especially as these entities are usually defined by arbitrary bureaucratic conventions, rather than by socially meaningful characteristics. This approach is unsympathetic to the idiographic tradition in human geography. But it does give reasons to believe that modest spatial zones – localities – may be meaningful units for research. This is because most people live, work, and form their immediate social relationships within a restricted geographical area. The 'local community [involves] sets of relations which are multiplex (neighbours who are workmates who are leisure-time companions, etc), where "everyone knows everyone else"'

(Lash & Urry 1987, p. 91). Its emphasis on social relations leads the restructuring approach to pay special attention to the locality. But at the same time, its emphasis on the capitalist character of the market (which is by no means local), means that this approach cannot be satisfied with a treatment of localities as autonomous units. No man, woman, no *place* is socially an island.

The first part of this chapter presents an historical overview of the restructuring debate. It begins with the social and political background, and the intellectual context within human geography. The debate gave rise to a body of claims which some regarded as restructuring theory. The elements of this theory are sketched out, followed by an account of its demise in favour of a less tendentious approach in the mid-1980s. The second part presents a résumé of this approach. At this point the reader should be warned that the account is necessarily selective. An approach is likely to be interpreted by different people in slightly different ways, and the details in my account may be contested by others. I hope, however, that the general analysis would be fairly widely accepted.

Historical overview

The social and political context

Like all systems of thought, the restructuring approach bears the marks of its intellectual, social, and political origins. It did not appear overnight, nor was it a purely cerebral development in an academic world of free-floating ideas. It was very much a product of specific observations and debates, undertaken by particular people influenced by particular political and intellectual ideas and interests. It is appropriate to begin with the political context, for the emergence of the restructuring approach was closely and visibly related to political developments set in chain in the late 1960s.

During the 1960s and 1970s employment in traditional industries in the metropolitan countries declined sharply, while new jobs were rapidly created in new industries. This process affected different countries, regions, and localities, and different groups within them, very unevenly. At the same time the political landscape also began to change sharply. In particular, traditional forms of socialist politics appeared to enter into decline, while new forms of dissent emerged. By the late 1970s capitalism 'seemed to have embarked on a radically new course in comparison with the economic and political structures that were set in place in the decades immediately following World War II' (Storper & Scott 1986, p. 361). These economic and political developments seemed to be intimately connected. The global restructuring of capitalism, it was argued, entailed not only the rapid industrialization of the Third World and deindustrialization of the First World, but also a series of radical alterations within the social structures of countries in Europe and the Americas. Capital was abandoning its historic centres in favour of newer, lower-cost sites (Frobel *et al.* 1982, Anderson *et al.* 1983). And this was subverting the political organization and leverage of the groups that had long held centre-stage in left politics and in social research; that is, mainly white, male, skilled manual workers (see Mingione 1983, Leys 1983, Panitch 1986, O'Connor 1984). The structure of the capitalist world economy was changing in a way which was undermining the sort of industry, and thereby the patterns of class formation and political mobilization, with which socialist and Marxist writers were familiar. A great deal of analytical and political rethinking was necessary (Hobsbawm 1981, Gorz 1982).

These changes had a marked spatial impact within the metropolitan countries. Economic decline was most severe in the traditional manufacturing heartlands of organized labour and left politics; the northern and Midlands conurbations in Britain, the Rustbelt in the US (Massey 1984, Peet 1983, Bluestone & Harrison 1982). At the same time trade unionism was extending to new groups of workers. In Britain trade unions added 3 million members in the 1970s. Most of the new unionists were in white-collar occupations, and for the first time in recent history many were women (who were often more militant than their male colleagues). At the same time, new kinds of political activity emerged with new spatial characteristics. The 1970s was characterized by a spate of campaigns over closures, jobs, environmental causes, and housing. These tended to be organized on a regional or local rather than a national basis. The womens' movement and the struggles of oppressed ethnic groups also tended to be organized around highly localized memberships and demands (Segal 1987, Buhle 1987). The orthodoxies of class analysis were challenged on a number of fronts; the de-alignment of working class voters; the growth of new social movements; the changing character of working-class communities, and the changing structural significance of the welfare state (Lash & Urry 1987, p. 211).

In this context it became particularly important to develop a radical understanding of the connections between economic and geographical change. Economic geography, and the questions of industrial location and labour markets in particular, took on a new political salience. The 1970s saw an outpouring of radical empirical and theoretical work in human geography in Britain, Europe, and the US. The restructuring approach can be understood as the outcome of a wide variety of such attempts to analyze the changing situation.

Meanwhile, locality-based empirical research was reviving, after some years of banishment to the academic shadows. Many of the most influential new studies were undertaken outside academe. The growth of new research within the state sector, or otherwise on state funds, reflected the new policy concerns of governments of all parties. In the US in the 1960s, and Britain in the 1970s, the problems of the declining regions were joined by the problems of the inner city, which attracted a new array of policy responses (Edwards & Batly 1978, Bluestone & Harrison 1982). Research studies were focused on small urban areas in order to inform the new spatial policies. Inner-city intervention in Britain was initially concentrated in the Home Office – the Ministry most directly concerned with social control (Britain's 'Ministry of the Interior'). The Home Office set up the Community Development Programme (CDP) to look at small spatially defined pockets of deprivation. But the CDP researchers argued that it was necessary to situtate the problems of these areas in the wider context of the capitalist economy. The problems of these areas 'were firmly tied to much more basic structural problems in society, and . . . the solution consists in fundamental and far-reaching political change' (Community Development Programme (CDP) 1975, p. 5).

Similar arguments were being made in a flurry of studies outside the state sector. The new politics of the 1970s was associated with a blossoming of local campaigns led by trade unions or community-based organizations. In many cases these groups sponsored research into their local economies. These were often action-research projects, prompted by a major factory closure, an environmental threat, or other local emergency. Local economic research was also stimulated by the availability of funds for local projects under policies such as the Urban Programme. In Britain, this kind of politically propelled research could be found in rural areas as well as the inner cities, in Wales and Scotland, and in most of the English regions (see, for example, Brown 1975, Merseyside Socialist Research Group 1980, Gaffikin & Nickson 1982, Lovering 1982). Like the CDP projects (by which they were often influenced), many of these studies tried to relate local problems to the wider restructuring of capitalist production and the role of the national state.

This 'turn towards the local' in policy and in research accelerated in the early 1980s. Local economic strategies became increasingly popular, favoured both by central government (following the election of governments committed ostensibly to the withdrawal of the state), by opposition parties, and by more localized organizations. In Britain Labour Party victories in the local elections brought a new official enthusiasm for local economic studies of a kind not far removed from those conducted by the CDP and local campaign groups. In the London Borough of Wandsworth the council tried to develop an economic strategy on the basis of a socialist analysis of the local economy (Davis 1979). The Wandsworth model influenced the forms of municipal socialism later adopted by the Labour-controlled Greater London Council until its abolition in 1986. Closely related strategies were adopted in the West Midlands and Sheffield (Boddy & Fudge 1984), and have since become a staple part of the diet of local politics in most cities and many smaller towns (Cochrane 1987).

The policies of new right governments in Britain and the US also fuelled strategies which emphasized decentralization, grass roots support, participatory planning, and localized economic strategies (Szelenyi 1984, p. 13, Massey 1981). In the US, where local government in general has more autonomy, there was a longer experience of local economic strategies, but the demand intensified. In some prominent inner-city cases, new strategies moved beyond the traditional boosterist model for local economic regeneration, to address the structuring of the labourmarket and the wider determinants of the location of industry (Tietz 1987).

But as the 1980s advanced the central state, and the political right, also 'went local'. The development of an 'Enterprise Culture' has been encouraged by the development of localized infrastructures of industrial promotion, business support, and urban regeneration measures. As more state and private organizations turned their attention to the local economy, a whole new research industry was spawned. An increasing proportion of local economic research was conducted in a conventional spirit, focusing on supply constraints, the flexibility of local labour and capital markets, and the need to attract firms. The radical connotations of local economic analyses and policies were diluted. It would be no exaggeration to say that local economic strategies have been largely hijacked by capital and the right.[2] But some initiatives continue to draw more or less explicitly on the perspective adopted by radical studies and later formalized in the restructuring approach.[3]

This political context has been every bit as important as academic debate in the development of the restructuring approach. This is partly for a straightforward sociological reason: many of the contributors to the restructuring debate were also participants in the political arena. Intellectual and political developments informed each other, notably through the channels provided by the Conference of Socialist Economists (in Britain) and the Union for Radical Political Economics (in the US), and to a lesser extent radical journals like *Antipode*, *Capital and Class*, and *Society and Space*. Against this background new academic work began to appear which addressed the problem of the local with both theoretical sophistication and a radical political edge. The restructuring approach in human geography centred on questions of the location of industry and regional development (Massey 1978b). But it was increasingly influenced by analytically and politically sympathetic arguments in other academic arenas, especially the debate over the nature of the local state (Cockburn 1977, Saunders 1981, Mingione 1983, Boddy & Fudge 1984), the rediscovery of the informal economy (Pahl 1984), and the exploration of gender relations (Hunt 1980, Hartmann 1978).

The restructuring approach and the critique of industrial location theory

While the broad political purpose behind the restructuring approach was to contribute to a new and pertinent socialist analysis, its analytical substance was shaped by an engagement with the dominant modes of enquiry in human geography. Since the 1960s mainstream theoretical explanations of spatial patterns had focused in particular in the location of industry. Positivism and the quantitative revolution in the social sciences had shifted the characteristic emphasis of geography and regional economists. Instead of producing detailed descriptions of the almost infinite variety of geographical outcomes, researchers were increasingly expected to explain this variety in terms of formal quasi-mathematical models.

The positivist approach rested on the assumption that it was possible to generate 'statements of a law like character which relate to phenomena that are empirically recognised' (Johnston 1986, p. 11). In human geography this took the form of spatial analysis in which theories were developed that generated predictions in line with observable events. The early work of neoclassical theorists of industrial location (such as von Thünen and Weber) was revisited. The recasting of human geography owed much to the domination of economics, and especially neoclassical economists, amongst the social sciences.

In the 1960s the neoclassical approach came to dominate virtually all specialisms within economics, including regional theory. The spatial structure of the economy was explained by applications of the general theory of markets. Neoclassical theory claimed to offer a complete explanation of production and exchange from the starting point of individual preferences and choices in response to price signals. These were essentially static theories resting on equilibrium models in which adjustment was achieved painlessly and instantaneously (through the fiction of the Walrasian 'auctioneer', who called out a succession of prices until a balance between supply and demand was established). From this unpromising start neoclassical economics claimed to be able to deduce a theory of the geography of production, by adding on assumptions about economies of scale and distance-related transport costs. Even more optimistically, theories of growth and geographical change could be deduced, by adding further assumptions about changing economies of scale and proximity through time (Kaldor 1970). The gist of these theories was that the market (strictly, the hypothetical perfect market), contained an inherent tendency to spatial homogeneity (for a neat review and critique from a position close to that of the Restructuring School, see Clark, *et al.* 1986).

The resurgence of neoclassicism led to a vast new geographical literature. The typical method in this work was to develop formal logical models which predicted spatial patterns (such as industrial location) from givens such as the spatial patterns of transport costs, productivity, and other factors. The corresponding style of emphasis in empirical work was to explain observable phenomena (as captured, for example, in statistics of regional employment) by showing that they corresponded more or less adequately to these theoretical models. A paradigmatic example would be the Newtonian gravity model – the assertion that interaction between two bodies depended on the degree of proximity (Johnston 1986, p. 37).

This whole style of analysis increasingly came under attack in the 1970s (of course, as is usual in any science, one could find earlier forerunners of these critiques, but their influence had been largely dormant). In economics, it was argued that the neoclassical system falsely equated logical (i.e. formal, theoretical) time with real historical time (Robinson 1978, Harcourt 1972). As a result neoclassical theories were unable to grasp the reality of change, or adjustment. There is no reason to believe that the adjustment process is inherently governed by a move towards equilibrium, and regional economics found itself without a firm basis. The complexities of the real world could not be understood from a viewpoint within pure theory of the kind offered by neoclassicism. It was theoretically, as well as empirically, necessary to investigate what kinds of institutions actually existed and how they worked.

An equivalent argument was developed in radical geography. The dominant positivism could produce fascinating maps, but at best these only showed correlations, and said nothing about causes. They failed to illuminate the underlying social relationships. But within the dominant scientific paradigm this kind of research appeared to work; it could generate reasonable statistical fits between theory and observations (and this helped attract research grants and promotion for its practitioners). So the radicals' first line of attack was to bypass the empirical issues and get at underlying questions of methodology. The restructuring approach began as a critique of positivism in industrial location studies (Massey 1974). The method of comparing theory with observations was not in itself objectionable, of course, but researchers should be aware that all observations are theory-determined, and the dominant theories were suspect. A descriptive focus on spatial processes was suspect because 'there are no such things as spatial processes without social content . . . what was really being referred to . . . was the spatial form of social causes, laws, interactions and relationships' (Massey & Allen 1986, p. 3). So the dominant approach was at best innocently misleading, missing some important social preconditions for spatial phenomena. At worst it was reactionary, an ideological disguise for the status quo. Alternative modes of explanation were needed which could show exactly how spatial patterns incorporated social relations.

This task required that research should begin not by describing observable spatial variations, but by grasping the social relations which underlay these variations; that is by 'breaking into the chain of causation at the level of the system as a whole' (Massey & Meegan 1985, p. 6). The particular conception of the system as a whole which the critics had in mind was Marxism. Society is capitalist society, based on the primary social division into a capitalist and a working class, operating under the priority of the profit motive (production for profit, not for need). Having demolished the established approach through its critique of positivism, the next task of radical theory was to develop alternative substantive theories which would incorporate Marxism's insights into 'the fundamental importance of the organisation of production in the creation and structuring of all social processes' (Johnston 1986, p. 121).

At the time, in the upsurge of confidence in Marxism following in the wake of 1968, it was implied that this could be done fairly straightforwardly. Marxism already offered the concept of uneven development.

The incompleteness of theories of uneven development and urban process

There was little discussion of geography in the literature of classical Marxism, and Marxist geography drew largely on later additions and interpretations (Quaini 1982). The term uneven development originally referred to the political state of development of class conflicts in different countries (Smith 1984, p. xii). But by the mid-20th century the term had been extended to embrace economic underdevelopment, both inter- and intra-national, as an effect of the extension of capitalist production (Brookfield 1975, Harvey 1982, Smith 1984, Massey 1978b, Browett 1984).

In this literature, the origins of uneven development lay in the contradiction inherent in capitalist production between use value and exchange value. Production may be motivated by profits (exchange values), but it necessarily involves the production of use values (goods and services). This means that specific capitals become committed to fixed investment in particular places (Smith 1984, p. 88). As a result they become partly immobilized. But capitalism as a whole is always finding new markets producing new products, and finding new ways of making them. Capital in general, therefore, tends to be industrially and locationally mobile. Established plants and established places will periodically be abandoned in favour of new sites, and this leads to perpetual 'development at one pole and underdevelopment at the other' (Walker 1978, Smith 1984, Harvey 1982).

An analagous style of argument was developed in Marxist approaches to the urban problem. The built environment was the physical framework for the production of profit and the reproduction of the labourforce (Harvey 1978, p. 114). The periodic eruption of class struggles requires capital periodically to switch investment from 'directly productive' circuits to less productive investment in working-class housing and social amenities. So 'spatial structures are created which themselves act as a barrier to further development in the form of immobile transport facilities and indeed the built environment as a whole' (Harvey 1978, p. 124). The built landscape therefore embodied the history of both accumulation and class struggle (Harvey 1978, p. 129). This notion was also inscribed in the new urban sociology developed by French Marxists. Cities were spatial containers for certain kinds of economic activity, especially 'collective consumption' (Castells 1977). The urban pattern therefore reflected the search by capitalist firms for locations which would enable them 'to reduce indirect production costs as much as possible' (Lojkine 1976, p. 134). But the intervention of the state, itself 'the reflection of class struggles' (Lojkine 1976, p. 142), conditioned this effect.

One consequence of these approaches was that new attention was drawn to the relationship between capitalist and non-capitalist processes. In the uneven development theory 'regions and nations are typically defined by the very barriers which *prevent* a full penetration and regional equalisation by capital: political boundaries, culture, monetary systems etc.'(Walker 1978, p. 28; emphasis added). The analysis of the urban process similarly pointed to ridigities imposed by the physical infrastructure on capital accumulation.

This sort of argument became popular, but on examination left much to be desired. Theory was pitched at a very general level, throwing little light on specific empirical cases. The causal relationships between productive capital and specific spatial forms, such as housing structures, and the wider social relations in particular localities remained obscure. This work was also generally silent on remedies, apart from an implicit or vaguely specified and ultimately millennial revolution.

These inadequacies can be traced to two major theoretical weaknesses inherent in the variant of Marxism on which they were based. The first weakness is still not widely recognized, especially outside the rather esoteric debate over the labour theory of value and the transformation problem (and even there it is often ignored). The problem is that the abstract categories of the Marxist economic analysis are necessarily at some remove from the empirical categories of real world investment, consumption, wages, and empirical social relations. This means that formal models of the kind that focus on the mobility of capital do not get to grips with the evident immobility of specific capitals, tied by the structure of the firms, state policies, and fixed investment. Similarly, the abstract concept of circuits of capital is a long way from the sectors of industry or categories of social expenditure which can be identified in published statistics in the real world. Marxist categories are defined by function; there is no obvious relationship which can be known in advance between these and the statistical categories of bureaucratic convention (for a similar point in relation to sociological categories, see Nichols 1979). This means that abstract analysis in the Marxist sense does not really say very much that is useful in researching concrete situations (for a contrasting view see Fine & Harris 1979). Ironically, while Marxism is generally unsympathetic to the method of ideal types favoured by Weberian sociology, a lot of Marxist analysis of the 1970s and 1980s actually has much in common with it.

The second weakness is more widely recognized. The emergent analysis emphasized the causal influences of the reproduction of labour power. The need to ensure reproduction committed capital to certain kinds of investment in particular places, either directly or through the state. But the explanatory power of this concept is severely limited. As new radical studies of the labour process appeared (Friedman 1977, Sabel 1982), and as feminist work exposed the power relations in the home and at work (Hartmann 1978, Walby 1986), it became increasingly clear that the reproduction of labour power is no simple matter. It is not a datum, but an outcome of a wide variety of processes. The income and conditions of any group of workers depend on a hugely complicated set of social relationships, so complicated that to assume that they can be understood primarily as necessary requirements of capital is simply reductionist (for a discussion see especially Urry 1981a).

But these reductionist pitfalls were not invisible to many who wanted to develop the insights of the uneven development and urban process theories, and an openminded spirit of enquiry emerged. A number of writers wanted to analyze industrial location and geographical change in terms that drew out the capitalist character of production, but did more than merely blame all geographical effects indiscriminately on capitalism (Harloe 1977, Sayer 1985). A more sensitive kind of analysis was needed, one which still started from the analysis of capitalist accumulation, rather than from a description of given spatial units (Massey 1978a, p. 116), but which could also provide new purchase on the specificity of individual nations, regions, and localities. In explaining specificity, this project might transcend the traditional dichotomy between idiographic and nomothetic geography (Massey 1984). This can be taken as the restructuring approach. It is essentially a long-term project to overcome the 'abstract and arbitrary' elements in Marxist writing on space in the 1970s (Massey 1978a, p. 108), without throwing out the insight of Marxism that the capitalist character of the economy exerts a pervasive influence on social life.

In theoretical terms, the central task in this project is to identify 'the *mechanisms* which produce specific spatial effects' (Massey 1978b, p. 50; emphasis added). 'Capital's response to spatial unevenness is itself a product of the interaction between the existing characteristics of spatial differentiation, and the requirements at any time of the dominant process of production' (Massey 1978a, p. 114). At the same time, it was important to acknowledge that 'the forms of spatial differentiation relevant to the process of accumulation are by no means confined to the purely "economic".' They also include 'the specific form taken by the class relation' (Massey 1978a p. 115). The maturing of the restructuring approach consists largely in the growing awareness of these non-economic dimensions.

The early restructuring literature tried to find these mechanisms by complementing an abstract Marxist analysis of capital accumulation with a set of lower-level theories (for clear examples, see Massey (1978a) and Peet (1983)). The choice of substantive theories at this second level was influenced by contemporary work on the product cycle, on corporate organization, and the labour process. Contemporary Marxist models of class structure were brought in as links in the chain from economic restructuring to local class formation (Massey 1984, Peet 1983).

In the first place there is capitalist production. This can be isolated in thought and examined in abstract, to yield knowledge of its inherent laws or tendencies (such as a drive to accumulation, global commodification, perennial class struggle, etc). Second, there are a series of more empirical or contingent processes through which these imperatives are mediated. By moving from the abstract analysis of capitalism through a set of medium-range theories of the labour process, technology, or the state, it should be possible to arrive at spatial tendencies and locality impacts. This unidirectional method of analysis – from abstract to concrete – was common throughout Marxist social science in the

1970s. It owed much to the contemporary popularity of structuralist forms of Marxism, and especially, although not exclusively, to the tradition associated with Althusser (Benton 1985).

The brief heyday of restructuring theory

This package of high-level and middle-level theories was presented by some as restructuring theory (although the term was rarely used at the time). Its main empirical propositions could be summarized as follows:

(a) The size of units of capital is rising, so the investment decisions of large firms play an increasingly central rôle in spatial patterns.
(b) Under the influence of a global change in patterns of competition, companies are restructuring production.
(c) This restructuring entails a deeper separation between the functions of conception and execution.
(d) This separation of activities is increasingly projected spatially. Basic assembly, requiring semi-skilled labour, and more complex production, involving skilled labour, tend to be allocated to peripheral regions and to urban conurbations respectively. Control functions and R&D are confined to the dominant metropolitan area (Massey 1978a, p. 116).
(e) This new spatial division of labour is superimposed upon a geography shaped by previous divisions of labour. As a result, combination effects are created giving rise to highly specific local or regional patterns in the social relations of production.
(f) These effects mean that locality-level change can feed back on to the wider pattern of investment. The spatial pattern therefore becomes both effect and influence on accumulation.

Different writers stressed different elements in this package, some offering it as something much stronger than an approach – a theory. It was implied that it was possible to move seamlessly from a highly abstract analysis of capitalism (drawn from Marx's *Grundrisse*), through theories of corporate forms and the urban system, to arrive at an explanatory theory of the geography of class conflict (Peet 1983). Some argue explicitly that there is a restructuring theory which is characterized by a set of 'master concepts that have explanatory intent' (Cooke 1987a, p. 74; see also Boddy 1986). One such concept is the claim that firms faced with heightened competition will rationalize workforces. Another is the claim that 'firms make skilled workers redundant at times of crisis since they are more expensive to keep than unskilled ones' (Cooke 1987a, p. 75) so the labourforce becomes deskilled.

Others were more cautious about the theoretical status of the collection of perspectives and substantive propositions gathered under the restructuring umbrella. Massey, in particular, stressed the nuances and complications in reality which the restructuring literature had yet to grasp adequately. She pointed to the problematic nature of capital – which can only partly impose its will since it must generally respond pragmatically to given circumstances (Massey 1983, 1984).

In fact, many of the medium-level theories incorporated into the stronger version of restructuring theory turned out on investigation to be inadequate. For example, the first three claims listed above are questionable. The size of firms appears to have stopped increasing (Lash & Urry 1987). The theory (drawn from Braverman) alleging an inexorable tendency towards deskilling in the labour process has been shown to be overstated (e.g. see Knights *et al.* 1985). The theory of class (drawn from E.O. Wright) has also lost credibility (e.g. cf. Wright 1978, 1984). And the attempt to pull these all together into a tight restructuring theory turns out to be little more than eclecticism. The strong version of restructuring theory was unsatisfactory and proved short-lived. Contentious empirical propositions (such as those concerning the internal organization of large firms, local business cultures, or deskilling in the labour-market), had been added on to a higher-level theoretical analysis of capitalist production, but not theoretically incorporated in any rigorous way. Ironically this theoretical looseness enabled the positivism which had been thrown out of the front door to creep in again through the back door, in the form of a search for regular spatial properties.

At one stage Massey, with Meegan, developed a 'repertoire of restructuring forms' (rationalization, technical change, and intensification (Massey & Meegan 1982). To the proponents of strong restructuring theory these were seen as having causal significance, explaining the spatial development of occupational patterns and the geography of class structure (e.g. Cooke 1983, 1984). In this vein, taxonomies of local labourmarket types have been constructed, some of which are extremely complex (Cooke 1983, Storper & Walker 1983). The implication was that identifying the local labourmarket type would have some explanatory value.

But it is one thing to claim that there are mechanisms linking underlying social relations to visible spatial outcomes, and another to suggest that these causal relations can be systematized in terms of a single empirical dimension such as labourmarket or production forms (Sayer 1985). And the search for a formula (such as labourmarket types) with which to interrogate empirical evidence sits uneasily alongside the antipositivist and, for some, realist intent of the restructuring approach. On investigation, it is hard to see how a taxonomy of restructuring forms can provide explanatory concepts, unless we give up the whole enterprise and go back to positivism and empiricism. These can only be examples of the diverse relations which may exist between a firm's economic adaptation and the spatial division of labour (Massey 1984). They exemplify how the restructuring approach has provided some useful new analytical boxes, making it easier to categorize aspects of reality. But this does not in itself provide causal explanations (Massey & Meegan 1985, p. 7)

The eclecticism of the strong version of restructuring theory also gave rise to theoretical imprecision. Massey's idea that rounds of investment interacted in some way with the residue of previous rounds offered an imaginative access to thinking about historical change. But if we examine this idea, the precise meaning of these interactions turns out to be less than clear. In the absence of any theories which could generate rules governing these combinations, the concept of spatial divisions of labour remains an 'heuristic device, a metaphor rather than a theory' (Warde 1985).

Perhaps the most curious weakness of strong restructuring theory was its tendency to eliminate its own theoretical object. If taken literally, it had the effect of reducing all spatial effects to non-spatial social relations. The spatial disappeared, being ultimately 'entirely social' (Duncan & Goodwin 1988, p. 53; Urry 1987). There was something wrong with a Marxist geography that obliterated space and denied geography any genuine existence (Massey 1986, p. 9; Harvey 1982). This reduction was inconsistent with the aim of the restructuring approach to situate and explain specificities, rather than eliminate them. The connections between an abstract domain of high theory at one pole, and a diverse collection of empirical studies at the other, remain incomplete (Massey 1984, p. 6; Smith 1985, Sayer 1985). This may explain why, in practice, empirical research tended to 'leave the theory behind' and get on with examining interesting observable details (Sayer 1985). And this, in turn, invited the criticism, not without some justification, that for all its pretentions the restructuring approach was a licence for a new empiricism, a new 'spatial fetishism' (Smith 1985).

The restructuring approach ten years on

In an assessment published in 1985 a number of writers argued that only half the argument on which the restructuring approach was based had been followed through (Massey & Meegan 1985, p. 4). There had been more emphasis on the way geography reflected (capitalist) social relations than on the way geography moulded them. But spatial outcomes do not merely reflect the prevailing constellation of social relations, as if those relations were projected on to a cinema screen of geographical space; spatial organization is also part of that constellation (Massey 1986, Soja 1985, Gregory & Urry 1985, Urry 1987). 'The social is spatially constructed too' (Massey & Meegan 1985, p. 6). This re-emphasis on a neglected part of the original project reflected wider developments in Marxist and non-Marxist theory, and the growing emphasis in empirical work on the social complexities of localities. One example is the Changing Urban and Regional System research initiative funded by the British Economic and Social Research Council. This has been an umbrella for a series of locality research projects inspired directly by the restructuring approach. These aimed too at tracing the connections between social forms and practices within defined localites, and the changing place of those localities in the wider economy and polity (for an overview, see Cooke 1989)

While new empirical work challenged the strong version of restructuring theory from one side, theoretical developments (and behind them, socio-political changes), challenged it from the other. The structuralism that haunted the versions of Marxism underlying the early restructuring work has generally lost favour. Most Marxists now accept the need for more sensitive and less totalizing conceptions of capitalist society. This reappraisal has drawn in particular on the work of Gramsci (see Hall 1977). The problem of ideology or civil society has drawn attention away from a narrow focus on questions of production and accumulation (Laclau & Mouffe 1983, Sayer 1985, Urry 1981a, Lash & Urry 1987). On the borders of Marxism, there has been a greater emphasis on the inadequacies of the Marxist theory of action and renewed attention to the Durkheimian normative dimension in social structure (Giddens 1984).

These issues were taken further in recent years through debates on realism and structuration. Philosophers and social scientists have returned to the central Marxist concern: the interaction between people – as knowledgeable but imperfectly informed actors – and the social spatial structures they inhabit and reproduce (Bhaskar 1979, Giddens 1984; for useful reviews see Thrift 1983, Cohen 1987). This has reopened a question at the heart of the restructuring approach, namely what exactly is the system as a whole with which this approach should start? The capitalist economy is not a mechanical apparatus, but is made up of active relationships between people. And these people are simultaneously men, women, whites, and blacks, living in particular places with particular modes of life. Economic activities in a given territory therefore have to be seen as bound up with the entire culture or civilization of the people living there (Thrift 1983, pp. 398–9). Capital accumulation and industrial location are only part of the story.

The debate on the spatiality of society has influenced the development of the restructuring approach, reinforcing those elements which pointed to social relations outside the immediate sphere of production. Amongst these, attention was directed especially to gender relations, since it was all too evident that the geography of production 'presented distinct conditions for the maintenance of male dominance' in different places (McDowell & Massey 1984, p. 128; Dex 1985, Hamilton & Barrett 1986, Crompton & Mann 1986, Walby 1986). There was also a revitalized interest in national specificities, since an important set of territorial boundaries is defined by political control, and the use of particular cultural symbols associated with nationhood (Sack 1986, Cooke 1989, Lovering 1988). This widening of focus has also meant paying more attention to cultural formations. The reproduction of labour power cannot be explained satisfactorily without explicitly taking into account processes which are not reducible to economy or state (Urry 1981b).

One result of the move away from a strong theory has been the recognition that geographical outcomes should not be seen as determinate or predictable. This means, in fact, 'that there can be no general theory of regional adjustment' (Clark *et al* 1986, p. 19), in the sense that both positivist and reductionist Marxists hoped for. A particular social practice in a particular locality cannot be explained by looking for a single cause or set of causes (for example, in the location of industry) (Urry 1987). A variety of political, cultural, and economic processes, operating at different spatial levels, should be expected to be implicated in any given outcome. Conversely, a given causal influence could give rise to a variety of local outcomes. For example, Massey showed that the spatial form of small local labourmarkets with dominant employers was related to militancy in one British locality (north east Lancashire) and acquiescence in another (south Wales) (Massey 1986).

One effect of the increasing recognition of complexity was to enable more fruitful discussions to take place between Marxist and non-Marxist geographers over both substance and method (e.g. see Massey & Meegan 1985, especially pp. 6–7). It became necessary to abandon the more tendentious theories on which the strong version of restructuring theory relied (Sayer 1985). The main casualties were the deterministic labourmarket and class theories drawn from Wright (Wright 1978). It is no longer reasonable to assume, with Braverman, that deskilling is an inexorable process and that there is some necessary *a priori* spatial distribution of deskilling, reskilling, and enskilling (Massey 1986). Above all, it is unreasonable to expect to read off local social structures and political characteristics from labourmarket categories (Wright 1984, Abercrombie & Urry 1981, Urry 1987). Not only was this a crude procedure, but if taken seriously, it also violated the assumption (close to the heart of the restructuring approach) that politics matters. Class is not just an offshoot of the labourmarket, it is also a matter of political culture and organization (Sabel 1982, Panitch 1986). If the restructuring approach was to remain a living project and not a rapidly dating collection of simplifications, it had to open out to allow for a variety of possibilities in the labourmarket and in class formulation. So room was made for conceptions such as the new service class, new social movements, and a recognition of the importance of empirical processes of social closure in segmenting labourmarkets (Goldthorpe 1982, Abercrombie & Urry 1981, Cooke 1983, Thrift & Williams 1987).

The unfinished and contradictory elements in Massey's seminal résumé and development of the restructuring approach (published in 1984 but written over several years), are witness to this continuing reappraisal. At some points she relies explicity on Wright's economistic class map, while at others she argues strongly that class relations, and especially gender relations, have a logic at least partly their own. This book can perhaps be regarded as standing at the transition from earlier more deterministic versions of restructuring theory to a more sensitive, if less tidy, approach.

The very openness of the restructuring approach means that different people may attempt to incorporate recent empirical and theoretical developments in it in quite different ways. Changing empirical patterns in the organization of industry, dominant political discourse, and aspects of culture have spawned a new debate over the meaning of postmodernism (Berman 1984, Jameson 1984). Some have argued that this has special implications for the restructuring approach. The salience of locality is, allegedly, becoming heightened as Fordist modes of organizing production give way to post-Fordist modes (Lipietz 1986, Cooke 1987c).

This issue is important because it is sometimes suggested that a locality focus is *per se* postmodern (Cooke 1987c, Graham 1988). The debate on postmodernism in human geography seems to conflate three quite separate issues. It is partly concerned to advance an anti-foundational position in social science, 'eschewing any absolute or categorical conceptions in moral discourse' (Lash & Urry 1987, p. 13). It is also associated with an emphasis on the experiential or life–world perspective as opposed to the systems perspective (Berman 1984; see Cooke 1987c). Finally, it tends to be associated with a denial of the validity of any form of class politics (Graham 1988, Beauregard 1988).

The claim that there is a connection between locality research and the postmodernist debate implies that there is some single agreed meaning to the concept of the locality. But the concept of locality is by no means a simple one (Urry 1987), and by the same token there are many different ways of undertaking locality research. It is possible to explore locality-level processes as an *aspect* of the 'post modern spatial paradigm' (Cooke 1987b). This enquiry could share the agenda of the restructuring approach. But it is also possible to conceive of a postmodern*ist* mode of locality research. The latter would probably not be compatible with the restructuring approach, and indeed might represent everything which that approach rejects – the celebration of fragments without regard to their context, even the very denial that context matters. The argument that a locality focus *must* become a licence for empiricism (Beauregard 1988) would be disputed from a restructuring perspective (Lovering 1988).

To choose a locality focus is merely to select a particular spatial frame of reference which there are reasons to believe will be particularly useful in examining the social processes of interest. Only if those processes are assumed to be inexplicable, not related in any way to each other or to systemic forces in which production occupies a key rôle, need locality-focused research become a postmodern rather than a restructuring project.

Conclusion

The restructuring approach is not a *theory*. I would argue that it is a package which includes: a philosophical position in regard to social scientific practice and the nature of explanation; a set of analytical concepts (tools) to sensitize researchers to certain kinds of phenomena; a set of claims about contingent historically specific processes which may cause spatial effects; and, most cautiously, a set of claims at the level of spatial outcomes. The next section summarizes the elements of this package (as I read them, in 1988).

The restructuring approach as a research programme for analyzing industrial and social change

Explanation and method

The specific perspective adopted here is realist. This is not uncontroversial, although it is widely shared to greater or lesser degree by many involved in spatial research. In a realist approach the task of research is to provide an account of the way observable events are explicable in terms of deeper social processes, which are themselves sustained by those events (Bhaskar 1979, Sayer 1982, 1984, Outhwaite 1987, Urry 1987). The restructuring approach attempts to explain geographical patterns as the visible manifestation of a set of less visible social relations. A study of a specific case (a locality, an industry, or even an individual) should be a window on to processes beyond. Although spatial research is concerned with the local, it should not be devoted to tracing geographical particularities for their own sake, although this is a temptation (Smith 1987, Duncan & Goodwin 1988, p. 51). What is at issue in research into geographical specificities is 'the articulation of the general with the local (the particular) to produce qualitatively different outcomes in different localities' (Massey 1984).

The restructuring approach is interested in localities in two senses. First, localities are places where wider processes manifest themselves. That is, locality events are understandable, in the language of realism, as contingent. General processes interact in a specific way in a particular place. Moreover, the way these wider processes interact locally may create combination effects which generate new causal entities. The locality, that is, may be a site of emergent causal powers (Savage *et al.* 1987, Urry 1987). What is particularly interesting about the latter is that they not only affect what happens in the locality (as stressed by Savage *et al.* 1987), but they also affect the wider systems of which the locality is a part. These emergent powers are all-important in understanding how localities both reflect and cause wider changes. An example might be the creation of a local business culture which channels finance, shapes markets, and generates training which in turn leads to the development of a distinctive local economy. This description would apply equally to the Welsh coal valleys earlier in the century, Silicon Valley in the 1950s and 1960s, or New England high technology in the 1980s.

The ideal that the task of research is to look for social causes of social events is rooted in the practical concerns of the Marxist tradition (Thrift 1983, Johnston 1986). Research should find *causes*, because this will reveal points of potential *change*; understanding society necessarily means knowing how it could be different.

Analytical concepts

As we have seen, a characteristic feature of the restructuring approach is its use of the concepts of a spatial division of labour, and the closely related notion of rounds (or layers) of investment (Massey 1979). The division of labour may refer to sectoral divisions (between industries), or to hierarchical divisions (between control and execution functions). These can be found both between firms (corresponding to the 'social division of labour' in Marx) and within large multiplant firms. The basis of the prevailing spatial division of labour will tend to vary historically. In the 19th century, for example, the spatial division of labour in the US and in Britain was largely a sectoral one. But in the late 20th century it is more usually intrasectoral. So places are known less for what they produce than for who is employed in them in specific stages of the production processes (Clark *et al* 1986, p. 23). Most localities contain a variety of firms in a variety of industries, and tend to represent a distinctive mix of places within different spatial divisions of labour. The pattern of eonomic places in a locality defines its place in the national and international spatial division of labour. It is quite possible for some localities to occupy several different places at once in the spatial divisions of labour – with, for example, low-level functions in one industry and higher-level ones in another.

These ideas offer useful ways to identify the different periods of history, and an important aim of the restructuring approach – as it is of Marxism – is to situate the present in history. But the concept of a spatial division of labour is still ultimately descriptive. Invoking this spatial division does not of itself explain the connections between a particular workforce and the wider system of production. The essence of the restructuring approach is its stress on the potentially two-way causal relationship between the local and the global. Local conditions may allow an industry to survive in the world economy; global economic pressures may destroy a local industry (Massey 1984). So research can usefully begin by situating a locality in terms of spatial divisions of labour, but the task of explanation will require a separate stage of work, which will be a different, more theoretical, kind. This means calling on other *kinds* of information, including historical studies (Sayer 1984).

A spatial division of labour in an industry is created with every round of investment. But the firm's investment decisions at any point in time are influenced by the existing spatial division of labour. They will invest to secure suitable, cheap, or scarce labour and this will be found unevenly across the world and within a country. So a new round of investment may involve drawing on different skills, or different kinds of workers. In such cases new rounds of investment will lay a different spatial division of labour on top of the residue of older spatial divisions. This captures two ideas: new spatial divisions of labour are created periodically, and the mix between new and old rounds has different implications in different places (Massey 1978a)

This way of placing sets of spatial patterns in chronological sequence – particular regional or local specialisms associated with particular historical phases in accumulation – clearly has something in common with the French Regulation School (see Lipietz 1986, or Dunford & Perrons 1986). If we find persistent spatial processes, such as the current cumulative cycle of growth in south east England or Sunbelt USA, the task of explanation can be helped by identifying the dominant prevailing rounds of investment. Persistent patterns of cumulative causation can be unravelled by examining the interplay between inherent geographical patterns and current investment decisions (Massey 1984, pp. 121, 142).

But each round of investment needs to be explained in terms of the contemporary complex of causes bearing upon capital's investment decisions. And these decisions are not simple things. They may reflect objective global trends in markets, but they may also incorporate highly subjective perceptions. The sociology of the particular investors concerned may be very important. The restructuring literature emphasizes the distinctivness of national capitals, state agencies, and even key individuals.

The local labourmarket occupies an important place in this approach (Urry 1981b). The labourmarket is a field of contact between the locality's place in spatial divisions of labour and the immediate pattern of social and political relations. Most of the population earns its living through the local labourmarket (which may be more or less local for different groups), and in this sense the labourmarket explains the level of prosperity and the kind of work relationships in the locality. At the same time the local labourmarket is the source of profit for capital. As such it is a key influence on investment decisions. The labourmarket is therefore one of the major channels through which a locality exerts its influence on the wider trajectory of accumulation. The local labourmarket offers a particularly useful case study through which to examine the articulation of the locality-level processes with global processes. Again, invoking the local labourmarket does not in itself supply explanations. To get at these, it is necessary to trace the social and political bases of the historically specific processes which work themselves out in the local labourmarket.

Claims as to spatially significant processes

The restructuring approach offers some guidelines in this search. It includes a number of substantive propositions – working hypotheses about the kind of processes at work on the side of capital and of labour which may be spatially significant.

ON THE PART OF CAPITAL

Companies restructure production to adapt, in a manner of their choosing, to competition in the product market. The restructuring approach maintains that capital generally has the geographical advantage over labour in that it is more mobile, and is becoming ever more so. But the way it uses this mobility is not mechanically determinate. It depends on the way companies decide to respond to the established pattern of immobilized physical capital, and in part on the response of labour (Storper & Walker 1981, 1983).

In countries like Britain and the US firms have a limited number of choices. These can be regarded as a repertoire of restructuring forms. For example, they can rationalize production (close plant), intensify production (reduce gaps in working time), invest in new techniques (increase productivity and/or produce new products) or they can merge with other companies (Massey & Meegan 1982, Cooke 1984, p. 4). Offices or factories may be cloned, so that identical plants are created. Alternatively, different plants may specialize in different parts of the production process. The entire process may be brought together on one site, or corporations may be split into a large number of separate companies, subcontracting to one another as autonomous firms.

Each of these ways of arranging the corporate organization of production could, under hypothetical circumstances, correspond to a distinctive spatial pattern (Massey 1984, p. 76). For example, different levels in the part-process model could be allocated to different regions (assembly in the low-wage regions, control and R&D in the metropolitan centre). But the repertoire of restructuring forms is not a cookbook for deriving spatial structures. It is useful to look for these forms when unravelling particular spatial patterns, but they cannot be used to predict spatial effects because there is no predictable connection between the restructuring forms within a firm and wider spatial patterns.

ON THE PART OF LABOUR

Labour's relative immobility means that localities have a special significance for labour. In the restructuring approach geography is seen as an influential element in the characteristics of labour supply. Some localities have a history of work in a particular industry or occupation, some have distinctive traditions of female participation in the labourmarket (or the lack of it). Some have accumulated an image for industrial militancy, others for compliance (either of which may be more or less based on fact). In some areas most women expect to go out to work like their mothers before them, in others they are more likely to think themselves lucky to get a job (McDowell & Massey 1984).

Spatial structures, therefore, have an influence on the potential options both to capital and labour. Investment patterns reflect both of these at once. It is to be assumed that capital will try, in general, to weaken workers resistance where this does not threaten other objectives. This may mean moving location, or it may mean staying put and changing work patterns or the workforce. This is why there is no determinate relationship between the form of restructuring adopted by capital and the spatial division of labour. Spatial outcomes are not reducible to any logic of either capital, or of labour alone: 'Spatial structures are established, reinforced, combatted and changed through political and economic strategies and *battles* on the part of managers, workers and political representatives' (Massey 1984, p. 85; emphasis added). In some cases local events might be determined completely and unavoidably by global changes (as where a specialization in an old and declining industry causes severe local job losses). In other cases local influences might affect the national, and even global, pattern of accumulation (as in the specialization of south east England in finance and military industry). In general we should expect the link between the local and the global to run both ways – down from the global, and up from the local. For example, investment decisions would be influenced by capital's perceptions of labourmarket conditions in the localities in which they had plants.

A set of claims as to empirical outcomes

As we have seen, a characteristic feature of restructuring theory in the 1970s and early 1980s was the claim that contemporary industrial change was giving rise to a new spatial division of labour. This implied a convergence between regions in terms of industrial structure, alongside continued divergence in terms of conception and execution in the production process (Massey 1979, Westaway 1974, Massey & Meegan 1982). These trends were related to the relocation of production associated with a round of investment in activities demanding lower skills. They were also related to the presence of a welfare state and the existence of national patterns of collective bargaining raising regional pay levels towards the national average. These generated a movement towards common patterns of consumption across the country.

This same round of investment was associated with a spatial recomposition of the working class. This gave rise to specific regional spatial coalitions (for example, in the campaigns in London Docklands, other inner cities, and the less advantaged regions where new regional alliances were emerging (Massey 1978a, Cooke 1984).

As the restructuring approach matured in the 1980s, placing less emphasis on predictions, it offered different, and more cautious empirical claims. In the 1980s spatial restructuring takes place within a new politico-economic context. For example, at the global level the flow of manufacturing capital out of the metropolitan countries has been accompanied by a significant return of some manufacturing to areas in those countries. The new international division of labour, wherein Third World countries increasingly specialize in manufacturing, is only part of the story. Labour-intensive investments in the old centres of manufacturing form an important part of the current round of investment (Morgan 1986, Sayer 1987, Davis 1987).

The growth of medium and smaller establishments and the decentralization of control functions in some companies has also demonstrated that corporate relationships are more complicated than the early restructuring literature suggested (Lash & Urry 1987). So the claim for an increasing spatial division of labour cannot be retained without qualification. It has also become apparent that the spatial differentiation between different labourmarkets needs to be given more attention. While semi- and unskilled workers are often effectively 'trapped by space', some élite labour groups are relatively mobile – even internationally – and are able to use space more as capital does, as a means of improving their bargaining position. This was pointed out by Massey in 1978, but its importance appears to be increasing. Local labourmarkets are in fact made up of layers with quite different boundaries, and the mix is an important dimension differentiating localities (for an illustration, see Boddy *et al.* 1986).

These developments in labourmarkets can be seen as part and parcel of the decline of the Fordist pattern of industry (Harrington 1987, Lipietz 1986). The spatial restructuring of capitalist production is taking place in the context of a changing macro-economic relationship between consumption and production. In retrospect, it appears that a major component of the round of investment of the late 1960s and early 1970s was related to the integration of the UK into the new international division of labour. In the 1980s it is not so easy to identify a single dominant round of investment, as different firms attempt to incorporate different segments of the national capital and labour into different international divisions of labour. One tangible effect is the emergence of 'the Third World within the First' (Davis 1987). The early restructuring theory also oversimplified the nature of the feminization of industry. In fact this has largely involved the growth of secondary labourmarkets employing women. Gender discrimination is being reconstructed in new forms, rather than being dismantled (McDowell & Massey 1984).

The restructuring approach was associated with the claim that regional and urban economies were disappearing. New sets of spatial divisions of labour were creating new ways of using space in which traditionally identified regions, and localities, were no longer the central 'containers' of accumulation (Lash & Urry 1987, p. 6). In recent years, the thesis of local economic disintegration has been tempered by recognition that there appear to be some regions where new localized economic systems are emerging. The prototypical examples are Silicon Valley, and the Third Italy (Scott 1988). As yet, there are few signs that this model applies widely, but these cases confirm that the restructuring approach is not able to generate firm empirical predictions. But neither, for that matter, is any other approach.

Conclusion

This hasty and selective résumé is intended to suggest that the restructuring approach is alive and well, even though it does not offer a set of neat theories or ready-made empirical claims. It is best seen as a *research programme* combining a distinctive and non-reductionist method with a preliminary set of analytical tools. Its purpose is to elucidate the interaction between capital's strategies and the socio-spatial pattern of production and other social relationships. It is part and parcel of this approach that pure theory can only go so far. Theoretical speculation can never generate all the types of discovery necessary to give knowledge of specific spatial outcomes. Formal analysis of the location of industry, for example, can only go so far as to point out possibilities, causal powers which may or may not be realized. Empirical work is therefore necessary, but not simply to apply theory or to test it; it is an essential input to the development of theory.

In particular, it is necessary to examine precisely how particular investments with particular spatial effects come to be chosen out of the range of possible alternatives. This means unearthing both the alternatives which presented themselves, and the decision processes which selected between them. Only empirical research can do this, and it must be of a particular kind. This can be elaborated with the distinction developed by Sayer between intensive and extensive research. Extensive research aims to detect regular patterns; it is familiar from orthodox science. Intensive research addresses the question, how does a process work in a particular case? It has erroneously been considered less respectable in conventional scientific discourse. But intensive methods aim to maximize the flow of information about a particular nexus of relationships, such as an investment decision (Sayer 1984, p. 223). There are no easy guides to extensive methodology, which can draw on semi- or unstructured interviews, informal sources of information, and other techniques. The point of such research is to study not only the objective facts such as, for example, the financial background to an investment decision. It is also concerned to draw out the rôle of the preconceptions (and even misperceptions) shared by the decision makers, and the social and cultural background to these. Not least, intensive research should also provoke the researcher to reappraise his/her assumptions about what the key processes really are. This is a matter of creative exploration, not a question of applying a pregiven theory.

But it is necessary to use both extensive *and* intensive methods. And at the extensive level, there should be no disdain for the extensive techniques of mapping, shift-share analysis, etc., even though these have traditionally been associated with positivism and have sometimes been looked down upon by radicals. This was due to a weakness in those forms of radicalism, rather than to the methods themselves. A research project in the restructuring approach should draw on quantitative data and statistical methods, but it should do so without thereby claiming to have found explanations. Intensive research should be coupled with extensive research to show the social relationships that underpin a particular spatial pattern at a particular time in history.

The world economy will continue to go through a series of restructurings, and civil society and national states are also undergoing massive upheavals. The restructuring approach is an attempt to make sense of this permanent revolution. It asks that research should lead both ways at once – unveiling empirical aspects of social reality, and developing new questions and chancing new answers on the theoretical level. Knowledge of this sort is not certain, much less predictive. But in the best sense it is a scientific attempt to grasp the sociospatial world we live in, and to inform a politics that might shift the trajectory of development towards less inhumane and hazardous paths.

Notes

1 This is not to prejudge the tricky question of what science is. For a recent discussion, in a vein which is sympathetic to the approach taken here, see Outhwaite (1987).

2 Calls for worker co-operatives, for example, now tend to be based on arguments for entrepreneurship rather than demands for a change in the balance of economic power.

3 Notable British examples include the South East England Development Strategy (SEEDS) project in England, and the two foci for local economic policies: the Centre for Local Economic Strategies (CLES), and the journal *Local Economy*.

References

Abercrombie, N. & J. Urry 1981. *Capital, labour and the middle classes*. London: George Allen & Unwin.

Anderson J., S. Duncan & R. Hudson 1983. *Redundant spaces in cities and regions?* Institute of British Geographers special publication No. 15. London: Academic Press.

Beauregard, R. 1988. In the absence of practice: the locality research debate. *Antipode* **20**, 51–9.

Benton, T. 1985. *The rise and fall of structural Marxism*. London: Hutchinson.

Berman, M. 1984. *All that is solid melts into air*. London; Verso.

Bhaskar, R. 1979. *The possibility of naturalism*. Brighton, Sussex: Harvester.

Bluestone, B. & B. Harrison 1982. *The deindustrialisation of America*. New York: Basic Books

Boddy, M. 1986. Structural approaches to industrial location. In *Industrial change in the United Kingdom*, W. Lever (ed.), 56–66. London: Longman.

Boddy, M. & C. Fudge 1984. *Local socialism*. London: Macmillan.

Boddy, M., J. Lovering & K. Bassett 1986. *Sunbelt city?* Oxford: Clarendon Press.

Brookfield, H. 1975. *Interdependent development*. London: Methuen.

Browett, J. 1984. On the necessity and inevitability of uneven spatial development under capitalism. *International Journal of Urban and Regional Research* **8** 155–76

Brown, G. (ed.) 1975. *The Red Paper on Scotland*. Edinburgh: EUSB.

Buhle, P. 1987. *Marxism in the USA*. London: Verso.

Castells, M. 1977. *The urban question*. London: Edward Arnold.

Clark, G. L., M. S. Gertler & J. Whiteman 1986. *Regional dynamics: studies in adjustment theory*. Boston and London: Allen & Unwin.

Cochrane, A 1987. *Developing local economic strategies*. Milton Keynes and Philadelphia, Penn.: Open University Press.

Cockburn, C. 1977. *The local state*. London: Pluto Press.

Cohen, I. 1987. Structuration theory and social *praxis*. In *Modern social theory*, A. Giddens & J. A. Turner (eds). Cambridge: Polity Press.

Community Development Programme (CDP) 1975. *The cost of industral change*. Home Office, London: CDP.

Cooke, P. 1981. Recent theories of political regionalism: a critique and an alternative proposal: *International Journal of Urban and Regional Research* **8**, 549–71.

Cooke, P. 1983. Labour market discontinuity and spatial development. *Progress in Human Geography* **7**, 543–65.

Cooke, P. 1984. *Global restructuring, local response*. London: Economic and Social Research Council.

Cooke, P. 1987a. Clinical inference and geographic theory. *Antipode* **19**, 69–78.

Cooke, P. 1987b. Research policy and review 19: Britain's new spatial paradigm, technology, locality and society in transition. *Environment and Planning A* **19**, 1289–301.

Cooke, P. 1987c. Local capacity and global restructuring; some preliminary results from the CURS research programme. Paper given to the Urban Change and Conflict Conference, University of Kent, Canterbury.

Cooke, P. 1989. Locality, economic restructuring and world development. Introduction to *Localities*, P. Cooke (ed.). London: Unwin Hyman.

Crompton, R, & M. Mann (eds) 1986. *Gender and stratification*. Oxford: Polity Press.

Davis, M. 1987. The streets of Los Angeles. *New Left Review* **164**, 65–86.

Davis, T. 1979. Employment policy in one London borough. In *Jobs and community action*, G. Craig, M. Mayo & N. Sharman (eds), 217–27. London: Routledge & Kegan Paul.

Dex, S. 1985. *The sexual division of work*. London: Wheatsheaf.

Duncan, S. & M. Goodwin 1988. *The local state and uneven development*. Oxford: Polity Press.

Dunford, M. & D. Perrons 1986. The restructuring of the post-war British space economy. In *The geography of deindustrialisation*, R. Martin & B. Rowthorn (eds), 543–605. London: Macmillan.

Dunleavy, P. 1980. *Urban political analysis: the politics of collective consumption*. London: Macmillan.

Edwards, J. & R. Batley 1978. *The politics of positive discrimination: an evaluation of the urban programme 1967–77*. London: Tavistock.

Fine, B. & L. Harris 1979. *Rereading capital*. London: Macmillan.
Friedman, A. 1977. *Capital and labour*. London: Macmillan.
Frobel, F., J. Heinrichs & O. Kreye 1982. *The new international division of labour*. Cambridge: Cambridge University Press.
Gäffikin, F. & A. Nickson 1982. *Job crisis and the multinationals*. Birmingham: Birmingham Trade Union Resource Centre.
Gamble, A. 1981. *Britain in decline*. London: Macmillan.
Giddens, A. 1984. *The constitution of society*. Oxford: Polity Press.
Giddens, A. 1984. *The nation-state and violence*. Oxford: Polity Press.
Goldthorpe, J. 1982. On the service class, its formation and future: In *Social class and the division of labour*, A. Giddens & G. Mackenzie (eds). Cambridge: Cambridge University Press.
Gorz, A. 1982. *Farewell to the working class*. London: Verso.
Graham, J. 1988. Post modernism and marxism. *Antipode* **20**, 60–6.
Gregory, D. & J. Urry (eds) 1985. *Social relations and spatial structures*. London: Macmillan.
Hall, S. 1977. The political and the economic in Marx's theory of classes. In *Class and class structure*, A. Hunt (ed.), 15–60. London: Laurence & Wishart.
Hamilton, R. & M. Barrett 1986. *The politics of diversity*. London: Verso.
Harcourt G. 1972. *Some Cambridge controversies in the theory of capital*. Cambridge: Cambridge University Press.
Harloe, M. 1977. *Captive cities: the political economy of cities and regions*. London: Wiley.
Harrington, M. 1987. *The next left: the history of a future*. London: I. B. Taurus.
Hartmann, H. 1978. The unhappy marriage of marxism and feminism. *Capital and Class* **8**, 1–33.
Harvey, D. 1978. The urban process under capitalism: a framework for analysis. *International Journal of Urban and Regional Research* **2**, 101–31.
Harvey, D. 1982. *The limits to capital*. Oxford: Basil Blackwell.
Hobsbawm, E. (ed.) 1981. *The forward march of labour halted*. London: Verso.
Hunt, P. 1980. *Gender and class consciousness*. London: Macmillan.
Hymer, S. 1974. The multinational corporation and the law of uneven development. In *Economics and world order: from the 1970s to the 1990s*, J. Bhagwati (ed.), 113–40. New York: Free Press/Macmillan.
Jameson, F. 1984. Postmodernism, or the cultural logic of late capitalism. *New Left Review* **146**, 53–92.
Johnston, R. J. 1986. *Philosophy and human geography*. London: Edward Arnold.
Kaldor, N. 1970. The case for regional policies. *Scottish Journal of Political Economy* **17**, 337–48.
Knights, D., H. Willmott & D. Collinson 1985. *Job redesign; critical perspectives on the labour process*. Aldershot and Brookfield, Vermont: Gower.
Laclau, E. & C. Mouffe 1983. *Hegemony and socialist strategy*. London: Verso.
Lash, S. & J. Urry 1987. *The end of organised capitalism*. Oxford: Polity Press.
Lee, R. 1987. Restructuring. Entry in *The dictionary of human geography*, 2nd edn, R. J. Johnston (ed.), 411. Oxford: Basil Blackwell.
Leys, C. 1983. *Politics in Britain*. London: Verso.
Lipietz, A. 1986. *Miracles and mirages*. London: Verso.
Lojkine, J. 1976. A marxist theory of capitalist urbanisation. In *Urban sociology:critical essays*, C. Pickvance (ed.). London: Tavistock.
Lovering, J. 1982. *Gwynedd in crisis*. Coleg Harlech, Harlech, Gwynedd: North Wales Employment Resource Centre.
Lovering, J. 1989. Postmodernism, marxism and locality research. *Antipode* **21**, 1–12.
Lovering, J. & R. Meegan (eds) 1989. *Economic restructuring*. Report on the findings of the CURS project. London; Unwin Hyman.
McDowell, L. & D. Massey 1984. A woman's place? In *Geography Matters!* D. Massey & J. Allen (eds). Cambridge: Cambridge University Press.

Martin, R. & R. Rowthorn (eds) 1986. *The geography of deindustrialisation.* Cambridge: Cambridge University Press.

Massey, D. 1974. *Towards a critique of industrial location theory.* Research Paper 5. London: Centre for Enviornmental Studies.

Massey, D. 1978a. Regionalism: some current issues. *Capital and Class* **6**, 106–26.

Massey, D. 1978b. Capital and locational change: the UK electrical engineering and electronics industries. *Review of Radical Political Economics* **10**, 398–54.

Massey, D. 1979. In what sense a regional problem? *Regional Studies* **13**, 233–43.

Massey, D, 1981. A politics of location. In *Socialism in a cold climate,* J. Griffiths (ed.), 144–60. London and Boston: Counterpoint/Unwin.

Massey, D. 1983. The shape of things to come. *Marxism Today* April, 18–27.

Massey, D. 1984. *Spatial divisions of labour.* London: Macmillan.

Massey, D. 1985. New directions in space. In *Social relations and spatial structures,* D. Gregory & J. Urry (eds), 9–19. London: Macmillan.

Massey, D. 1986. Industrial restructuring as class restructuring. *Regional Studies* **17**, 73–89.

Massey, D. & J. Allen (eds) 1984. *Geography matters!* Cambridge: Cambridge University Press.

Massey, D. & R. Meegan 1982. *The anatomy of job loss.* London: Methuen.

Massey, D. & R. Meegan (eds) 1985. *Politics and method.* London and New York: Methuen.

Merseyside Socialist Research Group (MSRG) 1980. *Merseyside in crisis.* 23 Glover St, Birkenhead, Merseyside: MSRG.

Mingione, E. 1983. Informalisation, restructuring and the survival strategies of the working class: *International Journal of Urban and Regional Research* **7**, 311–39.

Morgan, K. 1986. Re-industrialisation in peripheral Britain: In *The geography of de-industrialisation,* R. Martin & R. Rowthorn (eds). Cambridge: Cambridge University Press.

Murgatroyd, L., M. Savage, D. Shapiro, J. Urry, S. Walby & A. Warde 1984. *Localities, class and gender.* London: Pion.

Nichols, T. 1979. Social class: official, sociological and marxist. in *Demystifying social statistics,* London: Pluto Press.

O'Connor, J. 1984. *Accumulation crisis.* Oxford and New York: Basil Blackwell.

Outhwaite, W. 1987. *New philosophies of social science.* London: Macmillan.

Pahl, R. E. 1984. *Divisions of labour.* Oxford: Basil Blackwell.

Panitch, L. 1986. *Working class politics in crisis.* London: Verso.

Peet, R. 1983. The geography of class struggle and the relocation of United States manufacturing. *Economic Geography* **59**, 112–43.

Peet, R. (ed.) 1987. *International capitalism and industrial restructuring:* London and Boston: Allen & Unwin.

Pickvance, C. (ed.) 1976. *Urban sociology: critical essays.* London: Tavistock.

Quaini, M. 1982. *Geography and marxism.* Oxford: Basil Blackwell.

Robinson, J. 1978. *Contributions to modern economics.* Oxford: Basil Blackwell.

Sabel, C. F. 1982. *Work and politics.* Cambridge: Cambridge University Press.

Sack, R. D. 1986. *Human territoriality: its theory and history.* Cambridge: Cambridge University Press.

Saunders, P. 1981. *Social theory and the urban question.* London: Hutchinson.

Savage, M., J. Barlow, S. Duncan & P. Saunders 1987. Locality research: the Sussex programme. *Quarterly Journal of Social Affairs* **3**, 27–51.

Sayer, R. A. 1982. Explanation in economic geography. *Progress in Human Geography* **6**, 68–88.

Sayer, R. A. 1984. *Method in social science: a realist approach.* London: Hutchinson.

Sayer, R. A. 1985. Industry and space: a sympathetic critique of radical research. *Environment and Planning D, Society and Space* **3**, 3–29.

Sayer, R. A. 1987. Hard work and its alternatives. *Environment and Planning D, Society and Space* **5**, 395–9.

Scott, A. J. 1988. Flexible production systems and regional development. *International Journal of Urban and Regional Research*. **12**, 171–86.

Scott A. J. & M. Storper (eds.) 1986. *Production, work, territory: the geographical anatomy of industrial capitalism*. London and Boston: Allen & Unwin.

Segal, L. 1987. *Is the future female?* London: Virago.

Smith, N. 1984. *Uneven development, capital, nature and the production of space*. Oxford and New York: Basil Blackwell.

Smith, N. 1987. Dangers of the empirical turn: some comments on the CURS initiative. *Antipode* **19**, 59–68.

Soja, E. W. 1985. The spatiality of social life: towards a transformative retheorisation: In *Social relations and spatial structures*; D. Gregory & J. Urry (eds), 90–127. London: Macmillan.

Storper, M. & A. J. Scott 1986. Production, work and territory: contemporary realities and theoretical tasks. In *Production, work, territory: the gegraphical anatomy of industrial capitalism*, J. Scott & M. Storper (eds), 1–39. London and Boston: Allen & Unwin.

Storper, M. & R. Walker 1981. Capital and industrial location. *Progress in Human Geography* **4**, 473–509.

Storper, M. & R. Walker 1983. The theory of labour and the theory of location. *International Journal of Urban and Regional Research* **7**, 1–43.

Szelenyi, I. 1984. *Cities in recession*. London and Beverly Hills: Sage Publications.

Thrift, N. 1983. On the determination of social action in space and time: *Environment and Planning D, Society and Space* **1**, 23–57.

Thrift, N. & P. Williams (eds) 1987. *Class and space*. London: Routledge & Kegan Paul.

Tietz, M. 1987. Planning for local economic development. *Town Planning Review* **58**, 5–18.

Urry, J. 1981a. *The anatomy of capitalist societies*. London: Macmillan.

Urry, J. 1981b. Localities, regions, and social class. *International Journal of Urban and Regional Research* **5**, 455–74.

Urry, J. 1987. Society, space and locality. *Environment and Planning D. Society and Space* **5**, 435–44.

Walby, S. 1986. *Patriarchy at work*. Oxford: Polity Press.

Walker, R. A. 1978. Two sources of uneven development under advanced capitalism: spatial differentiation and capital mobility. *Review of Radical Political Economics* **10**, 28–38.

Warde, A. 1985. Spatial change, politics, and the division of labour. In *Social relations and spatial structures*, D. Gregory & J. Urry (eds), 190–212. London: Macmillan.

Westaway, J. 1974. The spatial hierachy of business organisations and its implications for the British urban system. *Regional Studies* **8**, 145–55.

Wright, E. O. 1978. *Class, crisis, and the state*. London: Verso.

Wright, E. O. 1984. *Classes*. London: Verso.

9 *Marxism, post-Marxism, and the geography of development*

Stuart Corbridge

Introduction

When David Keeble surveyed the literature on economic development in the mid-1960s he had little positive to report on the contributions of geography and geographers. Writing in *Models in geography*, Keeble complained of an 'apparent and remarkable lack of interest among geographers in the study of the phenomenon of "economic development"' (Keeble 1967, p. 243). Of the 251 major articles published between 1955 and 1964 inclusive in *Economic Geography*, 'only ten were concerned in whole or part with problems of economic development' (1967, p. 243). And writing in the *Annals of the Association of American Geographers*, 'the percentage falls still further, to 2.5 per cent (or six articles out of 242)' (quoted in Keeble 1967, p. 243). Worst still, these few contributions had little of interest to say. Only 3 of the 16 papers adopted a nomothetic approach to development; the rest displayed a tiresome concern for the local and the unique. The upshot is that Keeble's survey is 'perforce concerned primarily with the work of economists who, unfettered by an idiographic tradition, have at last moved to fill the wide intellectual void left open by geographers' (1967, p. 246).

Twenty years on there is less cause for disciplinary *Angst*. To be sure, government funding for geographical research in the developing world is depressingly low (especially in the United Kingdom: see Thrift 1985) and some colleagues need to be reminded that 'the world is our oyster' (Johnston 1984). It is also true that the renaissance of development geography is little indebted to the models of modernization theory. The typologies of Rostow and Lewis, North and Perroux, so much favoured by Keeble, enjoyed only a short ascendency in geography. Nevertheless, a renaissance there has been. Under the influence of Marxian social theory, development geography has found its voice – indeed several voices – on the key debates in development studies: on the constitution and dynamics of the world economy; on the concept of a Third World; on the scale and significance of industrialization in the periphery; on the relationship between resources and development; on questions of the state, gender, and political movements in the Third World; on the urban arena; on the persistence of peasantries; and on the issue of agrarian transformation in the periphery.

This is not to say that Marxism alone has raised its voice on these questions, or that its voice is always coherent or sufficient. In geography there remains a vibrant tradition of area studies (now being reclaimed for the new regional geography) and in development economics there is an equally vigorous, if rather chilling, counter-revolution in development theory and policy (Toye 1987, Meier 1987). It is to say that radical geographers have been at the cutting edge of new research in development geography since the early to mid 1970s. Most of the new models in development geography have come from this source and they continue to do so.

Given this blunt premiss, the chapter is organized as follows. First the emergence of radical development geography is charted, and the tradition of neo-Marxism around which it first crystallized is discussed.[1] Attention is paid to the models of Baran, Frank, and Wallerstein. The next section offers a critique of neo-Marxian development theory and presents a more classically Marxian interpretation of the dynamics of capital accumulation, realization, and uneven development. Prominent models here are the Warren model and the articulation of modes of production. The following section examines the metatheoretical commitments of Marxist development studies. It takes seriously the claim that Marxism is wedded to an unhelpful determinism and economism and that it mixes normative and positive discourses on development without due regard for logical consistency. The final section considers how Marxism is responding to these charges. New voices from analytical Marxism, post-imperialism, and the Regulation School are presented as three instances of a more tolerant post-Marxism. Among the defining features of post-Marxism are its disavowal of epistemological arguments and economism, and its sensitivity to the contingencies of capital accumulation in time and space. The latter quality makes post-Marxism especially attractive as a framework for future work in development geography. It also points up an important area of common ground between development studies and a reconstructed human geography.

The development of underdevelopment

A radical account of development and underdevelopment was first popularized in geography in the early 1970s. Work by Slater (1973, 1975) and Blaut (1973, 1976) and Harvey (1975) – following earlier work by Buchanan (1964, 1970), and in tandem with the essays by Cannon (1975) and McGee (1974) – pointed up the teleology, Eurocentrism, and apparent irrelevance of modernization theories. Instead of lauding the developmental impact of free trade, foreign aid, and direct foreign investment, attention now was directed to a secular decline in the commodity terms of trade of less developed countries [LDCs], to a widening development gap, and to the reproduction of neocolonial networks of dependence and exploitation. In place of studies which traced the diffusion of technologies down the central place hierarchy, geographers now concerned themselves with a process of spatial disarticulation which linked the production of regional imbalances in the Third World to the production of spatial divisions of labour in the metropolitan core (see also Massey 1984). In this endeavour geographers joined with Paul Baran and Gunder Frank in proposing a model of the development of underdevelopment which can be described by four linked propositions.

The underdevelopment model depends, first, upon an underconsumptionist account of the dynamics of capital accumulation. Paul Baran's work on *The political economy of growth* makes this point very clearly (Baran 1973, first published 1957). According to Baran, capitalism presents itself in two main forms. In its youth capitalism is competitive. At the time market forces are more or less untrammelled and local forces of competition are set free to produce an actual economic surplus which approximates a hypothetical potential economic surplus. The problem is that capitalism carries within it the seeds of its own destruction. Precisely because capitalism favours the strong, a process of monopolization occurs wherein economic production and exchange is centralized in a small number of oligopolistic concerns. These enterprises are able to generate an enormous economic surplus, but they are unable to distribute it beyond a small group of controlling capitalists. As a result, says Baran, monopoly capitalism is prone to crises of underconsumptionism which force its comptrollers to shore up demand by spending on arms, by deficit financing, and by extending the reach of capital into the periphery of the world system. The Third World, for Baran, becomes an investment outlet for metropolitan capitalism and a major and very cheap source of raw materials. For its part, the Third World underdevelops and begins to exhibit a peculiar 'morphology of backwardness'.

In the 1960s this insight was extended by Gunder Frank, who popularized an analysis of the development of underdevelopment. According to Frank, the development of capitalism in the core has from the very beginning depended upon the transfer of a surplus from the periphery. Says Frank:

> [From] the time of Cortez and Pizarro in Mexico and Peru, Clive in India, Rhodes in Africa, the 'Open Door' in China – the metropolis destroyed and/or totally transformed the earlier viable social and economic systems of those societies, incorporated them into the metropolitan dominated world-wide capitalist system, and converted them into sources for its own metropolitan capitalist accumulation and development. The resulting fate for these conquered, transformed or newly established societies was and remains their decapitalization, structurally generated unproductiveness, ever-increasing poverty for the masses – in a word their underdevelopment (Frank 1969, p. 225).

In this fashion, Frank detached an emerging neo-Marxism still further from its classical roots. For Frank (and later Wallerstein 1974, 1979, 1984), the economies of the periphery have been capitalist since they first produced for exchange in a world market. Again, it matters not that this production for exchange is carried on according to several different systems of labour control – for example, free wage labour, serfdom, and slavery. For Wallerstein, 'the relations of production that define a system are the "relations of production" of the whole system and the system at this point in time is the European world economy. Free labour is indeed a defining feature of capitalism, but not free labour throughout the productive enterprises' (Wallerstein 1974, p. 127).

A second element of the underdevelopment model describes the structure of metropolis/satellite relationships which facilitates the transfer of a surplus from the bottom to the top of the world system. Frank's early work describes this structure in terms both graphic and geographical. He tells of a chain of metropolis/satellite relationships wherein:

> At each stage along the way the relatively few capitalists above exercise monopoly power over the many below, expropriating some or all of their economic surplus and, to the extent that they are expropriated in turn by the still fewer above, appropriating it for their own use . . . at each point the international, national and local capitalist system generates economic development for the few and underdevelopment for the many (Frank 1969, pp. 7–8).

Later scholars added to this imagery a more precise account of unequal exchange. In Wallerstein's model of surplus transfer, actors in the core (the metropolitan capitalists) call on their state machines to manipulate an economic system geared otherwise to the geographical equalization of profits. In effect they use state power deliberately and persistently to weaken (underdevelop) the periphery – by conquest, by monopoly pricing, by protectionism, and so on; but not so the semi-periphery. Wallerstein implies that it suits the core states to preserve a (variable) semi-periphery as a sort of buffer between themselves and the periphery. More pointedly, Emmanuel (1972) provides a complex theory of unequal exchange which hinges upon the power of trade unions in the core to raise real wages in a manner not open to workers in the Third World. According to Emmanuel, this underpins an unequal exchange of goods in which the greater quantity of embodied labour time flows from the periphery to the core than vice versa. (For a critique, see Bettelheim 1972, Bacha 1978, Brewer 1980. For new theorizations, see Foot & Weber 1983, Hadjimichalis 1984, Roemer 1988.)

A third element of underdevelopment theory concerns itself with the effects in and upon the Third World of its dependent insertion into a capitalist world economy. Again, there are several variations upon a theme. In the work of Frank and Wallerstein the morphology of dependent social formations seems to be determined entirely by the logic and needs of metropolitan capitalism. It is not just that production for exchange takes place in a capitalist world system: for Wallerstein, the class systems of the Third World – or modes of labour control – take shape according to their ability to service this grand global machine. Thus: 'free labour is the form of labour used for skilled work in core countries whereas coerced labour is used for less skilled work in peripheral areas. The combination thereof is the essence of capitalism' (Wallerstein 1974, p. 127). In other accounts of *dependencia* (Dos Santos 1973, Sunkel 1973, Evans 1979), more attention is paid to the rôle played by *comprador* élites in enforcing local geographies of production.[2] Even in the work of Sunkel and Dos Santos, however, the determination of internal factors by external forces is never far from the surface. For Dos Santos, 'Dependence is a conditioning situation in which the economies of one group of countries is conditioned by the development and expansion of others' (Dos Santos 1973, p. 289).

A fourth proposition of underdevelopment theory suggests that industrial development within the periphery of the world system is at best unlikely and at worst unthinkable. Frank and Dos Santos are among those who have taken a stagnationist line (see also Kidron 1971, Sutcliffe 1972 and the first generation of radical development geographers), with Frank boldly declaring that 'the satellites experience their greatest economic development . . . if and when their ties to the metropolis are weakest' (Frank 1969, pp. 9–10). According to Frank the Third World must choose either socialism or barbarism: either a country breaks from the capitalist world system or it does not (see also Browett 1981, p. 160). In the work of Sunkel and others in the (reborn) ECLA tradition of dependency analysis, this stagnationism is neatly sidestepped. Sunkel prefers to speak of certain obstacles to development which are induced by class-dependent patterns of local market constriction. At this point, however, we are slipping towards more orthodox Marxian perspectives (the dividing line is sometimes very thin). Although Frank's early work is now regarded as an extreme case within the dependency paradigm, and thus of limited interest in itself, it remains true that underdevelopment theory 'exemplifies perfectly a form of analysis in common use' (Booth 1985, p. 762; see also Mouzelis 1988a, p. 27).

Marxist theories of development

A second account of the dynamics of global capitalism is rooted in the concepts of classical Marxism first developed by Marx, Lenin, and Luxemburg. By the late 1970s it was clear to many on the left (and not just the left: see Lall 1975) that underdevelopment theory was unable to account for certain important developments in the world economy, and not least the rapid industrialization of parts of the Third World. Some scholars judged this failure as emanating from the attachment of neo-Marxism to a circulationist, or neo-Smithian, conception of capitalist development. According to Laclau (1979), Brenner (1977), and Palma (1977) (and many others beside: Bath & James 1976, Slater 1977, Wolf 1982), neo-Marxism from Baran onwards has accepted (unwittingly) Adam Smith's thesis that capitalist economic growth depends only on the extension of an ever more efficient global division of labour. To this, the neo-Marxists have added a radical twist, in the form of a theory of unequal exchange. For Baran, Frank, and Wallerstein the extension of an uneven and exploitative world market makes possible the development of core capitalism through the underdevelopment of a capitalist periphery (which loses its economic surplus).

In the eyes of more orthodox Marxists this position is signally flawed. To begin with, the equation of capitalism with a system of production for world market exchange does not stand up. As Laclau explains, if this is all that capitalism amounts to, it must have shaped the lives of 'the slave on a Roman latifundium or the glebe serf of the European Middle Ages, at least in those cases – the overwhelming majority – where the lord assigned part of the economic surplus extracted from the serf for sale' (Laclau 1979, p. 23). Indeed, by this logic 'we could conclude that from the neolithic onwards there has never been anything but capitalism' (Laclau 1979, p. 23). More importantly, the world systems perspective fails to grasp the true uniqueness of capitalism as a system of qualitatively expanding commodity production based upon the prior separation of the workers from their means of production and of enterprise from enterprise. Lacking this insight, the neo-Marxists are condemned to draw two false inferences. First, they conceive of 'changing class relations as emerging more or less directly from the (changing) requirements for the generation of surplus and development of production, under the pressures and opportunities engendered by a growing world market' (Brenner 1977, p. 27). Instead of seeing in local class structures a context for the formation of a world market, the neo-Marxists present local modes of labour control as the functional outcomes of this grand world system. Second, the neo-Marxists are charged with mistaking the effects of an inflow of wealth from the periphery to the core. This will stimulate a systematic development of the core's productive forces

> only when it expresses certain specific social relations of production, namely a system of free wage labour where labour-power is a commodity. Only where labour has been separated from possession of the means of production, and where labourers have been emancipated from any direct relation of domination (such as slavery or serfdom) are both capital and labour-power free to make *possible* their combination at the highest level of technology. Only where they are free, will such combination appear *feasible* and *desirable*. Only where they are free will such combination be necessitated (Brenner 1977, p. 32; emphasis in the original).

This last statement directs us to what is distinctive about more orthodox Marxian theories of capitalist development. Put simply, we can say that classical Marxism recovers from Marx an account of the autocentric dynamics of capitalist accumulation. In place of the zero-sum logic of unequal exchange models, we learn now that capital is everywhere driven to exploit labour power, in part through the creation of those fixed capitals in which some see the trappings of development. It follows that there can be no question of capitalism promoting the development of underdevelopment (at least not in the sense that Frank understands it, and certainly not through local systems of production which are in fact non-capitalist). Marxists must instead explain the continuing underdevelopment of parts of the Third World in terms of a failure of capitalism to take root there (or to take root in forms which demand the production of relative surplus value). To date, this task has been undertaken in several ways, of which three stand out:

(a) Taking a lead from Rosa Luxemburg is a group of scholars concerned to theorize the articulation of modes of production. This tradition follows Luxemburg in suggesting that capitalism is driven to invade the non-capitalist world. This is not because the extraction of a surplus is a precondition for capitalist development in the core – *pace* Frank – nor must it lead, at once, to the promotion of capitalism in the periphery. Metropolitan capitalism expands because its domestic markets are incapable of realizing an expanded surplus. Once established in the periphery, capitalism comes into conflict with non-capitalist relations of production. According to Luxemburg (1972), capitalism must win its struggle with natural economy if it is to secure the liberation of labour power and its coercion into the service of capital. To this end metropolitan capitalism calls up the full force of colonial violence, using military might, oppressive taxation, and cheap imported goods to drive the peasants from the land and into the mines and plantations. Victory proves to be Pyrrhic, however, for once capitalism engulfs the periphery its external escape route is lost and the system is condemned to perish in the mires of overproduction and proletarian revolution.

Later theorists have sought to soften Luxemburg's conclusion while making use of her analysis. In the work of Rey (1971, 1973) attention is directed to a process of articulation wherein:

(i) an initial link is forged in the sphere of exchange, where interaction with capitalism reinforces the pre- (non)-capitalist mode of production;
(ii) capitalism takes root, subordinating the pre-capitalist mode of production but still making use of it;
(iii) there follows the total disappearance of the pre-capitalist mode of production, even in agriculture.

Others have moved still further from Luxemburg. In the 1970s a less mechanistic model of articulation was advanced which suggests that the preservation of pre-capitalist modes of production in the periphery can long continue to be in the interest of metropolitan capitalism. This is not just because the capitalists are content to exploit their partners through exchange, nor even is it because they face resistance in the periphery (although they surely do: Bayley 1988, Corbridge 1988b). It is because the reproduction of pre-capitalist systems removes from the agents of metropolitan capitalism the expense of providing real wages to a fully proletarianized labourforce. Instead, some of these costs are offset to the pre-capitalist sector where a small cash wage helps to supplement the traditional social wage of the villager. In this way the pre-capitalist society bears the major cost of reproducing the labourforces of capitalism. The articulation of modes of production ensures that there is 'a process of transfer of labour value to the capitalist sector through the maintenance of self-sustaining domestic agriculture' (Hoogvelt 1982, p. 179). It also ensures that Third World countries are marked by an extraordinary dualism of form, with pre-capitalist underdevelopment and capitalist development existing side by side in supposed symbiosis (Wolpe 1980, Sofer 1988; see also Gibson & Horvath 1983).

(b) A second strand of Marxist theory claims to take its lead from Marx and to stand in stark opposition to the later heresies of Lenin and the neo-Marxists. According to Warren (1973, 1980), the emergence of dependency theories in the 1960s must be traced back to a betrayal of classical Marxism first breached by Lenin in his pamphlet on imperialism (Lenin 1970). In 1928, says Warren, the 'traditional Marxist view of imperialism as progressive . . . was sacrificed to the requirements of bourgeois anti-imperialist propaganda and, indirectly, to what were thought to be the security requirements of the encircled Soviet state' (Warren 1980, p. 8). The Comintern now endorsed two theses that were only implicit in Lenin's original pamphlet: first, that imperialism had retarded the industrialization of the colonies; and, second, that as a consequence, the Soviet Union and the industrial bourgeoisies of the colonies were natural allies in the fight against imperialism. Warren rejects both claims. He reaffirms that, for Marx, capitalism, and indeed imperialism, is always progressive and is everywhere associated with an increase in democracy, individual freedom, scientific rationality, and undreamed of technological advance. As Marx himself puts it: 'the bourgeoisie cannot exist without constantly revolutionising the instruments of production, and thereby the relations of production, and with them the whole of society' (Marx & Engels 1967, p. 83). Nor does this stop at the borders of Europe and North America. Says Warren: 'Since Marx and Engels considered the role of capitalism in pre-capitalist societies progressive, it was entirely logical that they should have welcomed the extension of capitalism to non-European societies' (Warren 1980, p. 39).

Having thus reclaimed Marx, Warren concludes by demonstrating the vitality of capitalist development in the Third World. Warren acknowledges that the pace of development before 1945 was not especially fast. At this time the full flowering of peripheral capitalism was hindered by the politics of colonialism and imperial preference. Since 1945, however, Warren sees only progress in the Third World. He dismisses the view that GNPs have not grown rapidly or that income inequalities have widened significantly. He also rejects the claim that marginalization is endemic in the Third World and that people there have not gained in health, nutrition, and education. Most vigorously, Warren disputes the claim that metropolitan capitalism has acted to prevent the industrialization of the Third World. Looking in turn at statistics on national average rates of growth in manufacturing industry and on the percentages of national GDP earned by industry, Warren declares that 'the underdeveloped world as a whole has made considerable progress during the post-war period' (Warren 1980, p. 241). For Warren this is as it must be. His reading of Marx leads him to expect, and to welcome, the capitalist development of the Third World, if only as a precursor to global revolution and the transition to socialism.

(c) A third strand of Marxism stands firm against the conclusions of Warren. Like Warren (and *contra* Frank), a group of scholars concerned to theorize the internationalization of capital, or a new international division of labour (NIDL), accepts that capitalism is driven to expand into the periphery and that it is there promoting a selective industrialization. McMichael *et al.* (1978) are not alone, however, in dismissing Warren's claims that such industrialization is developmental, or even widespread (Slater 1987, Browett 1986, Peet 1987). They argue that Third World industrialization is still small in volumetric terms and that it consists, very often, 'in the simple elaboration of raw materials [or] the assembly of parts' (McMichael *et al.* 1978, pp. 110–11).

Later authors have developed this complaint. According to Frobel *et al.* p. 80), the recent industrialization of South-East Asia, and parts of Latin America, is an 'institutional innovation of capitalism itself'; the NIDL is not being established in response to the changing needs or strategies of Third World countries (for a critique, see Beenstock 1984, Harris 1986). Further, the emergence of a NIDL does not alter the fundamental structures of inequality which exist between core and periphery. The NIDL is said to be based on the exploitation of cheap (mainly female) labour (Ross 1983, Hamilton 1987); to be guaranteed by repressive Third World regimes (Lamb 1981; see also Sheahan 1980, Tokman 1985); to be directed by foreign transnational corporations (Landsberg 1979; but see

Dicken 1986); and to produce local enclave economies unconnected by positive multipliers to a still dependent periphery (Raj 1984). In short, the internationalization of capital is promoting growth without development; it is turning 'banana republics [into] pyjama republics' (Adam 1975, p. 102).

Marxism and development studies: the impasse

The discourse of neo-Marxism and classical Marxism proved valuable in the reconstruction of development geography in the 1970s and early 1980s. New questions were raised – on the historical reproduction of global structures of exploitation, and inequality, on the spatial organization of peripheral social formations, on the constitution and capacities of the state in developing countries, on the production of nature and environmental crises, on the political economy of trade and foreign aid – and new answers were proferred. But the answers were not to everyone's satisfaction, nor did they all survive the reaction to 'jumbo Marxism' which set in during the 1980s (Thrift 1983, p. 24; see also Duncan & Ley 1982, Benton 1984). In 1985, Booth published an important paper which described an impasse in Marxist development studies; more recently, Corbridge (1989), and Mouzelis (1988a) have each written of a crisis in Marxist (and non-Marxist) development studies.

Although Booth's critique of Marxist development studies departs significantly from the contours of the intra-Marxist debates of the 1970s and early 1980s, his point of entry is conventional enough. Booth confirms that neo-Marxian accounts of development and underdevelopment are flawed in three particular respects (and some versions more than others).

First, to the extent that neo-Marxism is wedded to an unhelpful circulationism, it is unable to distinguish between the stinging embraces of capitalist and non-capitalist social formations at different times and in different places. To read Frank and Wallerstein (in this respect the worst offenders) is to be referred to a game of musical chairs (Wallerstein 1984, p. 9) or the same figurative league table of nation-states. Some, few, countries move from periphery to core, and vice versa, but nothing of significance is changed by this; the game remains the same, the ladder is still in place.

Second, 'dependency theory was the child of its time, in both a passive and an active sense' (Booth 1985, p. 764). Booth complains that the discourse of neo-Marxism has reversed the optimistic logic of modernization theory while still locating, 'the apparently multiplying difficulties of the national development process . . . "outside" rather than inside the national society' (Booth 1985, p. 764). The difference is that the external is read positively by modernization theorists – it is the source of capital and enlightenment – while it becomes, for neo-Marxism, the active agent of underdevelopment.

Finally, the conceptual vocabulary of neo-Marxism is beguiling but leaves much to be desired. The concept of dependence has from the beginning been derided as too general, with critics charging that Canada and New Zealand are each classically dependent countries (by virtue of their position in the international division of labour; see O'Brien 1975). In some respects this is a naïve criticism, for the best authors within the *dependencia* tradition have taken care to specify the complex relations of power, production, and exchange which link the world economy with regional social formations. There is also the danger of assuming that real individuals consistently embrace all of the propositions associated with one model of development/underdevelopment. Such a view is surely encouraged by an essay like this, which at times puts the quest for pedagogic clarity before a proper attention to nuance and to exchanges within and across different paradigms (see Binder 1986). In other respects, the criticism is well-founded. In underdevelopment theory especially, and in some versions of world-systems theory, concepts are advanced which are little better than chaotic conceptions (Sayer 1984; see also Hecht 1986). Wallerstein's tripartite division of the world economy into core, periphery, and semi-periphery is a clear example of improper abstraction (Kearns 1988).

These criticisms enjoy wide support, but for Booth they are only part of the story. Booth accepts that neo-Marxism and classical Marxism differ in important respects. At a theoretical level the classical tradition is committed to a vigorous productionism which contrasts sharply with the exchange-oriented theorizing of neo-Marxism. Politically, too, there are departures. Several critics have suggested that autarchy and a form of Third Worldism are the logical end-points of circulationist thinking (Brenner 1977, Slater 1977). By contrast, 'Those writers who focus on imperialist-induced class structures within Third World countries [perceive] great scope for local struggles [and] for articulating defensive class alliances which may redefine and improve the links of dependency with the world capitalist system and which can construct paths to socialism' (Hoogvelt 1982, p. 172). At the level of metatheory, however, Booth sees mainly similarities in the 'Two Marxisms'. More exactly, he sees common failings. For Booth, the essential premiss of Marxism is its commitment to defining capitalism in terms of a set of necessary laws of motion which together work to produce a fixed set of spatial outcomes.

At first glance, this commitment would seem to be clearest in the models of neo-Marxism. Wallerstein signals the functionalism of this tradition when he declares that, 'free labour is the form of labour used for skilled work in core countries, whereas coerced labour is used for less skilled work in peripheral areas. The combination thereof is the essence of capitalism. When labour is everywhere free, we shall have socialism' (Wallerstein 1974, p. 127). It would be hard to find a more concise statement of the belief that class systems, or modes of labour control, are but the secondary results of the functioning of a world system. (It also suggests an eccentric definition of socialism.) But a similar commitment to 'system teleology' or to 'generic functionalism' (Booth 1985, p. 775) may be evident in classical Marxism. Once again, the defining features of a Third World social formation must be understood in terms of the needs or logic of metropolitan capital. We are faced here, says Booth, with an explanation devoid of real change. We must suppose instead a degree of 'teleological compatability' (Mouzelis 1978, p. 51), wherein the existence and preservation of peripheral pre-capitalist modes of production (PCMPs) (or not, as the case may be) is read off from the needs of metropolitan capital. If the PCMP survives (as in the bantustans) then that is evidence of its functionality for capitalism. If it does not (as in the plantations of Latin America), then that too is evidence of capitalism's functional requirements. In each case the possibility that a particular peripheral social formation might be the result of an unhappy compromise between two modes of production is swept away beneath the structural causality of capitalism's laws of motion.

These observations cut to the core of Marxist development studies. Booth does not suggest that the concept of a mode of production has been misused or used without sufficient care. 'What has been established instead is that the concept of mode of production is subject to multiple and in practice contradictory theoretical requirements which make it incapable of consistent application to the task of illuminating world development since the sixteenth century' (Booth 1985, p. 768; for elaboration, see Hindess & Hirst 1977). Booth also complains that Marxist development studies are beset by an unhelpful economism and by a tendency to methodological fiat.

There is merit in both these claims. The blatant economism of Frank and Warren has been remarked upon with regularity (Forbes 1984, Slater 1987). But economism is evident too in the modes of production literature. This is a conclusion which Mouzelis tries hard to resist. For Mouzelis: 'Althusser's insistence on the relative autonomy of the political and ideological instances . . . warns the student away from a mere reduction of political and cultural structures to the economic base' (Mouzelis 1980, p. 168; cf. Mouzelis 1988a, pp. 36–7). Others will dismiss this as special pleading. What the Althusserians have done is to disguise their economism. With the notion of relative autonomy the economy is subject to the feedback of other instances in the social formation, but only to a degree. Closer inspection reveals that the economy is determinant in the last instance, thus giving the lie to the claim that we have escaped the clutches of teleology. Only rarely do the Althusserians admit the possibility that the (economic and non-economic) conditions of existence of capitalism's relations of production will not be secured in the face of hostile political, cultural, or environmental action. In their curiously changeless theories capitalism is endowed with an endless and ageless capacity to secure its own perpetuation.

Booth is just as blunt on the epistemological shortcomings of Marxist development theory. He argues that, 'left-tending social scientists and activists have seen fit to close their minds to pertinent mainstream literature' (Booth 1985, p. 766; see also Corbridge 1986, Gould 1988). Booth attributes this blind spot to three moments of the radical discourse. There is, first, a penchant for theoretical arbitrariness. Booth illustrates this point with reference to concepts of super-exploitation and unequal exchange. Consider the claim that the goods and services of the small-scale (or pre-capitalist) sector, 'sell for less than their value . . . because the wages imputed to the members of the enterprise or the household are below what they would earn producing the same output under capitalist conditions' (Booth 1985, p. 771). Says Booth: 'The appeal to an abstract standard of equivalence representing a more "advanced" pattern is unwarranted' (1985, p. 771). The rôle of capitalism in the Third World is here opposed at a philological level. Patterns of development in the Third World are measured against an assumed and unrealizable equivalent in the already developed First World and then found wanting.

Such discursive tactics feed through to the strategy of 'bluff', the second moment of the radical discourse to which Booth objects. Bluff involves an attempt to mystify a process or an empirical observation by cloaking it in an unnecessary pseudo-science (Booth 1985, p. 771). In the modes of production literature, bluff is evident in accounts of the reproduction of informal sector enterprises. For Booth, it is enough to note that 'enterprises of different types move in and out of particular activities in line with expectations of profit and risk given the prevailing scale economies and so on. The possibility of superexploitation . . . does not enter into any of these decisions, and for the purpose of explaining actual behaviour we have no need of so complicated a hypothesis' (1985, p. 772; see also Bromley & Birbeck 1988)

Finally, there is the attendant rationalism of much modern Marxism. Booth shows how a resort to epistemological fiat has helped guard some Marxists from unwelcome empirical evidence, for example on the pace of peripheral industrialization or on state policies in the Third World. He also illustrates how a rationalist epistemology can inform an absentionist politics, or a politics which offers the Third World a crude choice between socialism and barbarism. Says Booth: 'In different but equivalent ways, both structural-functionalist theory and Marxism reify social institutions of a given type, placing them by metatheoretical fiat further beyond human control than they can be shown to be.' In each case the result 'is socially and politically corrupting' (Booth 1985, p. 775).

Marxism and post-Marxism

Booth's paper in *World development* has excited a spirited debate on the crisis in radical development studies and on possible departures from Marxism (see Vandergeest & Buttel 1988, Sklair 1988, Mouzelis 1988a, Corbridge 1989). It has served this purpose mainly because of its intellectual qualities, but also because it speaks to shared concerns. By the mid-1980s many students of the development process had succumbed to doubts about the rigour and practical relevance of neo-Marxism and classical Marxism. Against a background of reforms in China (and now in the Soviet Union), and with clear evidence of the rise of the NICs, new questions came to be asked of a discourse which seemed to oppose a developed capitalist core to an underdeveloped or misdeveloped capitalist/pre-capitalist periphery. The urgency of this task has been heightened by the counter-revolution in development economics and by the ascendency of the new right in the key offices of the World Bank and the IMF.

This does not mean that Booth's paper is without flaws. Critics might object that Booth lumps together some disparate voices within the Marxist camp, and that he marginalizes those who strike a less strident tone. (There is no reference to the tradition of Marxism as moral economy which informs the sophisticated political and cultural analyses of Scott (1985) or Taussig (1980).) Booth's paper might also be condemned for failing to find its own voice on the development process.

Be this as it may, the questions asked of Marxism by Booth and others (Bardhan 1986, Chakravarty 1987) cannot easily be gainsaid. The emerging critique of Marxist development studies suggests that a new *post-Marxism* may be required to illuminate the complex circuits of interdependency and political and cultural mobilization which mark the *fin de siècle*. I have argued elsewhere for an account of development/underdevelopment which is sensitive (a) to the constant yet shifting production of space under the rule of capital; (b) to the changing sites and temporalities of capital accumulation and crisis formation in the world economy; and (c) to the fragile (economic and non-economic) conditions of existence of national and international regimes of accumulation. Such an account would split open the determinism of those theories seeking to read off particular empirical developments from the 'logic of capitalism', and it would eschew forms of reasoning which conceive of capitalism as a totality with functional requirements and/or necessary laws of motion (Corbridge 1988a, pp. 64–5). Put another way, it seems likely that a new chorus in development studies – new voices rather than new models – will make itself heard as scholars address the following, related debates: (a) on concepts of causality, determination, and the conditions of existence of social and economic formations; (b) on the economy, its temporalities and spatial configurations; (c) on agency, power, and politics; and (d) on socialism and the proper rôle of normative discourse.

The rest of this chapter offers a brief review of each of these debates. The review is far from exhaustive, is often personal, and at times may call to mind a unitary post-Marxism when none such exists. I will come back to this in the conclusion. By way of preview, it will suffice to make three points. First, the metatheoretical concerns discussed here are linked intimately with the major substantive debates in development studies. Second, the forging of a new development studies is connected to a greater concern for the spatiality of social life (Soja 1980). Development studies has for too long operated with accounts of social and economic action which assign only a formal role to time and space (as for example in Wallerstein's appealing but ultimately anaemic metaphors of the core–state cycle and core, periphery, and semi-periphery). Third, progress is being made. Marxism is responding to its critics and a new and exciting synthesis of Marxian and non-Marxian social theory is beginning to appear.

Causality, determination, and conditions of existence

Questions of causality and determination are never far from the surface in social science. In the case of development studies the debate is more than usually complicated, with the main players (including Hindess and Hirst, Cardoso, Laclau and Mouffe, Geras, Booth) making their mark at different times, and on different stages, over the course of a dozen years. Nevertheless, the student who feels uncomfortable with this intellectual genealogy can still follow the debate. At stake is the way we write development studies.

The debate runs something as follows. Marxism has been painted as an evolutionary discourse which is steeped in 19th-century concepts and which is tied to various teleological models of the economy and society. Critics charge that Marxism, in its most vulgar forms, has advanced an untenable and mechanistic stages-model of global history which promises communism as the crowning glory of human endeavour. Such a view might conceivably be associated with Stalin and the Second International. More sophisticated Marxisms seek to disguise their intentions in the language of overdetermination and relative autonomy. We have seen, however, that such manoeuvres find no favour with Booth who would return us to the phrase 'determinant in the last instance'. In Marxist development studies these sins are said to be compounded by the assumption that the logic of capital accumulation must play itself out in a set of fixed spatial systems: core and periphery, north and south. The result, critics urge, is a penetrating but too formal discourse, which fails to identify those agents who must pander to systemic needs, which is uncomfortable with signs of rapid development in the Third World, and which is supportive of political programmes which are rarely feasible.

These charges have not gone unanswered. There is in progress a vigorous defence both of classical Marxism and of Althusser (Callinicos 1982, Elliott 1986, 1987, Harvey 1987). Its purpose is to affirm the truths of historical materialism even as it decries the virtues of an empirical turn and the prospect of a new 'true' socialism (Smith 1987, Wood 1986).

There are three moments to this defence. First, it is pointed out that Marxism is a good deal more textured and open-ended than its critics allow (Geras 1987). Second, there is a rehabilitation of what Watts has called 'the theoretical heart of Marxism': the labour theory of value, class analysis, the concept of relative autonomy and so on (Watts 1988b). Third, in the case of development studies, it is suggested that the main critics of a Marxist approach (Laclau & Mouffe (1985) explicitly, Booth implicitly) are indebted to the work of Hindess and Hirst and have followed these authors down a path of intellectual relativism and political nihilism. Much like the postmodernists, they are said to have embraced the 'absolutisation of language' (Anderson 1983). This is very much the view of Wood (Wood 1986) and it is the position taken by Geras in his important articles on post-Marxism (Geras 1987, 1988). Beneath the personal abuse the charge is that Hindess and Hirst have surrendered to a vacuous formalism. Having once removed Marxism from concepts of determination and relative autonomy, these 'post-Marxists' are left floundering in a world of absolute contingency and theoretical sophistry. The world of the post-Marxists is a world without a vantage point. Theirs is a social science without rhyme or reason, a history without cause or narrative.

It is hard to exaggerate what is at stake in these debates. Readers are encouraged to consult the works just cited and to make up their own minds. My own view is that the debate has become needlessly polarized. The main target of post-Marxism is Marxism as a closed and mechanistic discourse. Where Marxism exhibits these traits it is a legitimate target (*pace* Booth, Laclau, and Mouffe), where it does not (*pace* Geras) it must be called to account in other terms. Similarly, the main purpose of Hindess and Hirst is not to dismiss all notions of causality, nor do they invoke a world without vantage point. Their target is a general concept of causality which is guaranteed by epistemological protocols (Cutler *et al.* 1977, p. 128). Their quarrel is not with an empirical proposition such that the economy tends to be determining under the rule of capital; their objection is to concepts of the necessary primacy of the economy at the level of discourse. Their target is a world with one, privileged, vantage point.

This reading of Hindess and Hirst suggests a bridge to the accounts of structuration theory and theoretical realism. In each case attention is drawn to the necessary and contingent conditions of social structures and to the recursive dimensions of everyday life. We are also pointed forward. The work of Hindess and Hirst is proving attractive to post-Marxism because it offers a social scientific discourse which is wedded to causality and determination even as it opposed teleology and determinism.

In place of a mode of production as totality, Hindess and Hirst direct attention to the relations of production constitutive of a social formation and to their various and diverse conditions of existence. Hindess and Hirst accept that the reproduction of capitalist relations of production must presuppose the existence of private property rights and free wage labour and that these institutions in turn depend upon particular forms of labour discipline, accounting mechanism, legal practice, and so on. These are the 'definite conditions of existence' which speak to the ontological realism of Hindess and Hirst (Cutler *et al.* 1977, p. 172). At the same time, Hindess and Hirst deny that such conditions of existence must be produced 'for capitalism', or are produced in forms which are determined by the relations of production.

An example may help here. We can agree that the reproduction of capitalist relations of production depends in part upon the maintenance of a healthy labourforce. But this does not mean that the state will or must act to sponsor a system of health care, nor does it tell us whether the delivery of health care will be through private insurance schemes or through a system of socialized medicine, or through both.

The example of health care delivery has been carefully chosen. It should put to rest the suggestion that Hindess and Hirst are embarked on a course of political nihilism. The defence of the National Health Service in the United Kingdom is a major political issue but it resists the language of reformism versus revolution, socialism all or nothing. The existence of a socialist system of health care delivery within a capitalist social formation only underlines the difficulties of such a discourse. We see also why a post-Marxian concept of causality is of value to development geography. A focus upon the varied conditions of existence of capitalist and non-capitalist relations of production directs us to the recursive nature of social and spatial life and to the institutions through which the grand structures of the nation–state and world economy are reproduced.

Already there are signs of a third wave of radical geography which will avoid both a formless relativism and a formalistic structuralism. The promise of the new regional geography is to hold together the complex interplay of the local and the superlocal and to take seriously the constitutive rôle of class, of gender, of ethnicity, and of culture in the production of place (Forbes 1984). This is not a Third Worldist agenda, nor is it an invocation of new models in development geography, (cursed, as these have been, by a monocausal logic and/or by the ordinal ranking of factors).[3] The prospect, rather, is of new theory formation and of new causal narratives. It is an invitation to economic and social theory which recognizes the specificity of particular developing countries, but which resists an intellectual division of labour which seeks to cut off the Third World from its changing global context (see Pletsch 1981). Recent work by Watts (1983), by Carney (1988), and by Cristopherson (1983) only hints at what is to come.

The economy

The debate on causality and determination is linked to a debate on the economy *sensus strictus*. Marxism has been charged not just with economism, but with reducing the several circuits of the economy to a narrow productionism based on the logic of capital accumulation. Marxists respond to this charge in several ways. Some will deny that the economy can be described *sensus strictus*. The economy, for Marx, is said to be linked relationally to other instances of the social formation; in Althusserian parlance, it is overdetermined. Others have embarked on an ambitious reworking of *Das Kapital*. In geography, Harvey has provided an extraordinary account of *The limits to capital* which, if apocalyptic in its conclusions, is richly suggestive in its theory of crisis formation and displacement under the rule of capital (Harvey 1982; see also Smith 1984). Harvey's comments on fictitious capital and on the devalorization of capital through internationally transmitted inflation are of great value to development geographers struggling to understand the debt crisis and the crises of global Fordism (Corbridge 1988d). Still others have sought to combine the insights of Marxism with those of mainstream micro- and macro-economics, as in the work of analytical Marxism and the Regulation School respectively.

The work of the Regulation School may offer the clearest example of an economics informed by Marxism which yet avoids an unhappy essentialism and teleology. As one more instance of a prospective post-Marxism, the work of Aglietta and Lipietz and their colleagues at CEPREMAP (Paris), sits comfortably with the stance on causality taken by Hindess and Hirst and their followers. Its value in respect of a new development geography lies in three areas.

The work of the Regulation School is distinguished, first, by the challenge it presents to various 'ideologies of globalism' (Aglietta 1985). Aglietta insists that the world economy is theorized as a system of interacting national regimes of accumulation. This is an important point. Only a fool or a knave would deny that we live today in an interdependent world in which the economic powers of nation–states are being eroded and transferred to international capital. Nevertheless, what Petras & Brill (1985) call the 'tyranny of globalism' can be pressed too far, with the result that the changing constitution and dynamics of the world economy are lost amidst a welter of platitudes about core and periphery. As Lipietz explains:

> Something which 'forms a system' and which we intellectually identify as a system precisely because it is provisionally stable must not . . . be seen as an intentional structure or inevitable destiny because of its coherence. Of course it is relatively coherent: if it were not, we would have international conflict and there would be no more talk of systems. But its coherence is simply the effect of the interaction between several relatively autonomous processes, of the provisionally stabilized complimentarity and antagonism that exists between various national regimes of accumulation (Lipietz 1987, pp. 24–5).

The task is to hold these two levels together; to explore the redefinition of the other which each entails (Corbridge 1988e; cf. Holland 1987).

The Regulation School is marked, second, by its understanding of the process of accumulation and crisis formation under the rule of capital. In place of more orthodox Marxist formulations which stress the continuity of these processes, the Regulation School offers a set of meso-concepts which helps us to see the history of capitalism in terms of a theory of discontinuous equilibria (within which the regime of accumulation and the site of crisis formation changes periodically). A little detail is unavoidable here. In the work of Lipietz and Aglietta we are introduced to the concepts of a *regime of accumulation* and a *mode of regulation*. A regime of accumulation 'describes the fairly long-term stabilization of the allocation of social production between consumption and accumulation . . . [both] within the national economic and social formation under consideration and its "outside world"' (Lipietz 1987, p. 14). A mode of regulation 'describes a set of internalized rules and social procedures' which ensure the unity of a given regime of accumulation and which 'guarantee that its agents conform more or less to the schema of reproduction in their day-to-day behaviour and struggles' (Lipietz 1987, p. 14).

These concepts have been put to work to build up a four-stage model of capitalist development and crisis formation in the 20th century. (Needless to say, much is lost in this simple schema: see Aglietta 1982, Lipietz 1985, Noel 1987.) Until the early 20th century, the dominant regime of accumulation in the advanced capitalist countries was extensive. This regime centred upon the expanded reproduction of means of production and involved both a sharp international division of labour and a relative orientation to external markets. The corresponding mode of regulation was competitive, which means, in part, that national regimes of accumulation had to adjust to one another through international transfers of commodity money (and so by deflation). By the 1920s this combination of extensive accumulation/competitive regulation had entered a period of major crisis. According to Lipietz, the dominant regime of accumulation now shifted to a system of Fordism, centred upon the United States, which sponsored a growth in output beyond that which could be realized under a competitive mode of regulation. Put simply, a system of competitive regulation demands 'the *a posteriori* adjustment of the output of the various branches to price movements . . . and of wages to price movements' Lipietz 1987, p. 34). Accordingly, wages are able to rise only slowly – if at all – and capital falls into a crisis of overaccumulation.

After World War II a regime of Fordist, or intensive accumulation, came to be matched by a monopolistic mode of regulation. Within nation–states this mode of regulation, 'incorporated both productivity rises and a corresponding rise in popular consumption into the determination of wages and nominal profits *a priori*' (Lipietz 1987, p. 35). Internationally, a system of regulation emerged which acknowledged the United States as the new hegemon and which installed the dollar as the accepted international unit of account. This system proved stable so long as the US had a trade surplus with Europe and Japan, and so long as Europe and Japan had funds to buy American producer goods (Triffin 1960). Since the mid-1960s this equation has become less assured and we are living now through a second major crisis in 20th century-capitalism. The difference this time is that demand is holding up well – thanks to the international credit economy (Strange 1986) – but profits have fallen amidst generalized inflation and/or stagnation. The Third World, having in 60 years moved from colony to periphery to (in some instances) future core, now finds that a tentative regime of global Fordism is being curtailed by a US-inspired debt crisis (de Vroey, 1984).

The model of capital accumulation and crisis formation proposed by the Regulation School is enhanced, finally, by the philosophical stance of this School. Lipietz, especially, is scathing in his critique of theoretical 'finalism and functionalism'. With regard to regimes of accumulation and modes of regulation, he insists that:

> Whilst no immanent destiny condemns a particular nation to a particular place within the international division of labour, a provisional solution for the immanent contradictions of capitalism can at times be found (and I insist that it is a matter of chance discovery) in deviations and differences between regimes of accumulation in different national social formations. In such periods, a *field* of possible positions . . . does exist, but positions within it are not allocated in advance (Lipietz 1987, p. 24; emphasis in the original).

Put bluntly, the emergence of Fordism, and its extension as global Fordism, must not be seen as preordained solutions to capitalist crisis, neatly identified and invented by a controlling class of capitalists. They are rather – and as Hindess and Hirst might say – one of many experiments thrown up by capitalism, which survive only as successful mutants on probation.

Agency, power, and politics

A major strength of the Regulation School approach to political economy is that it directs attention to the political conditions of existence of a regime of accumulation. A mode of regulation is defined in a non-reductionist fashion with proper regard being paid to the varied moments of its constitution.

This raises an important issue. One of the most telling criticisms of Marxist development studies is that it so privileges structure over agency that questions of politics and power are reduced to simple analyses of the relative autonomy of the state or to representations of some deeper class struggle. In their review of *Marx, Weber and development sociology*, Vandergeest and Buttel suggest that: 'Marxist-influenced development sociology has been especially weak in building an understanding of the Third World state and its constitutive organizations – a lack which is particularly disturbing when we consider the overarching role played by the state in most of the post-colonial Third World' (Vandergeest & Buttel 1988, p. 689). Mouzelis makes a similar point. He argues that 'Marxism, having failed to elaborate specific conceptual tools for the study of politics, *builds the alleged primacy of the economic into the definition of the political*. In that sense it is unable to study the complex and *varying* relationships between economy and polity in a theoretically coherent and at the same time *empirically open-ended* manner' (Mouzelis 1988a, p. 37; emphasis in the original).

Such criticisms apply more to Leninism than to traditions of Marxism associated with Gramsci and the English historians, but that is not the central issue. It is clear that Marxism has enjoyed only limited success in its analyses of nationalism, of ethnoregionalism, and of the rise of Islam – three of the most pressing political issues in developing countries (see Anderson (1983) for a thoughtful review). The reasons for this are in dispute, but they relate to the limitations of an 'analysis of power as derived through economic advantage' (Vandergeest & Buttel 1988, p. 690). Giddens rightly insists that at 'the heart of both domination and power lies the transformative capacity of human action, the origin of all that is liberating and productive in social life as well as all that is repressive and destructive' (Giddens 1981, pp. 50–1; see also Mann 1986). A wider analysis of power would focus upon the capacity to direct, to oppress, to separate, to represent, to resist, to produce, to contest, and to destroy. In so doing it would direct attention not just to class and to economic position, but to the military, to the bureaucracy, to gender, to ethnicity, to culture, and to the capacity to script and create discourse.

The possibility of a post-Marxist account of power and politics is only just being thought through and it is faced by many problems. In the present context we can only list those intellectual currents which seem to suggest themselves as new models for development studies. They are four-fold:

(a) From Vandergeest and Buttel' we are offered a tradition of neo-Weberianism. This tradition aims to reclaim Weber from the Parsonians and to fasten, instead, on to Weber's discussion of bureaucratization, rationalization, and the state's monopoly of the means of violence. The trick is to detach Weber from his reliance upon ideal types and to return him to the comparative analysis of cultural and political institutions. The task also is to see that 'the relatively powerless always have some resources, or some strategic location from which they can influence or actively shape social processes. The task of a new sociology of development is to study relations of mutual dependence and access to resources as culturally defined, so as to know what is possible in a given situation' (Vandergeest & Buttel 1988, p. 690).

Vandergeest and Buttel deny that work within the neo-Weberian tradition must be without an agenda, or devoid of political praxis. Citing the work of Gaventa and the Highlander Center (Gaventa 1980), they argue that: 'When empirical study deals with power relations in the contexts of class, the state, cultural interpretation, and so on, the work quickly leads to strategies for the empowerment of the less powerful – strategies which emerge from the case itself, not from the dictates of teleological theory' (Gaventa 1980, p. 690). Vandergeest and Buttel are not slow, either, to claim for neo-Weberianism, a number of eminent scholars: Offe, Polanyi, Giddens, Bourdieu, Willis, and Tilly are all called to the cause.

(b) From Becker and Sklar we are offered a treatise on post-imperialism. Becker and Sklar define post-imperialism as 'an idea about the political and social organisation of international capitalism . . . [which] grew out of two bodies of thought: political theories of the modern business corporation and class analyses of political power in the "Third World"' (Becker & Sklar 1987a, p. ix).

Post-imperialism is associated with three or four main ideas. First, the rise of global corporations is serving 'to promote the integration of diverse national interests on a new transnational basis' (Becker & Sklar 1987b, p. 6) In so far as transnational corporations offer the Third World 'access to capital resources, dependable markets, essential technologies and other services', it follows that there is 'a mutuality of interest between politically autonomous countries at different stages of economic development' (Becker & Sklar 1987b, p. 6). This is Becker and Sklar's contribution to the debate on complex interdependency. It is a suggestive response both to Third Worldism and to Leninist theories of imperialism.

Second, the 'spread of industrialisation to all regions of the world' (Becker & Sklar 1987a, p. ix) has brought into being a managerial bourgeoisie. The managerial bourgeoisie defines 'a socially comprehensive category encompassing the entrepreneurial elite, managers of firms, senior state functionaries, leading politicians, members of the learned professions, and persons of similar standing in all spheres of society' (Becker & Sklar 1987b, p. 7). It is a class because it defends a position both against the proletariat and against the '"oligarchic" landed–financial–commercial dominant classes of yore' (Becker & Sklar 1987b, p. 7). Becker and Sklar argue that the rise to power of a managerial bourgeoisie is breaking forever the cast of imperialism. It ushers in a more nuanced politics.

Third, Sklar (1975, 1976) has put forward a concept of 'The doctrine of domicile'. The doctrine of domicile suggests that 'transnational business groups should and do undertake to adapt, to operate in accordance with the policies of states in which the subsidiaries are domiciled' (Sklar 1976, p. 9). Where this holds true, it follows that the managerial bourgeoisie can stand together despite its constitution as an alliance of privileged host-country capital and corporate foreign capital. It performs this manoeuvre in and through the politics of populism.

Finally, the doctrine of domicile directs attention to the variability of state/capital relations in the periphery. As such it places a particular emphasis upon the politics of the bargaining process. Although Becker and Sklar acknowledge the 'irreducible conflict of class interest between bourgeoisie and proletariat' (Becker & Sklar 1987b, p. 13), they doubt whether the living standards of the poor in the Third World will be improved by aggressive state action against foreign capital. On balance the system must be worked within. (For a critique of post-imperialism, see Frieden (1987).)

(c) From Mouzelis comes a plea for a 'non-reductionist Marxist theory of the polity' (Mouzelis 1988a, p. 40). Such a theory would create 'new conceptual tools which: (1) try to deal with the non-economic institutional spheres in a way that does not build into their very definition the type of relationship they are supposed to have with the economy; and (2) try to avoid economism without falling into the compartmentalisation of the political and economic spheres to be found in neo-classical economics and non-Marxist political science' (Mouzelis 1988a, p. 40).

More positively, Mouzelis points us back toward comparative history and sociology (Barington Moore is mentioned) and offers us the concept of a 'mode of domination'. A mode of domination consists 'of an articulation of specific political technologies (forces of domination) and specific ways of approaching such technologies (relations of domination) . . . [which] could, if theoretically developed, provide the conceptual means for studying the complex linkages between the economy and the polity in a logically coherent and empirically open-ended manner' (Mouzelis 1988b, p. 121).

(d) From critical theory we are pointed to discourse itself and to the enabling/oppressing powers of language and representation. Chatterjee talks of the 'cunning of reason' and demands that we recoil from those Enlightenment concepts which the West has thrust upon the world: concepts of development, of rationality, of modernization, of empiricism (Chatterjee 1986). Attention must be drawn instead to those linguistic tropes – allegories, metaphors, etc. – by which the other is made real and represented.

A classic example of such work is Said on orientalism (Said 1979). Said shows in detail how European culture produced and reproduced a concept of orientalism which served to mark down as exotic, inferior, and punishable an entire continent running from Egypt to Japan. Said contends that 'without examining Orientalism as a discourse one cannot possibly understand the enormously systematic discipline by which European culture was able to manage – and even produce – the Orient politically, sociologically, militarily, ideologically, scientifically and imaginatively during the post-Enlightenment period. . . . In brief, because of Orientalism the Orient was not (and is not) a free subject of thought and action' (Said 1979, p. 3).

The power of representation, the power to script, is part of the power to exploit and govern. Moreover, the power to script is not confined to words alone. It resides in a diverse series of texts, including the landscape. Work by Cosgrove & Daniels (1988), by Duncan & Duncan (1988), and by Duncan (1989) examines how an iconography of built form helps reproduce a dominant (if still contested) social and spatial order (see also Cuthbert 1987; on the discourses of geopolitics, see Agnew & O'Tuathail 1987, Agnew & Corbridge 1989).

It bears repeating that these four accounts of power and politics are not without difficulties of their own, nor are they mutually exclusive, nor can they be clipped together in some cosy post-Marxian synthesis. They are presented here as four moments of a critique of Marxism on power and politics which retains an affinity with Marxism. In this respect they are at odds with a libertarian account of power and politics now emerging in the counter-revolution in development theory and policy.

Socialism and normative discourse

Marxism's problematic perspective on power and politics is matched by problems in its own political agenda. An important subtext in Marxist development studies is the suggestion that *as* capitalism creates underdevelopment, *so* socialism will promise a better future. This claim is at best a double-edged sword.

It is clear that much of the strength of Marxism derives from its plausible critique of capitalism and (rightly) from its normative image of a better socialist future. The two elements are inseparable, a point underscored by the emphasis upon praxis in Marxist thought. It is because we can imagine a society in which food is distributed according to need that malnutrition and famine are so unacceptable. It is because we can hope to construct a society in which racial and gender distinctions are not asymmetrical that racism and sexism are so objectionable. It is because we see that capitalism is unstable, that it puts profits before people, and that it is associated with pollution and poverty, that we look kindly on a socialist alternative.

Nevertheless, to the extent that this future is simply assumed, and is assumed to be unproblematic, so the pretensions of a 'scientific socialism' are exposed. Far too often, Marxism has opposed capitalism-in-general (as bad) to socialism-in-general (as good). The tendency then is to legitimize a political discourse which opposes reform to revolution. It also leads to the strange spectacle of a Marxist development studies which apes modernization theory. In each case, there is a tendency to measure what goes on in the Third World against an idealized future state – capitalist nirvana/socialist utopia – and then to find it wanting. This tendency is most clearly displayed in neo-Marxist accounts of the NICs and of the new international division of labour. The suggestion that banana republics are becoming pyjama republics is typical of a pejorative socialist discourse which blinds itself to all signs of progress in the periphery (Bernstein 1982).

There are signs that a post-Marxist development studies will break with the agenda. Post-Marxism resists a static and totalizing vision of capitalism and/versus socialism. It claims a sensitivity to the diverse and fragile conditions of existence of a given regime of accumulation. As such, it is not likely to collapse the development experiences of a South Korea and an Argentine into one essential category of semi-periphery or 'pyjama republic'.

Post-Marxism is also marked by a methodological and political scepticism. It is not opposed to socialism and to socialist politics, but it is concerned to theorize each enterprise and to concern itself with the contours and contradictions of actually existing socialism. Giddens is right to insist that: 'Neither socialism generally, nor Marxism in particular, walks innocently in the world' (Giddens 1981, p. 249). He is also right to suppose that 'the principal contradiction of socialist societies . . . is between the planned organisation of production, mediated through the state, and the mass participation of the population in decisions and policies that affect the course of their lives' (Giddens 1981, p. 248). Evidence from China suggests both the virtues of a socialist food policy (Croll 1983) and the difficulties that a socialist country must face in promoting patterns of social and spatial equity (Paine 1981; see also Lai 1985, Massey 1987). The Chinese experience should make us wary of those socialist utopias which, on the basis of a purely relative conception of scarcity and assuming no opportunity costs, call to mind a new socialist being who is everything that men and women under capitalism are not: a caring, sharing, non-sexist polymath (Nove 1983).

Finally, there is within post-Marxism an insistent examination both of Marxist theories of exploitation and of the links between socialist politics and a theory of justice. This is not the place for a detailed discussion of the work of John Roemer and the analytical Marxists. (The reader is referred to Roemer 1986, 1988, and to Elster 1985; see also Carling 1986.) Suffice to say that Roemer has claimed to demonstrate (a) that the labour theory of value is logically untenable; (b) that exploitation of labour is only one moment in a chain of exploitative commodity relations and need not serve as an exploitation numeraire; (c) that the fact of exploitation is not itself sufficient to justify an ameliorative politics – one needs to show that exploitation in the technical, Marxian, sense is connected to a morally unjust distribution of skills and resources within a population (as opposed to a disposition to work more or less hard); and (d) that a just response to exploitation may include a traditional socialist politics, but will also attend to a wide range of rights, needs, and abuses of power. The fact that Roemer cites the theories of justice associated with Rawls and Dworkin shows how far removed from Leninism a post-Marxian politics might be (Rawls 1971, Dworkin 1986). It is also suggests a philosophical grounding for those works in development studies which are concerned with the provision of civil rights, basic needs, and welfare in the Third World (see Ward 1986, Bell 1988).

Conclusion: post-Marxism and development studies

This chapter has tried to provide an intellectual history of development geography which reaches back to the late 1960s and Keeble's chapter in *Models in geography*. Over the course of 20 years, development geography has been written around a series of models which trace their lines of descent to the neo-Marxism of the *Monthly Review* School and to various traditions within classical Marxism. (Many more geographies have been written without explicit reference to these models, but have internalized some of their claims none the less.) The signs are that development geography will now steer clear of the counter-revolution in development studies and will instead embrace a diverse post-Marxism. We can close this chapter by offering a brief definition of post-Marxism and by suggesting some lines of research for a post-Marxist development geography.

First, a word of caution. A definition of post-Marxism is offered here to point the way forward in development studies and to break with some aspects of the Marxian impasse. It is not my intention to label or falsely to divide. Many Marxists will feel comfortable with some parts of the post-Marxist agenda, as will many non-Marxists. Nor do I suggest that post-Marxism is in some simple way better than Marxism or that it has taken its place in history; the suggestion is that post-Marxism is indebted to Marxism and yet critical of its organizing concepts. These caveats entered, and being mindful that others have evaded the task of definition (see Geras 1987, 1988, Mouzelis 1988b), let us define post-Marxism in the following terms.

There are, first, the links to Marxism:

(a) Post-Marxism shares with Marx a materialist ontology and a commitment to causal analysis and a concept of determination.
(b) Post-Marxist accounts of the economy emphasize inequalities in the distribution of assets and power, and pay attention to contradictions in the process of accumulation.
(c) Post-Marxism accepts that people make history, but not under circumstances of their own chosing. (Clearly, some voices within post-Marxism stress this point more than others: compare the Regulation School and structuration theory.)

Then there are the departures from some traditions of Marxism.

(a) Post-Marxism is 'sympathetic to the idea that what is distinctive in Marxian theory is substantive, not methodological; and that as a science of society the methodology adopted by Marxists ought to be just good scientific methodology' (Levine *et al.* 1987 p. 68; this quotation does not imply an attachment to post-Marxism on the part of Levine, Sober, and Wright). Post-Marxism is opposed to the exclusivism to be found in some Marxism, and is opposed to a defence of Marxism on the grounds that its concepts are epistemologically privileged and/or are incommensurate with concepts emerging from non-Marxist traditions of social science (Wolff & Resnick 1987). Post-Marxism is committed to the *careful* wedding of concepts from Marxism and non-Marxism.

(b) Post-Marxism is opposed to propositions which speak of the necessary primacy of the economy, or of the economy's capacity to determine in the last instance. This objection is directed to the epistemological protocols which stand behind such a proposition. Post-Marxism does not deny the contingent dominance of the economy as an empirical proposition. It takes a similar attitude to the concept of relative autonomy.

(c) Post-Marxism is sceptical of the labour theory of value. It is likely to attach itself to a general theory of exploitation and class (cf. Roemer 1988).

(d) Post-Marxism is opposed to functionalist accounts of power, the state, and civil society. It is likely to attach itself to a general theory of power which draws insights from Marxism, from feminism, from discourse theory and from neo-Weberianism. The construction of such a general theory is in its infancy.

(c) Post-Marxism is unsympathetic to those dualisms which oppose reform to revolution and capitalism-in-general. Having regard for the diverse conditions of existence of both capitalism and socialism, post-Marxism is committed to a less certain politics which some will denounce as 'eclectic and discontinuous' (Watts 1988b), but which in fact is tied to concepts of moral justice and feasibility.

And so to development studies. Combining a 'post'-anything with an 'ism' is always open to objection and it is not my purpose to be programmatic. One virtue of post-Marxism is that it steers us away from manifestos and models, and away from the privileged worlds of epistemology and methodological fiat. But this is not an invitation to relativism or to some additive social theory in which factors a, b, c, . . . z are bundled together without regard for logical and empirical inconsistencies. To the contrary, the production of a post-Marxist development geography is distinguished by the demands it makes of its author.

A first task is to understand the constitution and dynamics of a changing world system and to pay close attention to its varied modes of accumulation and crisis formation in time and space. Such work is being done. Harvey's account of *The limits to capital* is a masterful work of Marxist theory which pays close attention to imperialism and inflation as two (false) fixes for the crisis of capital. (See also Smith (1984) on uneven development and Thrift & Leyshon (1988) on a new international financial system; see also the essays in Scott & Storper 1986.)

A second task is to examine the spatiality of the development process; to hold together the complex interplay of the local and the superlocal and to see in the joint production of place a continuing redefinition of the other. Such work is being done: see Armstrong & McGee (1985) on the production of urban geographies as theatres of accumulation; see Storper (1984) on social power and industrial decentralization in Brazil; see Shrestha (1988) on the dynamics of migration in peripheral areas.

A third task is to attend to the varied conditions of existence of capital accumulation; to take seriously the constitutive roles of class, gender, ethnicity, the discourse of geopolitics, and the environment, and to examine the bases of their transformation/reconstitution. Such work too is being done: see Blaikie (1985) on the political economy of soil erosion; see Carney (1988) on contract farming and gender relations; see Hecht (1985) on the environmental crisis in Amazonia.

A fourth task is to describe those patterns of political and cultural mobilization through which societies in the periphery seek to understand and contest 'the onset of modernity'. Such work is also being done: see Crush (1988) on the battle for Swazi labour; see Sutcliffe & Wellings (1985) on the geography of trade union activity in South Africa; see Watts (1988c) on the contested politics of place in The Gambia.

A fifth task is to understand the rôle of the state in the periphery; to examine its constitution and powers, and to investigate its capacity to mobilize and transfer resources. Such work is being done: see Harriss (1984) on the merchant state in South India; see Rakodi (1986) on the local state in Africa; see Watts (1984) on oil-based accumulation in Nigeria.

A sixth task is to investigate the territorial, economic, and political possibilities and contradictions of socialist development strategies. Such work is being done: see Thrift & Forbes (1986) on urbanization in war-torn Vietnam; see the essays in Forbes & Thrift (1987); see Corragio (1985) on territorial re-ordering in Nicaragua.

A seventh task is to connect an analysis of welfare provision in the periphery with theories of justice and spatial equity. Such work is being done: see Ward (1986) and Bell (1988); and see the critique of self-help housing by Burgess (1985).

A final task (sic) is to hold together production and empirical work in creative tension. Again, such work is being done: a new development geography is with us already.

Notes

1 The sections on the development of underdevelopment and on Marxist theories of development reproduce material from Corbridge 1988a.

2 The *dependencia* tradition is more sophisticated than some of its critics allow and it should not be equated with the discourse of underdevelopment theory (see Watts 1988a for a useful commentary; see also Palma 1977). At its best dependency theory offers a fruitful marriage of neo-Marxism and classical Marxism.

3 I am indebted to John Agnew for this phrase. My commentary on post-Marxism has benefited from exchanges with John Agnew, David Booth, Fred Buttel, Nancy Duncan, Nalanie Hennayake, and Michael Watts. To them, my grateful thanks. The usual disclaimers apply.

References

Adam, G. 1975. Multinational corporations and worldwide sourcing. In *International firms and modern imperialism*. H. Radice (ed.). Harmondsworth: Penguin.

Aglietta, M. 1982. World capitalism in the eighties. *New Left Review* **137**, 5–41.

Aglietta, M. 1985. The creation of international liquidity. In *The political economy of international money*, L. Tsoukalis (ed.), 171–202. London: Sage Publications.

Agnew, J. & Corbridge, S. 1989. The new geopolitics: the dynamics of geopolitical disorder. In *A world in crisis: geographical perspectives*, 2nd edn R. Johnson & P. Taylor (eds). Oxford: Basil Blackwell.

Agnew, J. & G. O'Tuathail 1987. Geographical order and domesticated space: towards a critical historiography of American geopolitics. Paper presented at the International Studies Association, Washington, DC (April).

Anderson, B. 1983. *Imagined communities: reflections on the origin and spread of nationalism*. London: Verso.

Anderson, P. 1983. *In the tracks of historical materialism*. London: Verso.

Armstrong, W. & T. McGee 1985., *Theatres of accumulation: studies in Asian and Latin American urbanisation*. London: Methuen.

Bacha, E. 1978. An interpretation of unequal exchange from Prebisch–Singer to Emmanuel. *Journal of Development Economics* **5**, 319–30.

Baran, P. 1973. *The political economy of growth*. Harmondsworth: Penguin.

Bardhan, P. 1986. Marxist ideas in development economics: an evaluation. In *Analytical Marxism*, J. Roemer, (ed.), 64–77. Cambridge: Cambridge University Press.

Bath, C. & D. James 1976. Dependency analysis of Latin America: some criticisms, some suggestions. *Latin American Research Review* **11**, 3–54.

Bayley, C. 1988. *Indian society and the making of the British Empire*. Cambridge: Cambridge University Press.

Becker, D. & R. Sklar 1987a. Preface. In *Postimperialism: international capitalism and development in the late twentieth century*, D. Becker & R. Sklar (eds). Boulder: Rienner.

Becker, D. & R. Sklar 1987b. Why postimperialism?. In *Postimperialism: international capitalism and development in the late twentieth century*, D. Becker & R. Sklar (eds), 1–9. Boulder: Rienner.

Beenstock, M. 1984. *The world economy in transition*. London: Allen & Unwin.

Bell, M. 1988. Welfare, culture and environment. In *The geography of the Third World: progress and prospect*, M. Pacione (ed.), 198–231. London: Routledge.

Benton, T. 1984. *The rise and fall of structural Marxism: Althusser and his influence*. New York: St Martin's Press.

Bernstein, H. 1982. Industrialisation, development and dependence, In *Introduction to the sociology of developing countries*, H. Alavi & T. Shanin (eds), 218–35. London: Macmillan.

Bettelheim, C. 1972. Theoretical comments. In *Unequal exchange*, A. Emmanuel. London: Monthly Review Press.

Binder, L. 1986. The natural history of development theory. *Comparative Studies in Sociology and History* **28**, 3–33.

Blaikie, P. 1985. *The political economy of soil erosion in developing countries*. London: Longman.

Blaut, J. 1973. The theory of development. *Antipode* **5**, 22–6.

Blaut, J. 1976. Where was capitalism born? *Antipode* **8**, 1–11.

Booth, D. 1985. Marxism and development sociology: interpreting the impasse. *World Development* **13**, 761–87.

Brenner, R. 1977. The origins of capitalist development: a critique of neo-Smithian Marxism. *New Left Review* **104**, 25–92.

Brewer, A. 1980. *Marxist theories of imperialism*. London: Routledge & Kegan Paul.

Bromley, R. & C. Birkbeck 1988. Urban economy and employment. In *The geography of the Third World: progress and prospect*, M. Pacione (ed.), 114–47. London: Routledge.

Brookfield, H. 1975. *Interdependent development*. London: Methuen.

Browett, J. 1981. Development, the diffusionist paradigm and geography. *Progress in Human Geography* **4**, 57–79.

Browett, J. 1986. Industrialisation in the global periphery: the significance of the newly industrialising countries. *Environment and Planning D, Society and Space* **4**, 401–18.

Buchanan, K. 1964. Profiles of the Third World. *Pacific Viewpoint* **2**, 97–126.

Buchanan, K. 1970. *The transformation of the Chinese earth*. Edinburgh: Bell.

Burgess, R. 1985. The limits of state self-help housing programmes. *Development and Change* **16**, 271–312.

Callinicos, A. 1982. *Is there a future for Marxism?* Atlantic Highlands: Humanities Press.

Cannon, T. 1975. Geography and underdevelopment. *Area* **7**, 212–16.

Carling, A. 1986. Rational choice Marxism. *New Left Review* **160**, 24–62.

Carney, J. 1988. Struggles over crop rights and labour within contract farming households in a Gambian irrigated rice project. *Journal of Peasant Studies* **15**, 334–49.,

Chakravarty, S. 1987. Marxist economics and contemporary developing economies. *Cambridge Journal of Economics* **11**, 3–22.

Chatterjee, P. 1986. *Nationalist thought and the colonial world: a derivative discourse.* London: Zed Press.

Corbridge, S. 1986. *Capitalist world development: a critique of radical development geography.* London: Macmillan.

Corbridge, S. 1988a. The 'Third World' in global context. In *The geography of the Third World; progress and prospect*, M. Pacione (ed.), 29–76. London: Routledge.

Corbridge, S. 1988b. The ideology of trial economy and society: politics in the Jharkhand, India, 1950–1980. *Modern Asian Studies* **22**, 1–41.

Corbridge, S. 1988c. The debt crisis and the crisis of global regulation. *Geoforum* **19**, 109–30.

Corbridge, S. 1988d. The assymetry of interdependence: the United States and the geopolitics of international financial relations. *Studies in Comparative International Development* **23**, 3–29.

Corbridge, S. 1989. Marxism and development studies: beyond the impase. *World Development.*

Corragio, J. 1985. Possibilities of a territorial ordering for the transition in Nicaragua. *Environment and Planning D, Society and Space* **3**, 191–212.

Cosgrove, D. & S. Daniels 1988. *The iconography of landscape: essays on the symbolic representation, design and use of past environments.* Cambridge: Cambridge University Press.

Christopherson, S. 1983. The household and class formation: determinants of residential location in Ciudad Juarez. *Environment and Planning D, Society and Space* **1**, 323–38.

Croll, E. 1983. *The family rice bowl.* London: Zed Press.

Crush, J. 1988. *The struggle for Swazi labour, 1890–1920.* Montreal: McGill-Queens University Press.

Cuthbert, A. 1987. Hong Kong 1997: the transition to socialism – ideology, discourse, and urban social structure. *Environment and Planning D, Society and Space* **5**, 123–50.

Cutler, A., B. Hindess, P. Hirst & A. Hussain 1977. *Marx's 'Capital' and capitalism today*, Vol. 1. London: Routledge & Kegan Paul.

Dicken, P. 1986. *Global shift: industrial change in a turbulent world.* London: Harper & Row.

Dos Santos, T. 1973. The crisis of development theory and the problem of dependency in Latin America. In *Underdevelopment and development*, H. Bernstein (ed.). Harmondsworth: Penguin.

Duncan, J. 1989. *The city as text: the politics of landscape interpretation in nineteenth century Kandy.* Cambridge: Cambridge University Press.

Duncan, J. & N. Duncan 1988. (Re)reading the landscape. *Environment and Planning D, Society and Space* **6**, 117–26.

Duncan, J. & D. Ley 1982. Structural Marxism and human geography: a critical assessment. *Annals of the Association of American Geographers* **72**, 30–59.

Dworkin; R. 1986. *Law's empire.* Cambridge, Mass.: Harvard University Press.

Elliott, G. 1986. The Odyssey of Paul Hirst. *New Left Review* **159**, 81–105.

Elliott, G. 1987. *Althusser: the detour of theory.* London: Verso.

Elster, J. 1985. *An introduction to Karl Marx.* Cambridge: Cambridge University Press.

Emmanuel, A. 1972. *Unequal exchange.* London: Monthly Review Press.

Evans, P. 1979. *Dependent development: the alliance of multinational, state and local capital in Brazil.* Princeton, NJ: Princeton University Press.

Foot, S. & M. Webber 1983. Unequal exchange and uneven development. *Environment and Planning D, Society and Space* **1**, 281–304.

Forbes, D. 1984. *The geography of underdevelopment.* London: Croom Helm.

Forbes, D. & N. Thrift (eds) 1987. *The socialist Third World: urban development and regional planning*. Oxford: Basil Blackwell.

Frank, A. G. 1969. *Latin America: underdevelopment or revolution?* London: Monthly Review Press.

Frieden, J. 1987. International capital and national development: comments on postimperialism. In *Postimperialism: international capitalism and development in the late twentieth century*, D. Becker & R. Sklar (eds). Boulder: Riennar.

Gaventa, J. 1980. *Power and powerlessness: quiescence and rebellion in an Appalachian valley*. Urbana: University of Illinois Press.

Geras, N. 1987. Post-Marxism? *New Left Review* **163**, 40–82.

Geras, N. 1988. Ex-Marxism without substance: being a real reply to Laclau and Mouffe. *New Left Review* **169**, 34–61.

Gibson, K. & R. Horvath 1983. Aspects of a theory of transition within the capitalist mode of production. *Environment and Planning D, Society and Space* **1**, 121–30.

Giddens, A. 1981. *A contemporary critique of historical materialism*. Vol. 1: *Power, property and the state*. London: Macmillan.

Gould, P. 1988. The only perspective: a critique of Marxist claims to exclusiveness in geographical enquiry. In *A ground for common search*, R. Golledge & P. Gould (eds), 1–10. Goleta, CA: Santa Barbara Geographical Press.

Hajdimichalis, C. 1984. The geographical transfer of value: notes on the spatiality of capitalism. *Environment and Planning D, Society and Space* **2**, 329–45.

Hamilton, C. 1987. Can the rest of Asia emulate the NICs? *Third World Quarterly* **9**, 1225–56.

Harris, N. 1986. *The end of the Third World: newly industrialising countries and the decline of an ideology*. Harmondsworth: Penguin.

Harriss, B. 1984. *State and market*. Delhi: Concept.

Harvey, D. 1975. The geography of capitalist accumulation: a reconstruction of the Marxian theory. *Antipiode* **7**, 9–21.

Harvey, D. 1982. *The limits to capital*. Oxford: Basil Blackwell.

Harvey, D. 1987. Three myths in search of a reality in urban studies. *Environment and Planning D, Society and Space* **5**, 367–76.

Hecht, S. 1985. Environment, development and politics: capital accumulation and the livestock sector in eastern Amazonia. *World Development* **13**, 663–84.

Hecht, S. 1986. Regional development: some comments on the discourse in Latin America. *Environment and Planning D, Society and Space* **4**, 201–9.

Hindess, B. & P. Hirst 1977. *Mode of production and social formation*. London: Macmillan.

Holland, S. 1987. *The global economy: from meso to macroeconomics*. London: Weidenfeld & Nicolson.

Hoogvelt. A. 1982. *The Third World in global development*. London: Macmillan.

Johnston, R. 1984. The world is our oyster. *Transactions of the Institute of British Geographers* **NS9**, 443–59.

Kearns, G. 1988. History, geography and world-systems theory. *Journal of Historical Geography* **14**, 281–92.

Keeble, D. 1967. Models of economic development. In *Models in geography*, R. Chorley & P. Haggett (eds), 287–302. London: Methuen.

Kidron, M. 1971. Memories of development. *New Society* **17**, 360–6.

Laclau, E. 1979. *Politics and ideology in Marxist theory*. London: Verso.

Laclau, E. & C. Mouffe 1985. *Hegemony and socialist strategy: towards a radical democratic politics*. London: Verso.

Lai, C.-F. 1985. Special economic zones: the Chinese road to socialism? *Environment and Planning D, Society and Space* **3**, 63–84.

Lall, S. 1975. Is dependence a useful concept in analysing underdevelopment? *World Development* **3**, 799–810.

Lamb, G. 1981. Rapid capitalist development models. In *Dependency theory: a critical reassessment* D. Seers (ed.). London: Frances Pinter.

Landsberg, M. 1979. Export-led industrialisation in the Third World: manufacturing imperialism. *Review of Radical Political Economics* **11**, 50–63.

Lenin, V. I. 1970. *Imperialism: the highest stage of capitalism.* Peking: Foreign Language Press.

Levine, A., E. Sober & E. O. Wright 1987. Marxism and methodological individualism. *New Left Review* **162**, 67–84.

Lipietz, A. 1985. *The enchanted world.* London: Verso.

Lipietz, A. 1987. *Mirages and miracles: the crises of global Fordism.* London: Verso.

Luxemburg, R. 1972. *The accumulation of capital.* London: Allen Lane.

McGee, T. 1974,. In praise of tradition: towards a geography of anti-development. *Antipode* **6**, 30–47.

McMichael, M., J. Petras & R. Rhodes 1978. Industrialisation in the Third World. In *Critical perspectives on imperialism and social class in the Third World*, London: Monthly Review Press..

Mann, M. 1986. *The sources of social power.* Vol. 1: *A history of power from the beginning to AD 1760.* Cambridge: Cambridge University Press.

Marx, K. & F. Engels 1967. *The Communist Manifesto.* Harmondsworth: Penguin.

Massey, D. 1984. *Spatial divisions of labour.* London: Macmillan.

Massey, D. 1987. *Nicaragua.* Milton Keynes: Open University Press.

Meier, G. (ed.) 1987. *Pioneers in development*, 2nd series. Oxford: Oxford University Press.

Mouzelis, N. 1967. *Modern Greece: facets of underdevelopment.* London: Macmillan.

Mouzelis, N. 1980. Modernisation, underdevelopment, uneven development. *Journal of Peasant Studies* **7**, 353–74.

Mouzelis, N. 1988a. Sociology of development: reflections on the present crisis. *Sociology* **22**, 23–44.

Mouzelis, N. 1988b. Marxism or post-Marxism? *New Left Review* **167**, 107–23.

Noel, A. 1987. Accumulation, regulation and social change: an essay on French political economy. *International Organization* **41**, 303–33.

Nove, A. 1983. *The economics of feasible socialism.* London: Allen & Unwin.

O'Brien, P. 1975. A critique of Latin American theories of dependency. In *Beyond the sociology of development.* I. Oxaal (ed.). London: Routledge & Kegan Paul.

Paine, S. 1981. Spatial patterns of Chinese development: issues, outcomes and policies, 1949–79. *Journal of Development Studies* **17**, 133–95.

Palma, G. 1977. Dependency: a formal theory of underdevelopment or a methodology for the analysis of concrete situations of underdevelopment? *World Development* **6**, 881–924.

Peet, R. (ed.). 1987. *International capitalism and industrial restructuring: a critical geography.* Boston: Allen & Unwin.

Petras, J. & H. Brill 1985. The tyranny of globalism. *Journal of Contemporary Asia* **15**, 403–20.

Pletsch, C. 1981. The three worlds, or the division of social scientific labor, circa 1950–1975. *Comparative Studies in Society and History* **23**, 565–90.

Raj, K. N. 1984. The causes and consequences of world recession. *World Development* **12**, 151–69.

Rakodi, C. 1986. State and class in Africa: a case for extending analyses of the form and functions of the national state to the urban local state. *Environment and Planning D, Society and Space* **4**, 419–46.

Rawls, J. 1971. *A theory of justice.* Cambridge, Mass.: Harvard University Press.

Rey, P. P. 1971. *Colonialisme, neo-colonialisme et transition au capitalisme.* Paris: Maspero.

Rey, P. P. 1973. *Les Alliances des classes.* Paris: Maspero.

Roemer, J. (ed.) 1986. *Analytical Marxism.* Cambridge: Cambridge University Press.

Roemer, J. 1988. *Free to lose: an introduction to Marxist economic philosophy*. Cambridge, Mass.: Harvard University Press.

Ross, R. 1983. Facing Leviathan: public policy and global capitalism. *Economic Geography* **59**, 144–60.

Said, E. 1979. *Orientalism*. New York: Vintage Books.

Sayer, A. 1984. *Method in social science*, London: Hutchinson.

Scott, A. & M. Storper (eds) 1986. *Production, work, territory: the geographical anatomy of industrial capitalism*. Boston: Allen & Unwin.

Scott, J. 1985. *Weapons of the weak*. New Haven, Conn.: Yale University Press.

Sheahan, J. 1980. Market-oriented economic policies and political repression in Latin America. *Economic Development and Cultural Change* **28**, 267–91.

Shrestha, N. 1988. A structural perspective on labour migration in underdeveloped countries. *Progress in Human Geography* **12**, 179–207.

Sklair, L. 1988. Transcending the impasse: metatheory, theory and empirical research in the sociology of development and underdevelopment. *World Development* **16**, 697–709.

Sklar, R. 1975. *Corporate power in an African state*. Berkeley: University of California Press.

Sklar, R. 1976. Post-imperialism: a class analysis of multinational corporate expansion. *Comparative Politics* **9**, 75–92.

Slater, D. 1973. Geography and underdevelopment – 1. *Antipode* **5**, 21–53.

Slater, D. 1975. Underdevelopment and spatial inequality. *Progress in Planning* **4**, 97–167.

Slater, D. 1977. Geography and underdevelopment – 2. *Antipode* **9**, 1–31.

Slater, D. 1982. State and territory in post-revolutionary Cuba. *International Journal of Urban and Regional Research* **6**, 1–33.

Slater, D. 1987. On development theory and the Warren thesis: arguments against the predominance of economism. *Environmental Planning D, Society and Space* **5**, 263–82.

Smith, N. 1984. *Uneven development: nature, capital and the production of space*. Oxford: Basil Blackwell.

Smith, N. 1987. Dangers of the empirical turn. *Antipode* **19**, 59–68.

Sofer, M. 1988. Core–periphery structure in Fiji. *Environment and Planning D, Society and Space* **6**, 55–74.

Soja, E. 1980. The socio-spatial dialectic. *Annals of the Association of American Geographers* **70**, 207–25.

Storper, M. 1984. Who benefits from industrial decentralization? Social power in the labor market, income distribution and spatial policy in Brazil. *Regional Studies* **18**, 143–64.

Strange, S. 1986. *Casino capitalism*. Oxford: Basil Blackwell.

Sunkel, O. 1973. Transnational capitalism and national disintegration. *Social and Economic Studies* **22**, 132–76.

Sutcliffe, B. 1972. Imperialism and industrialisation in the Third World. In *Studies in the theory of imperialism*, R. Owen & B. Sutcliffe (eds). London: Longman.

Sutcliffe, M. & P. Wellings 1985. Worker militancy in South Africa: a socio-spatial analysis of trade union activities in the manufacturing sector. *Environment and Planning D, Society and Space* **3**, 357–79.

Taussig, M. 1980. *The devil and commodity fetishism in Latin America*. Chapel Hill: University of North Carolina Press.

Thrift, N. 1983. On the determination of social action in space and time. *Environment and Planning D* **1**, 23–56.

Thrift, N. 1985. Taking the rest of the world seriously? *Environment and Planning A* **17**, 7–24.

Thrift, N. & D. Forbes, 1986. *The price of war: urbanisation in Vietnam, 1954–1985*. London: Allen & Unwin.

Thrift, N. & A. Leyshon 1988. 'The gambling propensity: banks, developing country debt exposures and the new international financial system. *Geoforum* **19**, 55–69.

Tokman, V. 1985. Global monetarism and destruction of industry. *Cepal Review* **23**, 107–21.

Toye, J. 1987. *Dilemmas of development: reflections on the counter-revolution in development theory and policy*. Oxford: Basil Blackwell.

Triffin, R. 1960. *Gold and the dollar crisis*. New Haven, Conn.: Yale University Press.

Vandergeest, P. & F. Buttel 1988. Marx, Weber and development sociology. *World Development* **16**, 683–95.

Vielle, P. 1988. The state of the periphery and its heritage. *Economy and Society* **17**, 52–89.

Vroey, de M. 1984. A regulation approach interpretation of contemporary crisis. *Capital and Class* **23**, 45–66.

Wallerstein, I. 1974. *The modern world system*, vol. I. New York: Academic Books.

Wallerstein, I. 1979. *The capitalist world economy*. Cambridge: Cambridge University Press.

Wallerstein, I. 1984. *The politics of the world economy*. Cambridge: Cambridge University Press.

Ward, P. 1986. *Welfare politics in Mexico: Papering over the cracks*. London: Allen & Unwin.

Warren, B. 1973. Imperialism and capitalist industrialisation. *New Left Review* **81**, 3–44.

Warren, B. 1980. *Imperialism: Pioneer of capitalism*. London: Verso.

Watts, M. 1983. *Silent violence: food, famine and peasantry in northern Nigeria*. Berkeley: University of California Press.

Watts, M. 1984. State, oil and accumulation: from boom to crisis. *Environment and Planning D, Society and Space* **2**, 403–28.

Watts, M. 1988a. The faces of famine: a response to Torry. *Geojournal* **17**, 145–9.

Watts, M. 1988b. Deconstructing determinism: Marxisms, development theory and a comradely critique of *Capitalist world development*. *Antipode* **20**, 142–68.

Watts, M. 1988c. Struggles over land, struggles over meaning: some thoughts on naming, peasant resistance and the politics of place. In *A ground for common search*, R. Golledge & P. Gould (eds), 31–50. Goleta, CA: Santa Barbara Geographical Press.

Wolf, E. 1982. *Europe and the people without history*. Berkeley: University of California Press.

Wolff, R. & S. Resnick 1987. *Economics: Marxian versus neoclassical*. Baltimore, MD: Johns Hopkins University Press.

Wolpe, H. 1980. *The articulation of modes of production*. London: Routledge & Kegan Paul.

Wood, E. M. 1986. *The retreat from class: a new 'true' socialism*. London: Verso.

Part IV
NEW MODELS OF THE NATION, STATE, AND POLITICS

Part IV
NEW MODELS OF THE NATION, STATE, AND POLITICS

Introduction

Richard Peet

Political geography grew out of the scientific geography of the late 19th and early 20th centuries. This view drew heavily on evolutionary biology, the leading natural science of the day, through the linking device of the organismic analogy – i.e. the idea that human societies were like natural organisms, evolving and becoming more perfect in relation to the physical environment. These were powerful, socio-biological ideas, apparently backed by the very latest in scientific thought and, more importantly, ideologically functional to the expanding Euro-American powers in the second imperialism of 1870–1914 (Dorpalen 1942, Peet 1985). Anthropogeographic ideas found their most powerful expression in the emerging schools of political geography in Anglo-America and geopolitics in Germany. The prominent British geographer Mackinder (1931, p. 326) was no stranger to organismic thinking, believing, for example, that a common blood flowed through the veins of generations of people living in the same natural region; however his widely known sayings were more narrowly geostrategic – for example his generalization that whoever controls the heartland of central Asia controls the world (Mackinder 1904). In Germany, Haushofer combined Mackinder's heartland idea with Ratzel's organismic notion of *Lebensraum* (a state's 'living space') in a world model of pan-regions dominated by Germany, Japan, and Anglo-America. Haushofer's ideas were influential in Nazi Germany and derivations (heartland–rimland) survived as explanations of US–Soviet rivalry in the Cold War (Taylor 1985, pp. 40–3).

But the philosophical bases of geopolitical thinking were changing. Environmental determinism was heavily criticized in the 1920s by Barrows (1923) and, more convincingly, by Sauer (1963, p. 359) on the grounds that 'natural law does not apply to social groups'. The original purpose of organismic thinking in the legitimation of imperial expansion was diminished (except in Germany and Japan) when the Treaty of Versailles carved the world into spheres of political influence. Organismic thinking did not disappear completely (e.g. Whittlesey 1939), but the main theoretical emphasis of political geography gradually shifted towards ideas drawn originally from Parsonian structural functionalism (e.g. Hartshorne 1950; for discussion see Kasperson & Minghi 1969, pp. 69–88). In the 1970s, systems analysis (also in a form related to structural-functionalism (Easton 1965, 1966)) was wedded to a welfare approach (Cox 1979), to form an eclectic theory of political space still eventually reliant on the analogy between biology and human geography. For example, systems theory was used to model the relations between political processes and space as a way of analyzing broad geopolitical processes (Cohen & Rosenthal 1971, Cohen

1973). However this new political geography of the 1970s was unable to find a unifying theoretical base in such formulations because, in Taylor's (1983) view, it tried to unify philosophically and politically incompatible perspectives. The reformist wing split into right and left camps, with the left increasingly adopting an explicitly Marxist stance. By the early 1980s we find calls for a reorientation of political geography towards Marxism or, in Taylor's (1981) case, towards Wallerstein's (1979) neo-Marxist world systems theory.

Thus Taylor (1982) makes the simple but essential point that political geography must be seen as part of a unidisiplinary study of society in general. In this view there is a basic unity of the political with the economic, yet the state is relatively autonomous, for example in terms of the variety of state responses to movements of the capitalist economy (see also Dear & Clark 1978). Within this approach, historical materialism directs analytical attention towards the dynamic of capital accumulation, in Taylor's case specifically to changes in the world economy. Taylor's (1985) original contribution involves linking the general world systems perspective to the specific case of political geography. World systems theory combines the French materialist history of the *Annales* School with Latin American dependency theory to see the world as a spatial system of core, semi-peripheries, and peripheries which expand and shift in space over long time periods (Wallerstein 1979). Under capitalism, surplus is extracted from the peripheries via an unequal exchange of commodities enforced by political-military power. In political geography the world economy is viewed through different geographical scales of analysis, world economy, nation–state and locality, with the nation–state as mediator between polar extremes. In this perspective, geopolitics and imperialism are reinterpreted in a dynamic model of the world economy; territory, state, and nation are reinterpreted in world systems terms as the spatial structure of the state, a theory of states in the world economy, and a materialist theory of nationalism; and the city is reinterpreted in locality terms as place of socialization into political culture. The end result is a political geography recast in world systems terms (see also Taylor 1988a).

As a result of the re-thinking of the early 1980s, typified by Taylor's argument, interests long hidden in embarrasment began to re-surface. The term geopolitics, for example, abandoned in disrepute after World War II, was revived in the late 1970s and 1980s (Hepple 1986). Part of this revival involves re-examining geopoliticians of the past a little more favourably. Paterson (1987, p. 112) reviews the earlier work and finds that much German geopolitical writing was 'academically unexceptional'. Bassin (1987a) distinguishes German geopolitics, which showed the influence of Ratzel's environmentalism, from Third Reich Nazism, which emphasized innate biological qualities – however, Ratzel is credited, still, with a scientific version of *Lebensraum* (Bassin 1987b). Other similar work looks at Mackinder (Blouet 1987), Haushofer (Heske 1987), and, from the Marxist side, Wittfogel (Perry 1988, Peet 1988), and Bowman (Smith 1984). Laundered of its original organismic connotations, geopolitics is now a part of the Marxist lexicon. Harvey's (1985) approach is to see capitalism internally enmeshed in contradiction and constantly involved in geographical shifts and restructuring, with crises becoming global in scope and geopolitical conflict part of the process of temporary crisis resolution. He sees the 1980s as a dangerously unstable decade in the historical geography of capitalism, marked by shifts in the international division of labour, the disintegration of regional class alliances, and the loss of coherence of state economic policies: such conditions make geopolitical conflicts appear inevitable. 'With this comes the renewed threat of global war, this time waged with weapons of such immense and insane destructive power that not even the fittest stand to survive' (Harvey 1985, p. 162, Harvey 1982, Ch.13). Similar concerns haunt the work of other geographers. Pepper & Jenkins (1983, 1984) urge the establishment of a reasearch agenda for political geography to include the geopolitics of the nuclear arms race (see also Pepper 1986). Openshaw & Steadman (1983) argue that the components and consequences of nuclear wars have a strong spatial dimension. They present detailed casualty estimates for two nuclear attack scenarios in Britain (Hard Rock – 11 million casualties and 'Hard Luck – 43 million) and predict the spatial distribution of casualties – 80–100 per cent in the main population centres, 0–19 per cent in northern Scotland. Wisner (1986) similarly proposes placing geography on the side of the weak by emphasizing the geographic consequences of military conflict. Such work is part of a broader trend to make the new political geography relevant to questions of war and peace (O'Laughlin & van der Wusten 1986). This requires more study of the actual geopolitical *practice* of governments, their politicians, and advisors (O'Tuathail 1986, Taylor 1988b).

Moving down the geographical scale, from global to national, the 1980s witnessed a vigorous debate on nationalism. This new interest has been kindled by two modern nationalisms – the new minority nationalisms of Western Europe and Canada (e.g. of the Basques, Scotland, Quebec) (Williams 1980, 1985) and the liberation nationalisms of the Third World (Blaut 1987). Smith (1985) has also recently looked at nationalism in the Soviet Union. In a powerfully stated book, Blaut (1987) argues the classical Marxist case that the crucial class struggle is for control of the state, with national struggle as one form of this contention for state power. Others, by comparison, criticize what they term economic-deterministic and reductionist theorizing on nationalism, arguing that there is a need to explore superstructural factors in interaction with economic factors in real historical and regional settings.

Cooke's chapter in this section continues this debate. In developing his approach, Cooke draws on Poulantzas, a structural Marxist interested in the state, who also had a strong theorization of space. For Poulantzas, pre-capitalist societies have 'spatio-temporal matrices' fundamentally different from those typical of capitalism. Pre-capitalism, for example, has open space devoid of national frontiers as we know them, whereas capitalism is characterized by the territorialization of space and national frontiers. Nations are made and remade through struggles to impose class-derived space–time matrices on territories. For Cooke, the logic of this argument is that national territories are *pre*-conditions for capitalist development rather than its result, lending support to a theory that is primarily political, rather than cultural or economic. Cooke thus turns to theories of modernity to gain freedom from the constraints of an economic-determinist mode of production analysis. Following Berman and others, modernity is the cultural experience associated with the Enlightenment; that is, a break with the past signified by the progress of reason as a mode of thinking and rationalization as a way of intervening in the world. For Cooke, the experience of modernity, among other things, directly underpins the emergence of nationalism. Hence Cooke's political model of nationalism has the following main elements: an educated class, imbued with modernity, with a capacity to interpret its cultural and territorial content; the development of civil institutions, notably political parties which can accommodate the popular masses of a territory; and apparatus (the state) to protect the nation and organize its internal life; and relations of spatially uneven development which focus political consciousness and practice on demands for national self-determination. For Cooke, then, the geography of the modern world, its economic patterns and political structures, derive ultimately from 'the contingent meeting of modernism and nationalism on the road from antiquity'.

The main body of work at this scale is concerned with what Johnston (1981) terms the central problem of political geography – the theory of the state. There was a considerable Marxian interest in the state in the 1970s (e.g. Holloway & Picciotto 1978). Drawing on this work but also using realist conceptions of state action, Johnston (1984) attempts to redefine political geography around the study of the state. In this Part, Johnston argues that the state is necessarily a territorial body associated with a clearly defined area. As a territorial unit, the state performs a variety of tasks in a number of class interests, giving it a foundation for independent action while still, of course, remaining within the constraints of its rôle in reproducing a given social order. The state's autonomy is constrained by sectional interests with unequal power and by its continuing relations with civil society – those relations are subject to crises of rationality, legitimation, and motivation. A more complex theory of the state also recognizes the many separate functioning units within it, to some extent in competition for relative power. And the multiple bases of state decision making have to be recognized by theory – here Johnston discusses corporatist models of the central state and pluralist models of the local state. These assume a democratic system in which the state is run by an elected government – so Johnston turns to the geography of liberal democracy. Using a world systems framework, he sees democracy as a characteristic which capital can tolerate in core countries; in the periphery, by comparison, social stability requires strong, often repressive governments. Thus modern political geography is society-centred in that it derives the need for the state from an analysis of society – in particular, in Marxist and neo-Marxist analysis, from the economic relations of capitalism. Future research will involve broadening the understanding of the territorial bases of state movements and deepening our knowledge of the relative autonomy of state action.

At this point the political geography of the state intersects with a wide range of other geographic work which also includes a strong emphasis on the rôle of the state. For example the connection between state policy and regional development is drawn by Johnston (1986a). Because such work is reviewed elsewhere in these volumes, we shall stay within the traditional sphere of political geography more narrowly defined. The defeat of the British Labour Party in the 1983 election, especially its virtual demise in southern England (except inner London), touched off a debate on the rôle of space in political allegiances and voting patterns. Johnston (1986b, 1987) argues that there are substantial differences in the vote for the Labour Party, both regionally and by size of settlement, which cannot be accounted for by differences in class structure alone. He thinks that political parties, rather than the media, are major agents of political socialization. Once a party is successful in mobilizing a substantial proportion of the electorate, processes of political socialization in the home and local institutions aid in the reproduction of a party's success. This pattern of reproduction is interrupted by changes in industrial structure, and erosion of local governments' power to provide high levels of collective consumption (housing, transport, etc.). McAllister (1987), by contrast, argues that those who think for themselves are not influenced by the socio-spatial context in which they live, but those who have their minds made up for them are so influenced. Savage (1987a) finds that local political traditions are continually re-formed so that the historical trajectories of localities are crucially important in explaining people's politics. Using a class analysis which stresses a practical politics centred in reducing the insecurity inherent in labour as a commodity, yet also stressing the autonomy of the political, Savage (1987b) outlines several working-class strategies (mutualist, economistic, statist) that emerge in different British localities. Other similar work (Savage 1989) analyzes the rise and fall of radical regions (Cooke 1985), while in a further reaction against economically deterministic explanation Rose (1988) sees local politics shaped by local cultural values and the local configuration of frames of awareness and communal sensibilites. The question of the relationship between place and politics has also recently been reviewed by Agnew (1987). This type of work is clearly influenced by the critical reaction to Althusserian determinism on the one hand, and by the localities movements on the other. Its major problem is that by deliberately embracing eclecticism it may sink into a disorganized morass incapable of being mentally retained as a set of generalizations.

By comparison with geographic work on the state, which now stretches through a considerable body of literature, Clark reports on the neglect of the geography of law. This is despite the existence of realist theories of law as a social practice reflecting and reproducing hierarchical societies, theories which would allow the geography of law to be linked with Marxian and neo-Marxian notions of the state. In the social sciences more generally, substantive theories place law into social and political context, providing a mode of analysis which also would allow an intellectually mature geography of law. Recent work interprets judicial action, for example in land use disputes, or capital–labour clashes over the relocation of production, in terms of the substantive basis of law in capitalist society. The political point is to place the geography of law into the critical context of the interactions between the legal and the social. The main academic point includes seeing the geography of law as part of a wider intellectual project – understanding the social structuration of the landscape of human experience.

Moving finally to the local scale, we find political geographic thinking transformed by work in the 1970s by Harvey (1973) and Castells (1977). Their reconceptualization of the city as locus of class struggles evoked a stream of research and writing at the juncture of urban sociology, urban geography, and political geography. For example, Cox (1978) argues that, particularly in North American cities, class struggle is transmuted into various forms of local political processes, like struggles between neighbourhood groups and among local governments (see also Cox & Johnston 1982). But the main focus within political geography at the urban level is the concept of the local state reviewed below by Fincher. Fincher's approach differs from that followed elsewhere in the book – she begins with three urban crises and moves into theory in search of explanation. These crises point to the ever changing conflict relations between different geographic levels of government and the class groupings formed around each level. Turmoil at the local level, she argues, has been interpreted by an expanding, if diffuse, analytical literature on a series of linked topics. A first theme involves making sense of the relations between the various levels of the state in different national contexts, focusing on the relative autonomy of local government, an important issue in places like Britain where serious political differences have occurred over the state's rôle in economic restructuring, service provision, etc. A second theme explicates the class relations of the communities of which the local state is part, and to which it contributes, with emphasis on the question of local control over local affairs. Fincher recommends that geographic research on the local state continue by looking at the relations between local economy and local politics, and the class and gender implications of local state actions. This research would, she concludes, inform more appropriate political responses to local state crises like those of the 1970s and early 1980s.

Similarly, Duncan *et al.* (1988; see also Duncan & Goodwin 1988) use the term local state institution to refer to all separate state bodies, organizations, and offices existing at a subnational level and attempt to explain why their behaviour varies. They relate the rise of the local state to the development of capitalism – local state institutions help dominant groups organize and manage the differentiated social system of capitalism, especially as local environments are developed, changed, and abandoned. Various levels of state institutions are needed to establish geographical stability through spatial fixes. State systems need specifically local extensions to organize and control spatially organized ecological systems and ways of living. However, they argue, local political variations cannot simply be 'read off' uneven economic development. Local state institutions stand in a mediating position between workplace and civil organizations, both of which are spatially constituted and combined in locally specific ways. Social relations in local state institutions, and hence local policy making, reflect these combinations – it is this heterogeneity, and the plurality of social conflicts, that are crucial to the distinction of the local state.

This brief review, and the chapters in Part IV, show several characteristics of the new political geography, which can be contrasted with the old. First, we find a far higher level of sophistication in the knowledge of the epistemological underpinnings of the theories employed – a linking of social philosophy with the politics of space. Second, as a partial result, the types of theories have changed, from organismic conceptions of society in the early 20th century, through structural functional theories, to Marxian and realist notions today – indeed, such is the dominance of political economic ideas, particularly in British political geography, that reactions against them (e.g. Nientied 1985) can sound like crank telephone calls. Third, again as a partial result of theoretical transformation, the topics constituting the core of political geography have changed, in part back to the fascination of the past with geopolitics and global strategies, in part towards new concerns like the local state. Finally, political geography has rediscovered the historical dimension so that its content is now viewed as integral to the dynamics and change that are our modern world. If political geography was once a 'moribund backwater' (Berry 1969, p. 450), 20 years later it is a lively eddy in the stream of political economy. There can be little doubt that Marxian ideas, concerns, and energies played a leading role in achieving this transformation.

Acknowledgement

Thanks to Peter Taylor for comments on this Introduction.

References

Agnew, J. A. 1987. *Place and politics*. London: Allen & Unwin.

Anderson, J. 1986. On theories of nationalism and the size of states. *Antipode* **18**, 218–32.

Barrows, H. 1923. Geography as human ecology. *Annals of the Association of American Geographers* **13**, 1–14.

Bassin, M. 1987a. Race *contra* space: the conflict between German *geopolitics* and national socialism. *Political Geography Quarterly* **5**, 115–34.

Bassin, M. 1987b. Imperialism and the nation state in Friedrich Ratzel's political geography. *Progress in Human Geography* **11**, 673–95.

Berry, B. 1969. Review of *International regions and the international system* by B. M. Russett. *Geographical Review* **59**, 450–1.

Blaut, J. M. 1987. *The national question: decolonising the theory of nationalism*. London: Zed Books.

Blouet, B. 1987. The political career of Sir Halford Mackinder. *Political Geography Quarterly* **6**, 355–67.

Castells, M. 1977. *The urban question*. London: Edward Arnold.

Cohen, S. B. 1973. *Geography and politics in a world divided*. New York: Oxford University Press.

Cohen, S. B. & L. D. Rosenthal 1971. A geographical model for political systems analysis. *Geographical Review* **61**, 5–31.

Cooke, P. 1984. Recent theories of political regionalism: a critique and an alternative proposal. *International Journal of Urban and Regional Studies* **8**, 549–72.

Cooke, P. 1985. Radical regions. In *Political action and social identity*, G. Rees (ed.). London: Macmillan.

Cox, K. R. (ed.) 1978. *Urbanization and conflict in market societies*. Chicago: Maaroufa Press.

Cox, K. R. 1979. *Location and public problems: a political geography of the contemporary world*. Chicago: Maaroufa Press.

Cox, K. R. 1986. Urban social movements and neighborhood conflicts: questions of space. *Urban Geography*, **7**, 536–46.

Cox, K. R. & R. J. Johnston (eds) 1982. *Conflict, politics and the urban scene*. New York: St Martin's Press.

Crouch, C. (ed.). 1979. *State and economy in contemporary capitalism*. New York: St Martin's Press.

Dear, M. & G. Clark 1978. The state and geographic process: a critical review. *Environment and Planning A* **10**, 173–83.

Dorpalen, A. 1942. *The world of General Haushofer: geopolitics in action*. New York: Farrar & Rinehart.

Duncan, S. & M. Goodwin 1988. *The local state and uneven development*. Cambridge: Polity Press.

Duncan, S., M. Goodwin & S. Halford 1988. Policy variations in local states: uneven development and local social relations. *International Journal of Urban and Regional Research* **12**, 109–28.

Easton, D. 1965. *A framework for political analysis*. Englewood Cliffs, NJ: Prentice-Hall.

Easton, D. (ed.) 1966. *Varieties of political theory*. Englewood Cliffs, NJ: Prentice-Hall.

Hartshorne, R. 1950. The functional approach in political geography. *Annals of the Association of American Geographers* **40**, 95–130.

Harvey, D. 1973. *Social justice and the city*. London: Edward Arnold.

Harvey, D. 1985. *The urbanisation of capital*. Oxford: Basil Blackwell.

Heske, H. 1987. Karl Haushofer: his role in German geopolitics and Nazi politics. *Political Geography Quarterly* **6**, 135–44.

Holloway, J. & S. Picciotto (eds) 1978. *State and capital: a Marxist debate*. London: Edward Arnold.

Hepple, L. W. 1986. The revival of geopolitics. *Political Geography Quarterly* **5**, 521–36.

Johnston, R. J. 1981. Political geography. In *Quantitative geography: retrospect and prospect*, N. Wrigley & R. J. Bennett (eds). London: Routledge & Kegan Paul.

Johnston, R. J. 1984. Marxist political economy, the state and political economy. *Progress in Human Geography* **8**, 473–92.

Johnson, R. J. 1986a. Placing politics. *Political Geography Quarterly* **5**, 563–78.

Johnston, R. J. 1986b. Places and votes: the role of location in the creation of political attitudes. *Urban Geography* **7**, 103–17.

Johnston, R. J. 1987. The geography of the working class and the geography of the Labor vote in England, 1983. *Political Geography Quarterly* **6**, 7–16.

Kasperson, R. E. & J. V. Minghi (eds) 1969. *The structure of political geography*. Chicago: Aldine.

McAllister, I. 1987. Social context, turnout and the vote: Australian and British comparisons. *Political Geography Quarterly* **6**, 17–30.

Mackinder, H. 1904. The geographical pivot of history. *Geographical Journal* **23**, 421–42.

Mackinder, H. 1931. The human habitat. *Scottish Geographical Magazine* **47**, 321–35.

Nientied, P. 1985. A 'new' political geography: on what basis? *Progress in Human Geography* **9** 597–600.

O'Laughlin, J. & H. van der Wusten 1986. Geography, war and peace: notes for a contribution to a revived political geography. *Progress in Human Geography* **10**, 484–510.

Openshaw, S. 1982. The geography of reactor siting policies in the U.K. *Transactions of the Institute of British Geographers* **7**, 150–62.

Openshaw, S. & P. Steadman 1983. The geography of two hypothetical nuclear attacks in Britain. *Area* **15**, 193–201.

O'Tuathail, G. 1986. The language and nature of the 'new geopolitics' – the case of US-El Salvador relations, *Political Geography Quarterly* **5**, 73–85.

Paterson, J. H. 1987. German geopolitics reassessed. *Political Geography Quarterly*. **6**, 107–14.

Peet, R. 1985. The social origins of environmental determinism. *Annals of the Association of American Geographers* **75**, 309–33.

Peet, R. 1988. Wittfogel on the nature-society dialectic. *Political Geography Quarterly* **7**, 81–3.

Pepper, D. 1986. Spatial analysis of the West's 'deep strike doctrines'. *Political Geography Quarterly* **5**, 253–66.

Pepper, D. & A. Jenkins 1983. A call to arms: geography and peace studies. *Area* **15**, 202–8.

Pepper, D. & A. Jenkins 1984. Reversing the nuclear arms race: geopolitical bases for pessimism. *Professional Geographer* **36**, 419–27.

Perry, P. J. 1988. Thiry years on: or whatever happened to Wittfogel? *Political Geography Quarterly* **7** 75–80.

Rose, G. 1988. Locality, politics, and culture: Poplar in the 1920's. *Environment and Planning D, Society and Space* **6**, 151–68.

Sauer, C. 1963. *Land and life: a selection from the writings of Carl Ortwin Sauer*. J. Leighly. (ed.). Berkeley: University of California Press.

Savage, M. 1987a. Understanding political alignments in contemporary Britain: do localities matter? *Political Geography Quarterly* **6**, 53–76.

Savage, M. 1987b. *The dynamics of working class politics: the Labour movement in Preston 1880–1940*. Cambridge: Cambridge University Press.

Savage, M. 1989. *What happened to red Clydesdale? Political parties in the local social structure*. Sussex Working Papers in Urban and Regional Studies.

Smith, N. 1985. Isaiah Bowman: political geography and geopolitics, *Political Geography Quarterly* **3** 69–76.

Taylor, P. J. 1981. Political geography and the world economy. In *Political studies from spatial perspectives*, A. D. Burnett & P. J. Taylor (eds), 157–72. Chichester: Wiley.

Taylor, P. J. 1982. A materialist framework for political geography. *Transactions of the Institute of British Geographers* **7**, 15–34.

Taylor, P. J. 1983. The question of theory in political geography. In *Pluralism and political geography: people, territory and state*, N. Kliot & S. Waterman (eds), 9–18. London: Croom Helm.

Taylor, P. J. 1985. *Political geography: world economy, nation-state and locality*. London: Longman.

Taylor, P. J. 1987. The paradox of geographical scale in Marx's politics. *Antipode* **19**, 286–306.

Talyor, P. J. 1988a. World systems analysis and regional geography. *Professional Geographer*, **40**, 259–65.

Taylor, P. J. 1988b. Geopolitics reviewed. *Seminar Paper* No. 54. Department of Geography, University of Newcastle upon Tyne.

Taylor, P. J. & G. Gudgin 1981. Geography of elections. In *Quantitative geography: A British view*, N. Wrigley & R. J. Bennett (eds), 382–6. London: Routledge & Kegan Paul.

Taylor, P. J. & H. Hadfield 1982. Housing and the state: a case study and structualist interpretation. In *Conflict, politics and the urban scene*, K. R. Cox & R. J. Johnston (eds), 241–63. New York: St Martin's Press.

Wallerstein, I. 1979. *The capitalist world-economy*. Cambridge: Cambridge University Press.

Watts, M. J. & T. J. Bassett 1986. Politics, the state and agrarian development: a comparative study of Nigeria and the Ivory Coast. *Political Geography Quarterly* **5**, 103–25.

Whittlesey, D. 1939. *The earth and the state: a study in political geography*. New York: Holt.

Williams, C. H. 1980. Ethnic separation in Western Europe. *Tijdschrift voor Economische en Sociale Geografie* **71**, 142–58.

Williams, C. H. 1985. The question of.natural congruence. In *A world in crisis?* R. J. Johnston & P. J. Taylor (eds). Oxford: Basil Blackwell.

Wisner, B. 1986. Geography: war or peace studies? *Antipode* **18**, 212–17.

10 *Nation, space, modernity*

Philip Cooke

Introduction

Following a period of quiescence there has been a revival of interest on the part of geographers in the questions and issues of political geography. An important source of this revival was the new literature on state theory published in the 1970s (Miliband 1969, Poulantzas 1973, 1978, Holloway & Picciotto 1978). This literature typically did not refer to space in its analyses for the most part, despite being rooted in a macrotheoretic discourse which professed to explain the economic development process under capitalism, deploying neo-Marxist insights to do so. An exception to the general rule that polities as well as economies are often perceived by their analysts to inhabit a dimensionless universe, is found in the work of Poulantzas (see, in particular, Poulantzas 1975, 1978). More will be said about this political theorist's analytical contribution later in this chapter; suffice it to say for now that his awareness of space as a theoretical construct deepened and broadened the perceptiveness of his theory of the state and, separately, the nation. In the process his work largely neglected by contemporary political geographers, speaks to the centrality of geopolitical thinking in furthering understanding of the modern world.

Despite the importance of state theory in reviving political geography as well as wider aspects of spatial studies, it is towards an exploration of theorizations of the nation, the relatively under-researched dimension of the nation–state couplet, that this chapter is directed. Geographers have been rather better at tackling the complex issues surrounding national movements and nations than many other social scientists, although, as we shall see, too much of this work has been descriptive in nature and focused upon the submerged non-state nations such as Brittany, Quebec, and Scotland rather than the variously powerful historic nations that became nation–states such as Germany, Japan, or the Soviet Union. This may be one product of a tendency for geography and geographers to think small, to underestimate the importance of the spatial dimension while protesting its omission by other social scientists. It may also be a product of geography's tradition of being theoretically underdeveloped by comparison with other social sciences. Whatever the reason, it is high time that geography made a more penetrating contribution to current thinking in national and international affairs, building on the subnational expertise that has developed, especially in political geography, in the past decade or so.

If one were broadly to categorize the geographical research on nations and national movements in the fairly recent past, it could be divided in two ways. Both are conscious of the importance of treating national struggles and national movements in their own terms rather than as displacements of cultural or economic struggles, and in this respect both have a degree of consistency with the argument that will be developed here. However, in my view neither takes the logic of their own analysis sufficiently far to register the fundamental importance of national movements as social agents that have exerted the most profound effect of all upon the social landscape in the modern era.

The first approach to thinking about national struggles and movements geographically is that of the radical geographers, often closely associated with the journal *Antipode*, who despite idiosyncratic inflections share a common view that nationalism is a form of class struggle. Taking a lead from the classics such as Marx and Luxemburg, they see the national struggle taking triangular form. Thus an *external*, occupying power (e.g. colonial or military) is opposed by an internal ruling class, the interest of which is to take over as the exploiting power in an independent nation–state. Also existing as an internal force is the presently externally exploited, subordinate class the interest of which lies in taking state power for itself, but which may combine with the local ruling class to destabilize the system as a prerequisite to achieving its key objective. This exogenous focus is most pronounced in the work of Blaut (1980, 1982) who makes the important point that imperialism should be seen as an extension of the national struggle of successful great powers, and that in this expansion lies one of the more pernicious expressions of institutionalized racism. The trigger for such expansionary drives, according to Blaut (1980), can be identified in the individual and class interest of the colonizers in easy accumulation of wealth. Resistance from the colonized peoples to such nationalist expansion is a defensive form of nationalism..

In a response to Blaut, Pringle (1982) generally agrees with the argument that an external perspective has to be brought to bear on the analysis of nationalism, and joins in with the condemnation of what both perceive as a Eurocentric tradition as represented in the work of Nairn (1981). This they see as unnecessarily diffusionist, not to say supremacist, in its assignment of primacy to European national struggles as the exemplars for subsequent ones in the colonized world. Where Pringle disagrees with Blaut is in his interpretation of the Marxist classics as giving unconditional support to all national liberation struggles (i.e. even if they are led by a putative ruling class). Drawing on contemporary Irish experience, Pringle aruges that its national struggle is retrogressive as it impinges upon the North because it is predicated on a permanent division within the working class along sectarian lines. It is thus non-progressive, even reactionary in its political objectives and, as a consequence, inconsistent with revolutionary strategy. This appears to agree with Anderson's (1980) 'one-nation' analysis of the Northern Ireland question, the argument being that to accept that there are two – one Catholic the other Protestant – is to fall into the trap set by both imperialists and capitalists with respect to Irish working-class solidarity. But Pringle condemns Anderson for his denial of the right for people to determine their own nationality, thus accusing him of a kind of territorial fetishism. This has some irony given Anderson's well known critique elsewhere of the tendency to spatial fetishism in much mainstream geography (Anderson 1975).

More recently, Blaut has returned to the fray with a theory of nationalism (Blaut 1986) which, critical of well known institutionalist theories such as those of Gellner (1983) and Smith (1971, 1979) discussed below, restates his previous materialist position on the exogenous drive of national movements. Importantly, for the thesis developed later in this chapter, he stresses once more the irremediably *political* (rather than, say, economic or cultural) determinants of such movements. But, somewhat questionably, he goes against an earlier argument (Blaut 1980) that nationalism is a modern political form, to assert that it is a timeless political tool, unaffected by stage or indeed type of societal development or mode of production. No evidence is offered to substantiate this assertion. Blaut recognizes the endogenous dimensions of national struggle, but treats these as subsidiary. Nationalism arises from the crises in surplus appropriation faced by a ruling class when the limits of exploitation are reached domestically, a condition necessitating exogenous, nationalist expansion. This equation of nationalism with imperialism is the abiding motif of Blaut's work and one which contains a fatal flaw for his analysis.

The problem is that while being critical of authors such as Nairn for the assumption that nationalism started in Europe and was then emulated elsewhere, Blaut's own analysis makes exactly the same assumption. In binding his theory of nationalism so tightly to a theory of imperialism he is forced to accept Europe as the prime mover in the development of nationalism, since it was the European powers of France, Spain, and England that invented the colonialism that subsequently, with the export of capital, became imperialism. Blaut may wish to argue that there is nothing intrinsic to imperialism to prevent it having a non-European provenance, and it would certainly be true that other continents have spawned imperialist powers, most notably Asia with the example of prewar Japan. But to argue both for a non-spatial and a non-temporal theory of nationalism seems difficult, to say the least. The position may owe something to Blaut's (1975) earlier interest in the origins of capitalism and a consequent confusion of the political and the economic as the two recede into the mists of time (for discussion of the 'European miracle' see Mann 1986).

The other strand of geographical work on nationalism is best represented in the work of Williams (1982, 1985), though it is also to be found in Kofman (1981, 1982, 1985) and the collection edited by Drakakis-Smith & Williams (1983). It is much concerned with the *endogenous* processes of national movements or the often synonymously referred to regionalist movements and their social and political bases. It is recognized, by and large, that they have essentially political origins, but these are traced back variously to what might be called slights commissioned by an insensitive and remote central state with respect to linguistic usage, regional planning policy, tourism, and so on. Such slights re-open wounds which regional or national minorities have long nursed, as a consequence of which politicization of such sentiment may ensue. Often the political leadership for such movements comes from the intellectual or bureaucratic strata who may themselves have a strong interest in expanding the demand for regionalist resource and occupational allocations. To the extent that such groupings are capable of mobilizing political support, concessions may be forthcoming, but they inevitably produce less by way of satisfying autonomist sentiment than they promise. This work is often supported with valuable empirical material, and a substantial archive of accounts of regionalist movements and their achievements now exists not only in the geographical but also in more specialist journals.

The predominant weakness of this work is its lack of well worked and consistent theorization. This is not to say that the contributions lack theoretical awareness; contributions often point to the weaknesses of crude consensus or core–periphery models of nationalist politics and adhere to a broadly defined conflict perspective. Perhaps the stage has been reached where the most significant step forward that can be taken with respect to this endogenous strand of literature in political geography is for the data to be sifted comparatively with a view to focusing and strengthening the theoretical underpinnings of the discourse. The urgency of undertaking this task lies in Agnew's (1981) injunction to explore not just the reasons why ethnic, linguistic, or socio-cultural motivations sometimes lead to the formation of regionalist or nationalist movements, but why, more often, they do not.

In this chapter I aim to derive a generalizable set of propositions which are sufficiently interlinked to constitute a model of the connection between national movements and nationalism. The former I take to mean the organized political base which exists to pursue the nationalist objective; it is usually a political party, though it may, exceptionally, not be. The latter, I take to be the political objective of achieving for a geographical area unified by territory, language, and culture, either full self-determination through achieving statehood or improved representation, possibly autonomy, within a multinational state. In undertaking the analysis I review theorized accounts of nationalism written from political, cultural, and economic perspectives. In evaluating this work I make no particular distinction between nationalism in long established nation–states and that in nations without states of their own.

The first section is a lengthy exegesis of the theory of nationalism developed by Poulantzas (1978). I argue that this is the fullest analysis yet available of the political theory of nationalism, one particularly suited to geographical analysis since it contains a most interesting theorization of space in the concept of the space–time matrix. Other recent political approaches to nationalism are also considered, notably those of Giddens (1985) and Skocpol (1977, 1979). Thereafter I provide a critique of cultural theories of nationalism after Gellner (1964) and Hechter (1975) and economic theories such as those of Nairn (1981) and Brustein (1981). Finally, I try to provide a theoretical underpinning for the general phenomenon of nationalism. This I do by reference to the work, in particular, of Berman (1983) and Habermas (1981) on modernity. I round off the last section by tying the different supporting arguments into a conceptual model of the political theory of national movements and nationalism which owes modified allegiance to the position outlined by Poulantzas in 1978. It is also consistent with the newer political geography of nationalism but seeks to balance the exogenous and endogenous polarities found in that literature.

The nation as a space–time matrix

The fact that capitalist space and time are not at all the same as their counterparts in previous modes of production implies that considerable changes have taken place in the reality and meaning of territory and historicity. These changes both allow and entail the constitution of the modern nation (Poulantzas 1978, p. 97).

It is instructive that Poulantzas should have given us his most fully worked out theory of space in the context of a discussion of the nation, rather than class, consumption, production, or the state *per se*. Modern geography was born with the emergence of the nation (Capel 1981, MacLaughlin 1986). The nation provides the co-ordinates within which modernity developed. Geography explains the processes that produce unequal distributions in degree and kind of development over space and draws appropriate political conclusions. The question of the reciprocal, constitutive relation between nation and space gains illumination from Poulantzas's (1978) discussion of the work of two earlier theorists who clearly influenced Poulantzas's central notion of the nation as a process, the Austro-Marxists Bauer and Renner (see Bottomore & Goode 1978).

The term nation is different from, and indeed predates, the concept of nation–state associated with the rise of capitalism in Europe between the 16th and 18th centuries. Nation refers to a sociospatial form which emerged when lineage society gave way to class-divided society (Anderson 1974a, 1974b). On this Poulantzas agrees with the Austrian School:

> Just as private ownership of the means of production and individual production develops out of the social system of primitive communism, and from this again, there develops co-operative production on the basis of social ownership, so the unitary nation divides into members of the nation and those who are excluded and become fragmented into small, local circles (Bauer 1978, p. 108).

This points to the political and processual character of the nation from the outset. The nation is constituted in terms of *inclusion* and *exclusion*, conditions which are objects of political struggle and which, in turn, supply the internal dynamic of the nation as a process rather than as a thing. Initially excluded from the nation are the peasantry and those who, to live, have to sell their labour power to the owners of private property. For Bauer, it is the educated classes not the popular masses that constitute the nation at an early stage of development because:

(a) the intelligentsia have the means to make, and subsequently remake, the nation as a totality based on the production of histories, the codification of a language, and the formation of a spatially extended system of education meant for the citizenry, those of similar upbringing to themselves;
(b) the popular masses are not citizens and thus have no access to the newly formed culture since they are not fully incorporated into the emergent education system. Their sense of belonging to the common, ancestral nation is weakened by temporal distance and spatial fragmentation. However, in time the national, cultural community extends to incorporate them as education becomes universalized.

Bauer assigns to the intelligentsia a crucial rôle in producing the nation as a cultural form. This is the keystone of the important culturalist strand of theorizing the nation proposed by Gellner (1964, 1983) and Smith (1971, 1979, 1982). Bauer's insight is reflected in Gellner's more recent analysis of nationalism in modern Africa in a way which underlines a certain transspatial, perhaps transhistorical, truth of the interaction between modernity and nationalism:

> The self-image of nationalism involves a stress on folk, folk-lore, popular culture, etc. In fact, nationalism becomes important precisely when these things become artificial. Genuine peasants or tribesmen, however proficient at folk-dancing, do not generally make good nationalists (Gellner, 1964, p. 162; quoted in Anderson 1986b).

Anderson (1986b) argues that the experience of modernity can shatter the 'small, local circles', as Bauer calls them, that remain relatively isolated and culturally self-sufficient within the interstices of a developed capitalist economy. The nation eventually comes to replace that lost sense of collective identification.

Nation, culture, development

To return to Poulantzas's analysis of the nation, a number of points should be noted. First, the relationship of nation to state is non-congruent in that the state may embody more than one nation, while nations may exist legitimately without their own state. Second, the trend of the modern nation–state is to seek to secure national unity even where it is multinational, thereby blocking the emergence of new states from within older formations. This tendency is most commonly explained by reference to economics. The state creates a space within which a base for production, exchange, and distribution is established, from which external trading and investment activities can be launched. Thus the nation becomes the internal market in which citizens and corporations exchange commodities as free and equal individuals, politically subject to the same state.

But why should the nation be the fundamental frame of reference for this process? The obvious answer is that the nation is far more than the container for economic activity that the above account suggests. As Bauer and others have long recognized, the ways in which the modern nation designates a common territory, language, and cultural history are themselves preconditions for the production, exchange, and distribution relations that come to take on specific forms. In other words, the logical priority is that of the nation acting as the co-ordinate for specific economic relations rather than those economic relations calling forth the nation as a specific kind of container.

This is an important point because it leads to the deduction that the trajectories of uneven spatial development result from cultural and territorial traditions, what Poulantzas calls 'the underlying conceptual matrices of space and time' (1978, p. 97) embodied in the modern nation and its specific power container, the modern state. Though this may sound like an idealist or crudely culturalist argument where certain shared values of a nation are expressed in its developmental forms, it is in fact a profoundly political analysis (for a similar analysis, see Mann 1986, pp. 500–17). Because the spatially combined and uneven development process – the differential rate and kind of development occurring over time across space, usually in combination with the residue of previous development processes – is intertwined with the social division of power within and between nations it is revealed in cultural, political, and then economic practices.

An example of what is meant by this will clarify the argument. Since I am talking here specifically about spatially uneven development rather than nations *per se*, my example will be drawn from the city–regional scale, though it will be clear that national forces underpinned the developments in question. The example is drawn from 19th-century Russia which was characterized, in the face of modernization everywhere to the west, by stagnation. Berman (1983) assigns key importance in the development of Russian politics and culture, even beyond the 1917 Revolution, to the absolutist politico-cultural will to develop as realized in St Petersburg (the modern Leningrad). It was founded as the new capital at the mouth of the Neva river in 1703 as a symbolic and real window on Europe and to escape the stifling constraints exerted on the nation by Moscow. Its population grew to 100 000 by 1723, 485 000 by 1850, and 2 million by 1914; it was on a par with Vienna as Europe's fourth or fifth largest city in the 19th century. The building methods were draconian, involving forced labour, nobles being made to invest in palaces there or forfeit their status, but, rather less tyrannically, professional educators, scientists, philosophers, and business-men being given patronage, as were architects, artists, and musicians. Thus culture was deployed politically in a forced march towards modernization, leaving the rest of the country underdeveloped as spatially uneven development of the most extreme kind was induced.

There followed periods of turbulent, quasi-revolutionary unrest often caused by the arbitrariness of such practices of modernization from above, but also a great flowering of modern literature from authors such as Pushkin, Gogol, Gorky, and Dostoevsky which took its inspiration from the modernity in a sea of tradition that the city represented. By 1905 St Petersburg was the major industrial complex of Russia with textile factories and heavy industry ringing the city, its workers were organized and frequently struck for better pay and shorter hours. Such demands were regularly met with extreme forms of suppression including indiscriminate firing upon protesting crowds, ultimately provoking the establishment in 1905 of the first Petersburg Soviet, enforcing the flight of the royal family, and eventually sparking off the events leading up to the Russian Revolution itself. The radicalism and cosmpolitanism of St Petersburg made it too dangerous to be the new Soviet capital, a rôle rapidly re-assigned to Moscow. But Leningrad remained a major axis of post-Revolutionary industrial development, a rôle which it has maintained to the present day.

What is revealed admirably in this example, is the way in which national political will, faced with evidence of relative cultural and economic decline, could create a new fault line in the spatio-temporal matrix of a vast empire. This fault line, created politically with a view to opening up the *Russian* nation to the culturally modern influences of the West, only thereafter developed as an economic growth pole. Even after the loss of the crucial trigger of imperial capital city status, it survived the effects of siege under two world wars and the economic planning of the Stalinist era.

The intriguing question arising from this account concerns the extent to which the St Petersburg experience is an extreme one. Although there is no space to develop empirical discussion of the intersection of the national as a specific space–time matrix with the local as a site of politico-culturally induced urbanization, it is likely that, in the modern era, under the impetus of the modern nation–state, it is the predominant form, certainly a distinctive category, in the spatially uneven unfolding of development processes across the world. Virtually all third World nations have uneven development of this kind as their major spatial motif (Chase-Dunn 1984). In some instances there are modern reversals of the Petersburg model where, instead of a city acting as a window on the developed nations, it acts as a putative beacon to supposedly benighted, usually state-socialist, nations. Examples of the latter form of uneven development would include West Berlin, Hong Kong, Taipei, Seoul, Saigon (for a time), and, possibly, Miami and even Los Angeles (see, for example, King 1976, Portes & Walton 1977, Soja *et al*. 1983; Agnew *et al*. 1984, Holton 1985, Thrift & Forbes 1986). Increasingly, developed nations are restructuring in response to world recession with their capital city–regions recovering sooner than their established industrial belts, as the cases of London, Paris, and Washington, DC testify.

Nation, state and space–time

Hence, the interaction between the modern nation, its state, and its unevenly developed space is central to our understanding of geographic process. To return to the example of St Petersburg, the national will to modernize (embodied in the monarch) could not have been fulfilled had the attempt been made in and around Moscow with its accretions of ideological, political, and economic power from feudal Russia. A new form of space had to be created, in considerable tension with the older social space, which facilitated the modernization process in terms of new philosophies, diverse cultures, cosmopolitan social relations, and new kinds of individual freedoms (interlarded, of course, with old, tyrannical power relations). After the 1917 Revolution, this now-not-so-new social space with its liberal culture was out of tune with the dictatorship of the proletariat, the command economy, and socialism in one country surrounded by a hostile world, hence it was demoted politically (though not economically) as Russia's space–time matrix was reco-ordinated in line with new Soviet rhythms.

For Poulantzas, pre-capitalist social relations, political power, and forms of state apparatus produce a spatio-temporal matrix that is as represented on the left side of Table 10.1, while those of capitalism are displayed on its right side. The term matrices is justified because each list describes both the spatial and the temporal dimensions of the mode of production in question. To take the spatial dimension of pre-capitalism first: it is *continuous*, *homogeneous*, and *symmetrical* in that it has no national frontiers as we understand them in the modern epoch, thus it is *open* space, the limit of which is only reached where this space (remembering that space is always social) meets its negation:

> its absolute reverse: namely the barbarians. But these barbarians are precisely a non-site: not only are they not a segment, however distinct, of a single space, they are the definitive end of all possible space; they are not a division of space but a without-space, not a no-man's-land but a no-land (Poulantzas 1978, pp. 101–2).

Table 10.1 Space–time matrices (after Poulantzas 1978).

Space-time matrices	
Pre-capitalist	Capitalist
Continuous	Serial
Homogeneous	Segmented
Symmetrical	Parcelled
Reversible	Cellular
Open	Irreversible
Repetitive	Cumulative

Space is homogeneous and symmetrical because its main centres are replicas of its primary centre rather than taking on an individuality derived from specific location. All Greek and Roman cities were modelled on Athens and Rome as, until recently, all Islamic cities were replicas of Mecca, hence such space is *repetitive*. The medieval period was even more spatially isotropic, certainly open despite the apparent closure implied by walled cities. Poulantzas argues that people were at their most mobile in that period – going on pilgrimages, crusades, as merchants, clerics, escapees from feudal constraints, and as part of the great peasant migrations of the era. In particular, space was patterned by Christianity, non-space was now occupied not by barbarians but by infidels. But space itself is *reversible*. For example, the papal seat moved from Rome to Avignon and the seat of the Holy Roman Empire, though centred on Constantinople, moved around constantly as did the monarchy and nobility of medieval societies.

Because in this era development in the modern sense does not occur, then time too takes on the same meanings as space. Time is literally a continuum in that the present always reproduces the past; in the fullest sense the past is the present since the socio-political anchors of space and time derive, ideologically and culturally, directly from an ancient cosmology. In terms of practice, production is determined by the repetitive symmetry of the seasons. Both cosmology and practice require that origins are deferred to for cognitive purposes, a factor which means that the present is reversible into the past, tradition is transcendent, particularly at points of collective remembrance – festivals, power transferences, and holy days, for example.

Capitalist space and time differ from this. The key difference is the appearance of frontiers, the *territorialization* of space as a precondition for modernity as represented in the release of labour from feudal ties, the emergence of an intelligentsia independent from church and sovereign, the circulation of literature, the growth and differentiation of workforces, the appearance of the modern, bureaucratic state, machine production, and large-scale industry. Above all, modernity meant the development of free cities whose very air seemed liberating by comparison with that of the countryside, as in the saying *Stadtluft macht frei* (see Berman 1983).

Territory is national; space is re-formed as a *series* of territories, it is cut up into *segments* of different sizes and *parcelled* out internally as property according not to an economic but a political logic. Territory not only is, in the sense of belonging to, the national, it constitutes it. National icons are territory, language, and culture (Sack 1980). As these are appropriated, reformulated, and represented politically with the indispensable assistance of the intelligentsia and the apparatus of the state, they delimit the modern nation and individuate its members as national subjects. There is a fixing of insides and outsides, and gradual moves towards setting in train what Derrida (1981) analyses as the boundary paradox whereby the pure inside must be protected from the impure surplus or excess of the outside. Thus, citizenship, the segregation of aliens and their exclusion from full involvement in national life, are features of the spatial power matrix which, Poulantzas reminds us, takes its purest form in the invention of the concentration camp.

But as space is segmented into territories with insides separated and protected from outsides, there remains the problem of how to unify or homogenize the interior, especially given that territorial boundaries may have become fixed and very largely *irreversible*, except by extraordinary resort to violence (including war), and thus include diverse national elements. This the state does through national law and sovereignty including the organization and location of state apparatuses, for example the expansion of the capital city, the distribution of power to localities, and the individuation of subjects. Here, suggests Poulantzas, is the modern nation–state producing the *cellular* internal structure all possess. From these fixed bases the nation–state seeks and organizes imperial expansion, assimilating and unifying other cultures as markets, territory, and property. A key instrument of this *cumulative* process is capital interacting with national–state power in competition with other national capitals (Skocpol 1979).

The temporal matrix under capitalism is equally serial, segmented, and parcelled. Time is divided between work time and free time, the circularity of the seasonal clock is replaced by measureable, controllable, cellular time involving calendars, clocking-on machines, stopwatches, factory hooters, school bells, and so on. Time becomes property as the employee exchanges portions (or parcels) of his or her life for a wage from the employer. These requirements mean that time must be unified over space, so that, for example, the trains can run on time (Kern 1983). But as well as this practical tendency towards universalization of time, it also becomes clear that the modernity and development which are unleashed by the forces embodied in the new space–time matrix signify change and difference from the past; time is thus rendered irreversible, practice more future-oriented, and experience cumulative rather than repetitive. The new experience of change which this temporal matrix opens up, also prefigures 'the imagined communities' (Anderson 1983) of the stateless nations forging their own states, and the disposessed masses their socialist utopias.

However, with respect at least to national movements, such imagined communities must have some pre-existing rationale, notably the possession of the key national markers of language, territory, and culture for the temporal experience of modernity to be transformed into a material, political force. Because the spatial expansion of the modern nation–state co-exists with its temporal expansion as capital's time-keeper in early modern industrial regions and cities, neonational movements often take contradictory form, embodying strong elements drawn from the desire to be modern and equally strong ones drawn from premodern cultural artefacts, notably folk culture (often rurally derived, albeit subject to modern revival).

There are two key features of the space–time matrix of the modern state which act as preconditions for political mobilization of the kind usually represented in national movements. The first is the propensity of the modern state in its expansionary, imperial mode (pushing its inside outwards as it enlarges the market for its capital), to try to make all space within its territory conform to the template of the nation–state ideal, that is a *common* territory, language, and culture. And yet, normally, it never quite fulfils that ideal. This is, in large measure, due to the second feature of the modern space–time matrix, the obverse of the first, namely that nations on the receiving end of the expansionary surge, lacking their own state for defence, are of course losing their collective memory and history. These twin pressures, both pointing towards oblivion, give rise to an opposite though by no means equal pressure of cultural and political resistance. The reason the modern nation–state does not affect that oblivion which is the logic of its dynamic is twofold. In the first instance states must unify territory by consent even though they begin with coercion. Such consent is achievable by respecting the self-definition of the locally dominant class in control of the dominated mass. Where such a negotiation is not possible, as for example with aboriginal groups who may have no recognizable class system, there tends not to be a territorially defined assimilation. The cases of native Americans and Australasians, but also numerous other ethnically but not nationally defined groups in Africa (e.g. the Berber) or Asia (e.g. the Montagnards of Indochina) may be understood in this way. The second reason why stateless nations can persist in the modern nation–state is that states are not omniscient; state members are reproduced and socialized beyond the state's own sphere, in civil society where knowledge inimical to the state's unificatory interest can circulate. Hence, consciousness of the past and present condition of the dominated territory may be kept alive informally and only later, where political mobilization takes place through the medium of the political party, formally by the nation–state itself.

Nation and class

This leads to the final element in the Poulantzian theory of the nation which picks up from the point made earlier about the class constitution of the nation. Given what has been said thus far about the relative modernity of the nation as such, and its interaction with the development of capitalist production, it goes without saying that the nation is constituted in terms of class division. As Poulantzas puts it:

> The modern nation is not then the creation of the bourgeoisie, but the outcome of a *relationship* of forces between the 'modern' social classes – one in which the nation is a *stake* for the various classes (Poulantzas 1978, p. 115); emphasis in original).

Each nation–state and stateless nation forms differently as a result of the particular historic settlements between their contending classes and their allies. So the particular ways in which the space-time matrices of nations are configured is contingent. A key element in the politics which develops around the concept nation is precisely the difference between the spatiality and temporality of the dominant and the dominated classes. This gives an extra dimension to the class form of national politics in both modern nation–states and non–state nations. In certain cases it may be the party representing the dominated classes which, in a nation–state, is the more successful at modernization and spatial redistribution – as has been the case in Scandinavia, possibly Britain, and now in many southern European countries. With respect to the mobilization of support for separatism and the establishment of a state in non-state nations, it is also possible for worker interests to be the dynamic force, as occurred in Quebec in the 1970s. More usually, the patriotic nation–state card is held most firmly by dominant-class parties, and it is normally the case that successful non-state national movements draw their support from both classes, the nation, perhaps temporarily, forming the bridge between them. The nation is made and remade by the effects of struggles to impose one or other class-derived space–time matrix upon its territory. And, finally, the logic of this overall analysis, placing the formation of national territories as a precondition for capitalist development rather than the reverse, implies that the nation *per se* need not be implicated in the demise of either that mode of production or the state which sustains it. While the state may or may not wither away at a future date, for there to be the triumph of international solidarity and, putatively therefore, an end to war, it would seem, as Poulantzas (1978, p. 118) suggests, that a national materiality must persist as the precondition of an inter-*national* materiality.

Another writer who argues in complementary fashion about the irremediably political nature of national movements is Giddens (1985, pp. 209–21). He argues that nationalism consists of four elements. The first is the nation's symbolic content, responsibility for which, like Bauer and Renner, Giddens assigns to an intelligentsia, particularly historians, who in seeking to describe the historical circumstances of a territory and culture, actually help constitute them as a nation (see also Hobsbawm & Ranger 1983). Language is a key marker in this process since it alone conveys the uniqueness of the collectivity in question. Second, such sentiment may be stimulated by disruption to the routines of modern life which normally provide psychic security; if anxiety is a consequence of such disruption national leadership may be sought in a leader figure. Thus, third, nationalism has an implicit tendency to be comingled with a cultural sensibility favouring sovereignty in the form represented by the administrative power of the modern nation–state. Where nation and nation–state fit neatly there is no political friction, but where, as is commonly the case, they do not and there is a coincidence between national sentiment and the regionalization of the negative aspects of uneven development under industrial capitalism, there are likely to be demands for administrative sovereignty in the region–nation in question. Hence, fourth, national movements are inherently political in that they seek to maximize territorial administrative power for a culturally uniform collectivity.

It should be said that, while broadly in line with the analysis favoured earlier, this is a much weaker and more negative view of nationalism than that of Poulantzas. Moreover, at a crucial point, that at which the account purports to explain the rise of oppositional movements, it is remarkably economistic and reductionist. It is far from empirically the case that uneven development is the main trigger for the mobilization of national movements; it may equally be connected to religious protest, demands for linguistic equality, or other cultural and/or more simply political demands. One can, in significant part, account for Irish, Welsh, Basque, and Catalan separatism respectively in these rather than in economic terms.

Finally, it is worth noting briefly the lineaments of a debate which bears, not entirely tangentially, on the discussion so far and helps clarify some of its points. The debate concerns world system theory (Wallerstein 1979). The key argument with respect to the modern nation–state is that it exists to facilitate the functioning of capitalism on a world scale. States, as it were, slot into the differential conditions of development found across the world and assist national economies to fit into the hierarchized world economic system. This enables the ruling classes of each nation to interact in the management of the system. Specifically, core nation–states exploit peripheral ones through their ability to enforce unequal exchange in the marketplace. Wallerstein's critic Skocpol (1977, Skocpol & Trimberger 1978) attacks the notion, that capitalism preceded nation–states, on theoretical and empirical grounds. She argues that this perspective completely overlooks the internal dynamics of nation–states which were crucial to the economic development of most national economies. The system of nation–states has its own motive force which is interdependent with but not reducible to the dynamics of the world economic system. The key factors in nation–state development, argues Skocpol, are political, amongst which military struggles play an important agenda-setting role (for discussion also see Anderson 1986a, pp. 217–30). Moreover, Skocpol (1979) stresses the competitive nature of the multistate system of nations as the key mechanism of such militarization (see also Mann 1986).

Cultural and economic theories of nationalism

Thus far I have argued, drawing on a long-established and still dynamic theory of the national, that it is to be understood as, principally, a political phenomenon though one which is intimately interrelated with cultural, linguistic, spatial, and territorial, economic, and even psychological issues. In arguing this I have been pressing further along a line of analysis I developed first in the paper which sought to explain contemporary national movements within existing modern nation–states (Cooke 1984a), but which connected to other work seeking to explain *inter alia* socially and spatially uneven development at the local and regional levels and, specifically, the phenomenon of what I called 'radical regions' (Cooke 1983, 1984b, 1985a, 1985b). My general conclusions were that such phenomena could not be understood by primary reference to processes of economic uneven development, but nor could their political manifestation and organization be divorced from the ways in which the social division of labour had been developed unevenly over space. Moreover, in explaining why apparently similar economically unevenly developed spaces had projected contrasting political profiles, I placed considerable emphasis on cultural factors. These included the existence of egalitarian or inegalitarian, inclusive or exclusive practices deriving from previous social, political, and ethnic experiences usually linked to pre-industrial social relations. But, once again, it will be noted how even when discussing culture it is impossible to divorce it from the other key dimensions of social practice – economic and political particularly. In this section I want to explore the cultural and the economic explanations for nationalism and show why they are inferior to the political one presented in the previous section, but nevertheless argue that both types of theory connect to crucial segments of reality without which the political theory of nationalism and nationalism itself could not function. This discussion will be a prelude to the last section which will seek to show that what underlies and stimulates movements as political phenomena is the experience, individual and collective, of modernity.

Cultural theories

If we look at culturalist explanations of nationalism first, it is readily apparent that such an approach has attracted some of the foremost writers on national movements such as Deutsch (1966), Hechter (1975), Gellner (1964, 1983) and Smith (1979, 1982). Moreover, there are distinct differences of approach from within the culturalist frame adopted by these authors. Deutsch, for example, places the stress upon the rôle of communications as the principal means for developing the national component of the nation–state. This is achieved as development (in the economic sense) enables the communicative apparatus of the state to disseminate shared understandings and so assist in the process of generating a common sense of historical, moral, cultural, and political identity on the part of members of the collectivity. In other words, modern media, controlled or at least influenced by the nation–state, are principally responsible for developing solidaristic attachment to a specific culture and territory through the use of a dominant language to convey a sense of membership of the national community.

Clearly, the modern media are extremely powerful instruments for conveying such messages, and their rôle cannot be excluded in any analysis of the processes of cultural reproduction which are an essential part of nationalism. However, there are two obvious weaknesses of this thesis as a general explanation: first, in the past nations have been formed in the absence of the kinds of communicative apparatus available to modern states. Moreover, such nations, notably England and France, have remained highly – one might say unhealthily – exclusive in terms of their national self-definition over the years. Second, there are numerous examples of national movements achieving varying degrees of success in the absence of advanced means of communication, indeed in the anti-colonial situation, in the teeth of the possession of such instruments by their very antitheses, the colonial power. Nevertheless, the fact that control of the means of communication is seen as one of the key objectives when national struggle involves the violent overthrow of a colonial or a neo-colonial power, is indicative of the perceived centrality of communications to national mobilization.

Hechter's (1975) thesis is more all-encompassing than that of Deutsch, whom Hechter sees as adhering too rigidly to a simple cultural diffusion model (Hechter 1975, p. 25), itself based on untenable neoclassical economic assumptions regarding inter-regional equalization and integration through factor mobility. Hechter's own approach cannot be accused of that since it is based on an explicit rejection of equilibrium explanations of any kind. Rather, he starts from the position that resources are distributed unequally and that such socially uneven development can be spatially expressed where it coincides with cultural differences. Such differences can, and in many cases do, occasion a *cultural division of labour*. Such a division of labour is based, economically, upon the greater degree of modernization present in the core or dominant nation possessed of state power, unlike the dominated one. At some unspecified point hostility towards the dominant power may give rise to political responses which embody both the nation-forming work of the intelligentsia and the formation of the political party or parties pressing for independence. Where such national movements and their cultural base exist within the territorial boundaries of a dominating nation–state, which is itself culturally and linguistically different, then such dominated spaces constitute what Hechter calls (as earlier did Dobb 1963) *internal colonies*.

The internal colonial model may be summarized therefore in the following: first, spatially uneven development of industry creates modernized and less-modernized social groupings. Their power inequalities are crystallized by the initial advantage enjoyed by the first recipients of modernization. Second, the more advanced social group takes control of the state apparatus, pursuing policies which further entrench the power of its social space at the expense of subordinate, excluded social groups. Third, where ethnicity is involved this stratification system becomes the cultural division of labour, where rôles are assigned in the social structure on the basis of cultural markers (ethnicity, territory, language, religion, etc.). Thus, fourth, the cultural division of labour becomes an extra marker for distinctiveness and an objective symbol of the unfair treatment meted out to both individuals and the collectivity. Finally, such morally indefensible conditions may give rise – perhaps in circumstances where a symbolic act which epitomizes that objective inequality is performed by the nation–state – to political mobilization, provided conditions for the communication of political information (associations, clubs, a party) exist. Or, alternatively, where they do not exist such means of interest communication and representation may be *formed* by a similar process.

This model seems to explain a great deal of modern neo-nationalism, for example, the Greek, Scandinavian, contemporary Flemish, Quebecois and Celto-Brittanic movements with their varying degrees of shift away from an apparently internal colonial status. Yet it does not satisfactorily explain Catalan and Basque national movements, where the cultural division of labour (though not the political) works to the advantage of the minority cultures from an economic point of view, nor that of the admittedly much smaller Val d'Aosta province in Italy which is the richest in the whole country. The problem with Hechter's otherwise subtle and generally well theorized analysis is that, on the one hand, minority culture is equated too uncritically with economic backwardness, and on the other, it has absolutely no purchase on the nationalisms of the successful modernizers – England, France, Germany, etc. – which, as Giddens (1985, p. 213) notes, can scarcely be explained solely in terms of economic backwardness either. In brief, therefore, Hechter's theory rests on a misconception of the relationship between cultural minority status and economic backwardness and an inappropriate reduction of the latter to the former in his quest for the origin of national movements.

Gellner's position is somewhere between that of Deutsch and Hechter in that he is concerned to establish the link, which Deutsch fails to do, between the diffusion of communication and the development of a national consciousness. He does this by reference to culture as the mediating and determining factor in that process as follows:

> If a man is not firmly set in a social niche, he is obliged to carry his identity with him, in his whole style of conduct and expression: in other words his 'culture' becomes his identity. And the classification of men by 'culture' is of course the classification by 'nationality' (Gellner 1964, p. 157).

So culture is the process whereby individuals receive a kind of personality impress which marks them off as different from those in receipt of a different imprinting experience. The process is activated by mass education, mass literacy, and the homogenization of culture. These function in the interests of modern industrialism which requires an educated, mobile workforce, and the modern nation–state which seeks compliance from its subjects. Language is the key mediator and marker of the national territory. But Gellner's analysis is, if anything, more problematic than Hechter's mainly because he tends to take a top–down view of the nation as nation–state. Thus he fails adequately to deal with the many examples of national movements whose territory is linguistically divided over space and in which, for those movements, the provision of a common language is not the solution but precisely the national problem. Moreover, he has a very reductionist view of the relation between cultural difference and the birth of new nations, simply asserting at one point that where territory contains an immovable cultural frontier it will automatically result in two nationalisms. However, this seems not to have happened where old, non-state nations are bisected by a new nation–state boundary as, for example, with regard to both Basque and Catalan groupings in Spain and France, or the Kurds in Iran and Iraq, or even for Catholic Irish in Northern Ireland for whom such more recent boundaries are less real in some ways than the 'imagined communities' to which they feel they belong.

The last culturalist theory to be outlined here is that of Smith (1979, 1982). He criticizes all the foregoing cultural theories for being too wedded to economic development as the key influence on the emergence of national movements, and it is certainly true that all are more economically reductionist in their chronology of national movements than, say, Bauer, Poulantzas, or Giddens – an irony given their privileging of the cultural sphere in general terms. Smith's framework addresses four questions:

(a) What is the social base of the ethnic nationalism?
(b) How and why is its intelligentsia politicized?
(c) Why separatism rather than other political routes?
(d) What are the conditions of success for contemporary national movements?

Smith's point of departure is modern bureaucracy which is interventionist, powerful, and efficient. It needs boundaries and territory within which to operate and a modern type of person as its agent, the modern intelligentsia with an understanding of historical forces for communal change. Out of this process emerges cultural nationalism. This becomes political when ethnic minorities, competing for jobs in the cities, experience discrimination and exclusion. The turn to separatism can be explained in terms of the political and economic cycle. As the latter enters decline, so the labourmarket dries up and there is a questioning of state failure. The issue is whether to seek more autonomy and a better deal for the minority nation or whether that should be sought in a new, separate nation. That dilemma may be solved by the action or inaction of the nation-state. The conditions for a successful separatism would seem to include the support of a powerful neighbour (possibly a superpower), and with the balance of forces as delicately poised as it is currently separatism is less likely to be achieved than limited autonomy. So this theory has a cultural dimension to it, but it also has an equally strong political analytic element in its content. As such, it is more in line with the here-favoured approach to explaining national movements. However, its economic content is extremely weak, strikingly reductionist – to speak of economic cycles interacting with political cycles to cause nationalist revivals is alarmingly crude, given the care with which other parts of the analysis are prepared. But perhaps the biggest flaw is its emphasis on the bureaucracy as the leading edge of national movement. It is a common misconception of culturalist theorists that ideas as translated into book form cause revolutions whether national or social, or both. The educated bureaucracy, skilled in administration and accultured to historic national sentiment therefore appear – as Max Weber balefully thought – to be the leading edge in such struggles. The reality is often different. Most national movements do not consist of bureaucrats but workers, the unemployed, small professionals, and the petit bourgeoisie. Such movements often press, not unsuccessfully, *for* a bureaucracy of their own rather than the remote one located elsewhere which rules the life of the non-state nation. National movements are led by culturally informed, economically sensitive *political parties* who struggle to create spaces into which national demands are channelled. It is instructive that none of the theorists of the cultural persuasion, and few of any persuasion, pay serious attention to this precise instrument of national movements. In the next subsection I shall examine the work of some economic theorists of nationalism to see whether they can do any better.

Economic theories

The most fully developed of the theories of nationalism which give primacy to economic development processes as the ultimate stimulus is that of Nairn (1981). He is quite clear and undoubtedly correct in answering his own question about the origins of the phenomenon:

> How may we describe the general outlines of nationalist development, seen as 'general historical process'? Here, by far the most important point is that nationalism is *as a whole* quite incomprehensible outside the context of that process's *uneven* development (Nairn 1981, p. 96; emphasis in original).

Later he draws on Gellner's characterization of nationalism as a phenomenon intermingled with the uneven diffusion of industrialization or modernization and underlines its economic dimension by reference to the ways in which free trade (which England dominated) embodied a kind of economic imperialism.

This, in turn, spurred Germany, Italy, and other European countries to develop the sort of national consciousness which would result in nation–state formation, thus enabling them to compete economically with English industry. Hence, the general historic process is clearly, in Nairn's (and, one might add, Gellner's) view a fundamentally economic one, where the industry of one country develops more rapidly than that of another, a factor which tows other parts of the socio-economic system of that country into modernization too, eventually.

Inevitably, nationalism is a bourgeois phenomenon; the industrial, commercial, and professional classes have a clear interest in overturning the absolutist, archaic, and aristocratic regimes in their empires, city–states and prefigurative nations. But within the bourgeoisie it is the intellectual strata that provide the unificatory ideas through their history writing, philosophy, and involvement in the burgeoning school and university system. However, the way in which this system develops is left unexamined by Nairn, as it is by his mentor Hobsbawm (1962), although both note how schools and universities tend to be the most ardent champions of nationalistic ideals. As we saw earlier, Smith (1979, 1982) has a good explanation for this in the struggle by the nascent service class to create a space for itself within the class structure; nationalism is a powerful weapon in that struggle, which can both unify divergent class interests but also distance the intelligentsia from capital and labour, with their often strong internationalist motivations, and thereby secure the patriotic middle ground for themselves.

Nairn hints at a grasp of the political dynamic of national movements in his discussion of the relations between the intelligentsia and the popular masses. Intellectuals draw heavily upon the folk culture of the still relatively unmodernized popular classes (peasants, artisans, workers) for ideological purposes, but also need their political power to mobilize civil society and rid the state of its absolutist rulers. Unfortunately, Nairn does not go on to look at the precise mechanisms – the political parties – which are the necessary but not sufficient means of bringing about successful national *rapprochements*. This has rather destructive effects upon his analysis and prognosis of 'the Break-up of Britain', especially where that analysis is at its most fine-grained – as it is in his discussion of the neo-nationalist movement in contemporary Scotland. He struggles for many pages with an unsuccessful analysis of the reasons for Scotland's nationalistic slumbers in the 19th century despite its apparent possession of most of the necessary socio-economic (but not political, i.e. no nationalist part or nationalist strand within a dominant party) preconditions for a national movement. Nairn says Scotland was too culturally, philosophically, and economically advanced, but the reality is that it was politically reactionary. Quoting Lord Cockburn, Nairn answers but does not analyze the absence of political nationalism:

> If Scotch Jacobinism did not exist, Scotch Toryism did, and with a vengeance. This *party* engrossed almost the whole wealth, and rank, and public office, of the country, and at least three-fourths of the population (Cockburn 1856, quoted by Nairn 1981, p. 119; emphasis added).

More recently, it could be added, the failure of contemporary Scottish nationalism, again given propitious economic conditions in the form of dramatic economic weakness and the prospect of an economic saviour – oil –

being claimed not for Scotland but for the UK, can be traced not to the lack of a party but its failure to make the link between the interests of the intellectual and professional strata in an independent state and those of the working-class Labour voter in the benefits of such a state. Far from recognizing this failure of politics, the party in question actually expelled the factions most alert to the need for such a link.

Finally, Nairn's tendency towards an economic determinist analysis often leads him to overlook the ways in which nationalist movements gain force from being recipients of political decisions poorly adapted to meet the requirements of the nation in question. This responsive politics, as it may be termed, has much in common with the ways in which social movements form at a more local level than that of a nation. The concept of urban social movement introduced by Castells (1977) has been usefully transferred into the regional field by Hadjimichalis (1985). Here the argument, in a nutshell, is that uneven development leaves some parts of countries (or parts of cities) economically and socially disadvantaged. The nation–state, responding to pressure from regional interest-representatives (regional capital or labour organizations, groups of MPs, etc.) produces typically hamfisted policies – often of a modernizing kind – which exacerbate the problems of disadvantage as perceived locally. For example, large construction programmes may introduce migrant labour from elsewhere with few jobs going to locals, or cities may be made into growth poles at the expense of rural areas, or mass tourism may be induced with, once again, relatively few of the economic benefits entering local pockets and, worse, significant cultural, social, or political damage to fragile local structures being a result. Such *regionalization* policies may result in a growth of *regionalism* as a political response. Usually this does not lead to separatist demands, but it may result in some pressures for greater decision-making autonomy. Where the region in question is a non-state nation, the effect may be similar but more extreme in that pressures moving beyond those likely to be satisfied by degrees of autonomy may be unleashed.

A similar weakness is implicit in the approach adopted in the work of Brustein (1981) who seeks to explain regionalist and nationalist autonomist and separatist pressures in terms of the ways in which different regional modes of production are distributed over space, and result in spatially uneven development. Such regional modes of production are determined by the particular form of property rights found in different spatial locations. These influence the form of organization of production, which, in turn, determines settlement patterns, social structure, and relative resource advantage over space. Thus, territorially specific social groupings perceive political interests in terms of their regional mode of production. This implies that voters support parties most likely to further their own material interest. Brustein's is a fairly straightforward economic determinist theory of party allegiance which works for those aspects of voter-intention which are governed by economic concerns, but not for those which are not. It could be argued that national movements are less motivated by narrow economic issues than most political movements in that they frequently have concern for matters cultural, social, and linguistic which have very little directly to do with economic advantage.

In general, therefore, the conclusion drawn here is that while the economic theorists of national movements offer an insight, often theoretically quite well worked out, into the connections between uneven economic and spatial development and territorial political mobilization around the concept of nation, there are too many theoretical, hence, ultimately, empirical lacunae in this approach to make it other than a useful adjunct to a more generalizable theory of nationalism. Probably the major flaw in each of these approaches is the assumption that what happens in the economy ultimately determines what happens in society and politics. The contention developed here in contradistinction to the above thesis is that the political constitution of the nation is a necessary first step to the processes which produce spatially uneven development. This generalization applies to the first nations, the absolutist states of England, France, and Spain whose political unification had definitive implications for their subsequent spatial development. It applies equally to the later nation–states, such as Germany, Italy, and Japan for whom economic development (catching up) would have been impossible without the spatial delimitation and internal unification of the nations in question. Lastly, it applies to those even-later nations and nation–states who have struggled successfully or unsuccessfully for independence as a means towards the achievement of their own variety of internally influenced spatially uneven development and against the effects of that distortion of their economic geography which colonialism, internal or external, has imposed upon them. For the latter, whether in India (see Hobsbawm 1968), Ireland (see Perrons 1986), Flanders (see Mandel 1963), or Wales (see Cooke 1980) was normally secured by the discriminatory application of political power. This prevented development of one kind, often by exclusionary laws, and enforced development more suited to the interests responsible for the emergent social and spatial division of labour in the dominant nation–state.

Thus the last remaining questions concern first, of what shape is the model that is being here proposed as a generalizable account of the development of nationalism? And second, which are the forces responsible for unleashing what has clearly become one of the most powerful social instruments available to humankind, namely the division of the Earth's surface into nations for the appropriation of its resources with uneven effect over space and time? In order to arrive at an answer to the first question it will be necessary to tackle, fairly briefly, the second one and this I attempt now.

Modernity and the politics of national movements

Throughout this chapter, the terms modernization and modernity have been used in passing as reference has been made to theorizations of nationalism from the political, cultural, and economic analytical wings. Of those whose work has been considered, probably only Poulantzas is wedded so closely to a neo-Marxist perspective as to be unwilling to deploy the language of the intellectual enemy, the theorists of modernization who used structural–functionalist sociology and neoclassical economic theory to explain underdevelopment, and, viewed from a radical perspective, justify it. However, of late a different and radical (though by no means exclusively so) theorization of *modernity* as a totalizing experience has emerged and for the reason to be discussed below, I propose to incorporate aspects of this theorization into the model to be elaborated subsequently.

One of the best reasons for deploying a theorized concept such as modernity in this context is that it enables the model so developed to be distanced somewhat from the constraints of modes of production theorizing of the onset of capitalism with its over-restrictive privileging of the economic sphere over all others in social analysis. I do not want to develop a model of nationalism which privileges the sphere of politics in a language which inevitably tends to drive the discourse towards a privileging of economics. A secondary reason for deploying modernity as the discourse of the model is that it is able, unlike Marxism, to relate aspects of the social development process to the self-development of the individual without reducing one to the other, or more importantly, either to the economic development process. It therefore suits the analysis better, but also I believe it can offer an explanation for the political turn which results in nationalism, and in most, though not all cases, in nation–states, through offering a sound analytical basis for the political phenomenon of the specific national movement.

So, what does the theory of modernity tell us? As expressed in the work of Berman (1983) and others such as Elias (1978, 1982) and Habermas (1981) modernity is the cultural experience associated with the break in the past signified by the onset of the Enlightenment. This, in turn, is inextricably mixed with the progress of reason as a mode of thinking about the world, and rationalization as a means of intervening in it. The appearance, in a dominant cultural posture, of rationality and rationalization are, precisely, first order abstractions signifying the second order abstractions of modernity and modernization. Berman takes this analysis further in his discussion not of a two-pronged conceptualization of modernity, but of a three-pronged one. In Berman's terms, *modernity* is the experience of reason, *modernization* is the objective external product of the application of reason to the world – the unleashing of new economic forces, social change, mass migrations from countryside to cities, etc. – while *modernism* is the cultural vision of reason as expressed in the arts, literature, music, and architecture.

Now, the important factor in this formulation – apart from its capacity to integrate, in objective terms, the subjective experiences, the structural processes, and the creative energies – is that it acts as an invitation to think precisely about what modernity is. Modernity is, above all, the consciousness, that deciding on courses of action – individual or collective – on the basis of rational discourse with oneself or others than on traditional recipes is both a liberating and a frightening experience. As Berman puts it in a much cited quotation:

> To be modern is to find ourselves in an environment that promises us adventure, power, joy, growth, transformation of ourselves and the world – and, at the same time, that threatens to destroy everything that we have, everything we know, everything we are. Modern environments and experiences cut across all boundaries of geography and ethnicity, of religion and ideology . . . . But . . . it pours us all into a maelstrom of perpetual disintegration and renewal, of struggle and contradiction, of ambiguity and anguish. To be modern is to be part of a universe, in which, as Marx said, 'all that is solid melts into air' (Berman 1983, p. 15).

This seems to me to capture much of what nationalism consists in: its Janus face, looking backwards in order to proceed into an unknown future; its tension between tradition and modernity, the disrupting effect which stimulates often great, certainly new art forms; the need to intellectualize about change, loss, and opportunity. The experience of modernity, principally unleashed through practices, cultural, social, political, and economic, which took as their starting point the generalization of Enlightenment modes of thinking is what, amongst the other things it did, directly underpinned the emergence of nationalism. Defining territory, seeking to universalize linguistic and cultural usages within it, delimiting interiors and exteriors, segmenting space and so on, are precisely social means of controlling the feared disintegration of individual and collective identity, protecting a sphere for the pursuit of opportunity, and saying a qualified 'Yes' to modernity.

So, to move towards a reprise and summary of the key points in the model adumbrated in the preceding pages, a model which stresses the generalizability of a fundamentally *political* theory of nationalism and the movement which sustains it, the following are elemental:

(a) the availability to the educated classes of a mode of reasoning, and/or evidence of its individual and social implications, which may be referred to as modernity;
(b) the capacity in the intelligentsia to modernize society to: apprehend the nature of modernity; produce literary, historical, and philosophical interpretations of its cultural and territorial content; and politicize the dominant class to effect the development of the space–time matrix to control it;
(c) the development of institutions in civil society, notably appropriate political parties, such that the class divisions of modern, capitalist territory can be accommodated without the exclusion of the popular masses;
(d) the articulation of an objective to secure a coherent cultural, linguistic, and institutional apparatus – the state – to protect the nation in its political, economic, and military dealings with other states, and to organize its internal political and economic life;
(e) the emergence of relationships of spatially uneven development which, given the cultural, linguistic, and territorial preconditions of the nation, focus political consciousness and practice upon political demands for national self-determination in this and related spheres.

This formulation, I believe, captures the main co-ordinates of the concrete concept of the nation and the dynamic force of the national movement without which a nation cannot exist. It is applicable to historic nation–states and to those historic and non-historic nations which are continuing to seek to fulfil the political process of development to either full nationhood and statehood or full nationhood in the context of a multinational state. It avoids the weaknesses of attempting to reduce and 'read off' national development from psychological, cultural, or economic indicators which have in the past only produced non-generalizable deductions, and proposes – for what is incontrovertibly a political process – a political theorization of the national question. Such a proposal, rather surprisingly, has received remarkably little serious theoretical or empirical attention until relatively recently, in particular until the advent of Poulantzas's theoretical insights in the late 1970s.

Concluding remarks

Geographers have for too long been neglectful of political geography and this neglect may have cost them dear in terms of the development and prestige of the subject. In this chapter I have sought to show that the main lineaments of the modern world, in which I take it all geographers are closely interested, derive from the response of all societies to the experience of modernity which took a general form, that of nation-building. But I go further, following Poulantzas (1978) in arguing that the geography of the modern world, its economic patterns (though not necessarily all its economic processes) and its political structures all derive ultimately from the contingent meeting of modernism and nationalism on the road from antiquity.

Nationalism is responsible for the structure of virtually all of the non-natural environment of the Earth's surface. There can be no question that the varying distributions of social attributes of modern populations derive from the emergence at different times and in different spaces, of nations. More than that, the constitution of nations involved a fundamental change in the nature of space, and time, once modernity had made its appearance. Nations introduced frontiers, cores, and peripheries, reshaped the social and economic landscape, and restructured the relations within and between these new social spaces.

It is my view that by focusing on this rather neglected corner of until recently, a relatively neglected subdiscipline of geography, political geography, a surprising richness of insight into the nature of the modern world and the nature of geography can be harvested. This chapter touches the surface, by offering a new and almost completely untried model of geographic process viewed from the political vantage point of the nation-building process. It may be that critical thinking about what I have written will enhance the theoretical development of geography, a discipline which, I hope, will not continue to underestimate its importance to social science.

References

Agnew, J. 1981. Structural and dialectical theories of political regionalism. In *Political studies from spatial perspectives*, A. Burnett & P. Taylor (eds). Chichester: Wiley.

Agnew, J., J. Mercer, & D. Sopher (eds) 1984. *The city in cultural context*. London: Allen & Unwin.

Anderson, B. 1983. *Imagined communities*. London: Verso.

Anderson, J. 1975. *The political economy of urbanism: an introduction and bibliography*. London: Architectural Association, Department of Urban and Regional Planning.

Anderson, J. 1980. Regions and religions in Ireland: a short critique of the two nations theory. *Antipode* **12**, 44–53.

Anderson, J. (ed.) 1986a. *The rise of the modern state*. Brighton: Wheatsheaf.

Anderson, J. 1986b. On theories of nationalism and the size of states. *Antipode* **18**, 218–32.

Anderson, P. 1974a. *Lineages of the absolutist state*. London: New Left Books.

Anderson, P. 1974b. *Passages from antiquity to feudalism*. London: New Left Books.

Bauer, O. 1978. The concept of the 'nation'. In *Austro-marxism*, T. Bottomore & P. Goode (eds). Oxford: Clarendon Press.

Berman, M. 1983. *All that is solid melts into air*. London: Verso.

Blaut, J. 1975. Where was capitalism born? In *Radical geography*, R. Peet (ed.). London: Methuen.

Blaut, J. 1980. Nairn on nationalism. *Antipode* **12**, 1–17.

Blaut, J. 1982. Nationalism as an autonomous social force. *Science and Society* **46**, 1–24.

Blaut, J. 1986. A theory of nationalism. *Antipode* **18**, 5–10.

Bottomore, T. & P. Goode (eds) 1978. *Austro-marxism*. Oxford: Clarendon Press.

Brustein, W. 1981. A regional mode-of-production analysis of political behaviour. *Politics and Society* **10**, 355–98.

Capel, H. 1981. Institutionalization of geography and strategies of change. In *Geography, ideology and social concern*, D. Stoddart (ed.). Oxford: Basil Blackwell.

Castells, M. 1977. *The urban question*. London: Edward Arnold.

Chase-Dunn, C. 1984. Urbanization in the world system. In *Cities in transformation*, M. P. Smith (ed.). Beverly Hills: Sage Publications.

Cooke, P. 1980. Dependency formation and modes of integration: the historic and contemporary case of Wales. *Papers in Planning Research* 15. Cardiff: University of Wales Institute of Science and Technology.

Cooke, P. 1983. *Theories of planning and spatial development*. London: Hutchinson.

Cooke, P. 1984a. Recent theories of political regionalism: a critique and an alternative proposal. *International Journal of Urban and Regional Research* **8**, 549–71.

Cooke, P. 1984b. *Region, class and gender: a European comparison*. Oxford: Pergamon.

Cooke, P. 1985a. Class practices as regional markers. In *Social relations and spatial structures*, D. Gregory & J. Urry (eds). London: Macmillan.

Cooke, P. 1985b. Radical regions? Space, time and gender relations in Emilia, Provence and South Wales. In *Political action and social identity*, G. Rees (ed.). London: Macmillan.

Derrida, J. 1981. *Dissemination*. London: Athlone Press.

Deutsch, K. 1966. *Nationalism and social communication*. Boston: MIT Press.

Dobb, M. 1963. *Studies in the development of capitalism*. London: Routledge & Kegan Paul.

Drakakis-Smith, D. & S. Williams 1983. *Internal colonialism: essays around a theme*. London: Institute of British Geographers.

Elias, N. 1978. *The civilizing process*. Oxford: Basil Blackwell.

Elias, N. 1982. *The civilizing process*. Vol. 2: *State formation and civilization*. Oxford: Basil Blackwell.

Gellner, E. 1964. *Thought and change*. London: Weidenfeld & Nicolson.

Gellner, E. 1983. *Nations and nationalism*. London: Weidenfeld & Nicolson.

Giddens, A. 1985. *The nation–state and violence*. Cambridge: Polity Press.

Habermas, J. 1981. Modernity versus postmodernity. *New German Critique* **22**, 3–14.

Hadjimichalis, C. 1985. Regional crisis: the state and regional social movements in southern Europe. In *The crises of European regions*, D. Seers & K. Ostrom (eds). London: Macmillan.

Hechter, M. 1975. *Internal colonialism*. London: Routledge & Kegan Paul.

Hobsbawm, E. 1962. *The age of revolution: Europe 1789–1848*. London: Weidenfeld & Nicolson.

Hobsbawm, E. 1968. *Industry and empire*. Harmondsworth: Penguin.

Hobsbawm, E. & T. Ranger 1983. *The invention of tradition*. Cambridge: Cambridge University Press.

Holloway, J. & S. Picciotto 1978. *State and capital*. London: Edward Arnold.

Holton, R. 1985. *Cities, capitalism and civilization*. London: Allen & Unwin.

Kern, S. 1983. *The culture of time and space 1880–1918*. London: Weidenfeld & Nicolson.

King, A. 1976. *Colonial urban development*. London: Routledge & Kegan Paul.

Kofman, E. 1981. Functional regionalism and alternative regional development programmes in Corsica. *Regional Studies* **15**, 173–81.

Kofman, E. 1982. Differential modernism, social conflicts and ethno-regionalism in Corsica. *Ethnic and Racial Studies* **5**, 300–12.

Kofman, E. 1985. Regional autonomy and the one and indivisible French Republic. *Environment and Planning C, Government and Policy* **3**, 11–25.

MacLaughlin, J. 1986. State-centred social science and the anarchist critique: ideology in political geography. *Antipode* **18**, 11–38.

Mandel, E. 1963. The dialectic of class and region in Belgium. *New Left Review* **20**, 5–31.

Mann, M. 1986. *The sources of social power*. Vol. 1: *A history of power from the beginning to A.D. 1760*. Cambridge: Cambridge University Press.

Miliband, R. 1969. *The state in capitalist society*. London: Weidenfeld & Nicolson.

Nairn, T. 1981. *The break-up of Britain*, 2nd edn. London: Verso.

Perrons, D. 1986. Unequal integration in global Fordism: the case of Ireland. In *Production, Work, Territory*, A. Scott & M. Storper (eds). London: Allen & Unwin.

Portes, A. & J. Walton 1977. *Urban Latin America: the political condition from above and below*. Austin: University of Texas Press.

Poulantzas, N. 1973. *Political power and social classes*. London: New Left Books.

Poulantzas, N. 1975. *Classes in contemporary capitalism*. London: New Left Books.

Poulantzas, N. 1978. *State, power, socialism*. London: New Left Books.

Pringle, D. 1982. Marxism, the national question and the conflict in Northern Ireland: a response to Blaut. *Antipode* **14**, 21–32.

Sack, R. 1980. *Conceptions of space in social thought*. London: Macmillan.

Skocpol, T. 1977. Wallerstein's world capitalist system: a theoretical and historical critique. *American Journal of Sociology* **82**, 1075–90.

Skocpol, T. 1979. *States and social revolutions: a comparative analysis of France, Russia and China*. Cambridge; Cambridge University Press.

Skocpol, T. & E. Trimberger 1978. Revolutions and the world-historical development of capitalism. *Berkeley Journal of Sociology* **22**, 101–13.

Smith, A. 1971. *Theories of nationalism*. London: Duckworth.

Smith, A. 1979. *Nationalism in the twentieth century*. Oxford: Martin Robertson.

Smith, A. 1982. Nationalism, ethnic separatism and the intelligentsia. In *National separatism*, C. Williams (ed.). Cardiff: University of Wales Press.

Soja, E., R. Morales & E. Wolff 1983. Urban restructuring: an analysis of social and spatial change in Los Angeles. *Economic Geography* **59**, 195–230.

Thrift, N. & D. Forbes 1986. *The price of war: urbanization in Vietnam 1954–1985*. London: Allen & Unwin.

Wallerstein, I. 1979. *The capitalist world economy*. Cambridge: Cambridge University Press.

Williams, C. (ed.) 1982. *National separatism*. Cardiff: University of Wales Press.

Williams, C. 1985. When nationalists challenge: when nationalists rule. *Environment and Planning C, Government and Policy* **3**, 27–48.

11 *The state, political geography, and geography*

R. J. Johnston

A major component of the spatial organization of the Earth's surface is its division into 150 or so sovereign states, each with a well defined (though perhaps contested) territorial reach. This division is commonly used in social sciences as a basic set of units of analysis – the world is described, and accounted for, in terms of similarities and differences among its states. That division is often accepted as unproblematic: it is natural to use countries as reporting units (despite some debate over whether countries are regions (Johnston 1984d)). The validity of this approach has rarely been questioned by political geographers, among whom the need for, and hence existence of, states has normally been taken for granted (Johnston 1980a, 1981a, 1981b). Further, the links between political geography and economic and social geography have rarely been explored, so that until recently the rôles of the state in everyday life, and the importance of state territory in that, have not been central geographical concerns.

Attention has recently been directed towards the study of the state, however, and to incorporating political geography within the broader corpus of the discipline. In part this reflects the growth of the institution and apparatus of the state; it has been enlarged into a phenomenon that cannot be ignored in analyses of the contemporary scene (as many of the other chapters in this book indicate). There are studies not only of what the state does, but also of why: to understand state action, it is argued, we must understand why we have states.

The issue of why we have states could be dismissed as of little relevance to the study of human geography. However, according to some arguments (e.g. Mann 1984, 1986) the state differs from other institutions *because it is necessarily a territorial body* associated with a clearly defined area. For this reason, geography is crucial to the understanding of the state – and, of course, the state is crucial to the understanding of geography.

The focus of this chapter is a society-centred view of the state; it identifies the state as necessary to the operations of society (especially capitalist society, which receives most attention) and develops an understanding of state operations accordingly. This does not imply a deterministic approach; the autonomy of the state (strictly speaking, of those who run the state) is recognized as a funtion of its particular situation in the societal superstructure. Other approaches – state-centred rather than society-centred – are possible. For example, Lovering (1987; see also Giddens 1985) suggests an internal/external conflict between racial groups (citizens/non-citizens; us/them) which is distinct from the capital/labour conflict of the capitalist mode of production, which involves territorial boundedness, and which provides the basis for a nation-state system that may be separate from the class conflict (see also Lovering 1986). For these, he argues that 'the nation-state has causal powers logically independent of those of capital or classes' (1986, p. 35). The two overlap and real world events take place where capital/labour and citizen/alien conflicts intersect. Mann (1986) presents a somewhat similar conception, identifying four types of power – ideological, economic, military, and political – which represent different ways of pursuing human goals. They intersect as power struggles; each comprises a power network and at any one time and place one of those networks dominates social organization. Thus in some situations, military power will dominate the structure of a particular society, whereas in others economic power may. (And, of course, the four are not as readily isolated empirically as they are theoretically.)

A full analysis of the state and political geography requires equal treatment of the state-centred and society-centred approaches. Here, the latter gets virtually all of the attention, in a brief overview that focuses on why we have states (why political, military, and ideological power must be linked to economic) in modern societies and why those states are necessarily territorial institutions. Thus the first section considers the need for the state in conflict-ridden capitalist societies and the territorial requirements of the rôles that it must play. After this, attention turns to the autonomy of the state, and then to the concept of the state apparatus and to the running of the state. A final brief section deals with the pattern of states and geopolitics.

Towards an understanding of the state

Why do we have states? A variety of answers to this question is available, representing not only different academic interpretations of the state but also separate ideological justifications for its existence and actions (the two are often linked). Many (as summarized in Johnston 1982) are little more than descriptions of state functions without analyses of why those functions are performed: thus the state may be presented as a protector, a neutral arbitrator, a facilitator, and a cohesive force, as an investor, and as a bureaucracy – all of which descriptions are true, but why the state acts in those ways is not explained.

Theories of the state can be grouped into three general categories (see, for example, Alford & Friedland 1985). The first – *the pluralist grouping* – focuses on the state as a locus of decision making in the context of expressed demands. The second – *the managerialist grouping* – situates it as a separate institution within society, acting independently of other institutions. And the third – *the instrumentalist grouping* – presents the state as part of the control mechanism established by the dominant class within society. These three are not reviewed in turn, but the chapter presents a picture of the state that requires elements of each.

Conflict and capitalist societies

One of the most compelling arguments presented as a rationale for the state is that without it a society founded on and driven by the forces of competition would not survive. The state provides the necessary regulation of these forces, thereby ensuring that individual self-interest does not defeat the collective interest; realization of this leads to the acceptance of the state by all.

A simple illustration of this argument is Hardin's (1968) classic paper which uses the example of overgrazing on common land. It makes clear the case that although restraint is in everybody's interest, it is in no one person's interest to exercise self-restraint if there is no guarantee that all others will too. Lacking such a guarantee, the common good will only be achieved by an external body that imposes restraint on all; such a body is the state, and eventually world government. Thus the need for public policy making via the state, a body whose power to impose solutions on all within its territory is accepted (see also, for variants on this case, and arguments against it, Rawls 1972, Brams 1975, Taylor 1976, Laver 1981, 1986, Clark 1986; a contemporary example, the regulation of shop opening hours in the UK, is provided by Blomley 1986).

Hardin's illustration was developed with regard to the pressure on environmental resources, but it can be applied to a great range of conflicts in which the pursuit of individual self-interest, as the only rational action in the absence of restraints, is ultimately self-defeating in that the conflict destroys the system (the resource base in Hardin's example). Conflict is inherent to the capitalist mode of production – although it is usually unequal conflict, and not a conflict of equals, as in Hardin's example – and thus capitalism is inherently unstable, containing within itself the seeds of its own destruction.

Only the briefest sketch of that inherent conflict and its many different components can be presented here (based largely on the full and clear presentation in Harvey 1982, and his succinct summary in Harvey 1985c). Capitalism is driven by the need to accumulate wealth, realized through profit taking. Profits are achieved through selling the products of labour in competition with other products; price is fundamental to that competition, so survival means that labour costs must be continually reduced. Eventually this leads to crises of overproduction (or under consumption); the greater efficiency of labour means that more can be produced than markets can absorb, and thus profitablity declines. This decline exacerbates the conflict between capital (the investors and profit takers) and labour (the profit makers, who seek to increase their proportion of the profits through increased real wages and living and working conditions).

The capital/labour conflict is divided into a large number of subconflicts, between different sectors of production, different organizations within and across those sectors, and different places. In addition there are conflicts within capital and within labour, again spatially disaggregated, and attempts to resolve some of these involves the switching of resources (Harvey 1982), with capital much more mobile than labour (Peet 1983, 1986). In total, this suggests five types of conflict:

(a) between capital and labour in one place;
(b) between capital in one place and capital in another;
(c) between labour in one place and labour in another;
(d) between different segments of capital in one place; and
(e) between different segments of labour in one place.

Some of these conflicts, notably (a), (b), and (d) occur in the sphere of production; others, in particular (c) and (e), occur there and also in the sphere of reproduction, as different segments of labour compete for both privately and publicly produced goods and services. Containing such conflicts involves the state, with individuals and the various interest groups accepting the existence of such a body (i.e. they give it legitimacy) and using it to press their particular claims.

The state and conflict

The above discussion indicates that capitalism is built on conflict between and within its two main classes – capital and labour – and that it is inherently unstable because of the nature of its dynamic forces. If the conflicts were not contained, that instability would be even greater. The state has thus evolved as a necessary institution to capitalism, to contain conflict and regulate the instability. The conflicts are, in part, between people in places, and for this reason the state is necessarily a spatially identified institution (Mann 1984).

The ways in which a separate body – the state – contains and regulates conflict and competition are many. It performs particular rôles, however, and their identification has recently claimed academic attention. Three rôles are suggested by several writers (e.g. O'Connor 1973), and Clark & Dear (1984, p. 43) have argued that they can be ordered according to their importance.

The first rôle, they claim, is *securing social consensus*, whereby all residents of a state's territory accept certain rules for the operation of society. Without such acceptance, there is no order, stability, or security, and thus no incentive for the investment of capital. It is necessary for the state to create those conditions through, for example, rules relating to ownership of property to contracts, to inter-class relations, and to inter-personal relations. Only with such rules, accepted (often implicitly, and with the threat of coercion if necessary) by all, will production and exchange – the basis of profitability – be undertaken and reproduction assured (see Johnston 1984a).

Second, the state must *secure the conditions of production* through the provision of a co-ordinated infrastructure within which production and exchange can take place; this infrastructure is both physical (e.g. communications systems) and abstract (e.g. a monetary system). Its rôle is thus to guarantee the conditions for profit making, thereby advancing the interests of capital over those of labour. If necessary it must invest, directly or indirectly, in production, and also in reproduction of the labourforce – not just in its physical reproduction but also in its intellectual reproduction, through the creation of necessary skills.

Third, the state must *ensure social integration* by ensuring the basic welfare of all, especially that of the exploited groups within society (labour) who gain least from it and who are therefore most likely to attack the system and harm (if not destroy) it. This usually involves what is widely known as the welfare state – direct state provision or subsidy for the means of consumption and reproduction – as well as an ideological function: creating an acceptance of inequality.

These three rôles can be separated and identified theoretically. Empirically, they may be intertwined (Taylor & Johnston 1984). For example, education can be involved in all three. Social consensus can be promoted through the educational system, which instils the disciplines of working in a capitalist society and ensures acceptance of the implicit social contract. Further, it is one way of securing the conditions of production, since it provides the skills necessary for the conduct of the various tasks undertaken by labour. And it is also involved in securing social integration, through the ideological function, for example (instilling a national – i.e. state territory – identity).

The state as place

Why, in undertaking these three rôles, is the state a place, an institution with a defined territory over which its sovereignty is (generally) accepted both by the residents of that territory and by the residents and governments (those controlling the state) of all others? 'According to Mann (1984, p. 185) the state is 'a place, an *arena*, in which the struggles of classes, interest groups and individuals are expressed and institutionalised.' Mann's argument can be interpreted as a particular example of the concept of territoriality. This was first popularized by Ardrey (1969), who argued that it is an innate characteristic of humans to organize their lives and their societies in clearly defined territorial bases (see also Pickles 1985), thereby using territory to achieve what he identifies as the three basic human needs: identity, stimulation, and security. A territorially defined institution, i.e. a state, can provide all three. Sack (1983, 1986) has pursued a theory of human territoriality further, without exploring the phenomenological issues raised by Ardrey and Pickles. Social organization involves the exercise of power, and territoriality is a strategy for implementing control, he argues, because of its particular characteristics. Thus, for example, territory can be used to promote social consensus ideologically, by classifying people according to where they live, using that membership to develop an identity with the state, and promoting the state ideology through various means, such as iconography – the state flag, anthem, monarchy, etc.

The use of territorial strategies for the exercise of power extends beyond the important ideological rôle, however, for two reasons. First, many of the functions of the state, especially with regard to its rôle of securing the conditions of production almost certainly could not be achieved unless they were contained within a defined territory. The provision of physical infrastructure to promote all forms of communication must be for a defined area; similarly the laws that govern competition, must refer to a territorial unit, as must, for example, the value of the currency. Secondly, in order to compete, people will usually seek strength in alliance with others. Such alliances need not be spatially defined, but very many are, because association with a definable place not only provides a clear identity for the alliance but also allows the use of the institution associated with that place – the state – as an agent of the alliance. Thus, for example, the state in enacting treaties with other states and in providing representation there advances the interests of local against foreign capital. Similarly, in operating tariff barriers it protects both local capital and local labour. Indeed, the state as territorial unity makes possible the regulation of all five types of conflict identified above. As Harvey (1985a) expresses it:

> regional class alliances, loosely bounded within a territory and usually (though not exclusively or uniquely) organised through the state, are a necessary and inevitable response to the need to defend values already embodied and a structured regional coherence already achieved. (p. 151).

The territories within which such alliances are created may not be particularly suited to them, being the residuals of pre-capitalist modes of production, perhaps, or the outcome of unsuccessful inter-state conflict. But as containers they provide a valid shell for inter- and intra-class combination, although as discussed below, attempts may be made to change their spatial form.

The state in non-capitalist societies

Other modes of production preceded capitalism and had their own requirements for a state or similar institution. There are only a few, very small remnants of such societies now in existence, and they will not be considered here. Alongside contemporary capitalism, however, there is a set of societies – variously termed communist, socialist, Second World etc. – which promotes an alternative mode of production. What is the rationale for the state in such contexts?

The existence of such societies poses problems for many analysts with regard to their links with the capitalist world economy. To some they are entirely separate from it, because they operate on very different principles in which profitability and accumulation play no part (see, however, Leeming 1986, Shaw, 1986). To others (such as Chase-Dunn 1982) they are firmly linked to capitalism, whereas others (such as Szymanski 1982) see them as increasingly insulated from it. Whatever the empirical situation, however, theoretically they are entirely separate, and in the long term do not need a state.

The classic works of communism see a communist society as the successor to capitalism. The latter has solved the problems of production, so that human ingenuity has been harnessed to ensure the means of reproduction for all. But this is achieved at the cost of a very unequal distribution of those means. To remove that inequality, the means of production must be appropriated by the state and placed in common ownership. The result is the dissolution of a class-based society and its replacement by a communist one. State socialism is an intermediate stage between capitalism and communism, therefore. Further, as the latter is achieved so the state will wither away, since the absence of conflict will remove its rationale.

Far from withering away, however, the state has become increasingly powerful in the non-capitalist countries, for two main reasons. First, the classic transition is not being followed, for the countries that are now socialist were not formerly prosperous and capitalist. Thus there the problems of production have not been solved, and the state's rôle is to produce solutions, by organizing a social and economic transformation. As Davis & Scase (1985) express it, the only difference between capitalist and state socialist societies is in the location of accumulation as the result of investment. In state socialist societies

> All but a very small fraction of this investment is monopolised by the state and it is guided by a centralised planning process instead of by individual capitalists' search for profit (Davis & Scase 1985, p. 75).

Thus the rôle of the state is to plan and carry through an investment strategy which will achieve high standards of living for all. Such planning and its implementation in the use of scarce resources requires a large bureaucracy and in state socialist, as in capitalist, societies this bureaucracy both has autonomy and is self-seeking. This provides the second reason why the withering away of the state has not occurred, and may well not occur; it is not in the interests of those who control the state. Thus the state is necessary to the goals of the mode of production, at its current level of achievement; the autonomy that it has means that it may well sustain its own necessity.

The autonomy of the state

The view of the state outlined here is of a necessary component of the superstructure of the capitalist mode of production, an institution without which the inherent strife within and volatility of capitalism would almost certainly ensure its rapid demise. The state is needed to promote and legitimate capitalism, not in abstract but in empirical terms to the residents of a defined, and defended, territory. It is, as Taylor (1982) has argued, the spatial unit which links the individual's scale of experience, the localities in which all live and learn about the world, to the scale of reality, the global world economy. The state, he claims, is the scale of ideology.

To some critics, this view of the state is a very partial one, since it implies that the state is merely the agent of capital. They see it as part of the portrayal of people within capitalism as little more than cultural dupes, bearers of a structure and servants of a disembodied economic determining force (see Duncan & Ley 1982; van der Laan & Piersma (1982) express this view in a slightly different way in a wider-ranging critique of models of man). Marx's argument that 'The executive of the modern state is but a committee for managing the common affairs of the whole bourgeoisie', is used to substantiate this argument; it proposes a theory of the state which is instrumentalist, a form of determinism which the critics find unacceptable in its treatment of individual agents.

The autonomy of the state has been a focus of much academic debate, therefore. Those who seek to counter the instrumentalist charge focus on two aspects of the issue of autonomy. First, they point out that in order for the state to perform its necessary rôles within capitalism it must *give the empirical appearance of autonomy*. Its rôle is to promote accumulation and to legitimate capitalism generally within its own territory: thus, unless it is to rule by coercion rather than consensus, it must not appear to be linked to the interests of any specific group within society (within capital as well as between capital and labour), otherwise its neutrality in, for example, the enforcement of contracts and the resolution of conflicts would be queried and the social contract involved in the consensus that it builds and maintains would be in jeopardy (see Johnston 1984a). The state must appear to be an independent (or neutral) agent in order to undertake at least some of its rôles. It is continually being called upon to resolve conflicts and to decide between alternative courses of action – it may have to decide, for example, what segments of industry (and therefore what particular interests) to boost and what to run down, in policies designed to promote its interpretation of the general good. Its decisions must be taken within the constraints of the general rôles outlined above, but how it interprets those rôles in particular circumstances depends on how those involved in running the state – i.e. government, comprising politicians and bureaucrats – determine the best course of action, in the context of pressures being placed upon them by interested parties (e.g. the nature of welfare state policies, which differ substantially between states).

The second argument focuses on the *foundation of state autonomy* which is a function, according to Mann (1984), of its necessary spatial elements – its territoriality. Mann notes that many writers argue for the necessity of the state and for the multiplicity of functions that it is called upon to perform, but ignore a third element – what he terms its 'territorialised centrality' (1984 p. 194). He notes that states are not alone in exercising economic, military (or physical), and ideological power; these are used in some form in all social relationships.

> The power of the state is irreducible in quite a different *socio-spatial* and *organizational* sense. Only the state is inherently centralised over a delimited territory over which it has authoritative power. Unlike economic, ideological or military groups in civil society, the state elite's resources radiate authoritatively outwards from a centre but stop at defined territorial boundaries. The state is, indeed, *a place* – both a central place and a unified territorial reach (Mann 1984, p. 198; emphasis in origianal).

And from this he deduces that 'Territorial-centralization provides the state with a potentially independent basis of power mobilization being necessary to social development and uniquely in the possession of the state itself' (1984, p. 200). So that even though a state may be established or structured to promote the interests of particular groups, its very establishment and structuring as a state make it autonomous of those groups. The state is needed to do a great variety of things which could not be undertaken by forces within civil society. Those forces do not lack any control of the state, but they lack the defining characteristics that separate it from them. Thus the state – i.e. those who control it – has a foundation for independent action. There are limits to what can be done, of course, because state actions may be against the interests of members of civil society. A majority may support in principle what the state is doing, but its ability to act may be undermined by the power of the minority that disapproves; the result will be either an alteration in state policy (as when the IMF forces changes in fiscal policy: Johnston 1982) or a change in the personnel who control the state.

Sectional interests constrain the autonomy of the state, therefore. Of even greater potential import is the relationship between the state and civil society as a whole. This is expressed in general crises of the state, of which three types of crisis are usually recognized (after Habermas 1976). Rationality crises occur when the state fails to promote accumulation, and thereby loses the support of capital: their solution may involve either activities designed to discipline the state (e.g. the withdrawal of investment, leading to a change in economic policy, perhaps after an election precipitated by that withdrawal) or its replacment through some form of *coup d'état*. Legitimation crises occur when the state is unable to maintain proletarian support, usually because of high levels of unemployment and poor welfare services; resolution may involve yielding to proletarian pressure (either voluntarily, as with a return to democracy, or forced, again with some form of *coup d'état*), or repression (as in Hungary in 1956, Czechoslovakia in 1968, and Poland in 1984). Motivation crises occur when both types coincide.

State apparatus and the control of the state

The multiplicity of functions which the modern state (capitalist and non-capitalist) undertakes within its defined rôles means that it cannot be treated as a unity. Rather it is a complex of separate functioning units, each to some extent autonomous of the others and in competition with them for relative power. (On bureaucratic power, see Downs 1967.) This complex is usually referred to as the *state apparatus*, defined by Clark & Dear (1984, p. 49) as 'the set of mechanisms through which state power is exercised and state functions realised'. Within it they identify separately: (a) the subapparatus, the state agencies, organizations, and institutions where the functions are carried out; and (b) the para-apparatus, auxilary agencies established by the state but separate from it. They suggest that there are eleven separate subapparatuses (whose functions, or part of them, may be undertaken by para-apparatuses) in the following areas, the first three of which relate to the rôles of the state identified above:

(a) creating and maintaining social consensus – political, legal, and repressive;
(b) securing the conditions of production – public production, public provision, and treasury;
(c) ensuring social integration – health, education, and welfare, information, and communication and media; and
(d) controlling the executive – administration, and regulatory agencies.

Each is the focus of much separate study. Here attention will not be directed towards these particular elements but towards one spatial component that is identified as a subapparatus performing several rôles – the local state.

The local state

A major element of the state apparatus is what has become widely known in recent years as the local state. The term is rejected by some, because it implies that local states – units of local government or administration, whether general or particular – are autonomous agents in the same way that states are. They do not deny the empirical autonomy of local units, within the constraints set by superior bodies (Johnston 1984b), but argue against any real autonomy; hence they argue against the term local state because of its false implications.

Why have local government and/or administration? In some countries, the answer is in part that the local administrations predated the capitalist state, but this does not explain why they were not removed. Clark (1981, 1985) has argued convincingly from the US case that indeed to a considerable extent those boundaries (notably those of the States, ostensibly with some sovereign powers under the federal constitution) have been overridden, though not removed, in order to secure the conditions of production.

The rationale for local administration is usually presented in terms of one or more of the following: (a) the liberty function, which divides powers between local and central, providing a counter to the latter and the possible development of autocracy – it thus helps to avoid the empirical appearance of what is presented in the instrumentalist theory; (b) the participation function, which allows a substantial proportion of the population to be involved in the state apparatus; and (c) the efficiency function, which uses local needs as the determinant of the type and level of service provision (Johnston 1979; note Tiebout's (1956) classic paper which provided a theoretical justification, based on neoclassical economics, for a fragmented pattern of local governments within individual urban areas; see also Whiteman 1983, Johnston 1984b, 1986a). Of the three rôles of the state, these functions are at present related much more to the securing of social consensus and integration than to securing the conditions for production.

This was not always the case in the past, and the development of many towns has been strongly influenced by the boosterism policies of their local governments. Many involved in local politics and administration wish that were possible today, and seek to develop policies that will promote their towns against others (see Boddy & Fudge 1984). But central governments, notably in the UK, identify many of the fiscal problems of the modern state with high levels of local government spending and are constraining it, seeking to release market forces and to reduce local state activity as influences on locational choice within those forces. Participation involves (at least implicity) acceptance of the state activity participated in, so that local government activity is a form of co-operation. Provision of services according to local demands avoids potential legitimation crises. But if local governments were involved in securing the conditions of production in a major way, this would set parts of the state against each other – with the possible conclusion suggested by the tragedy of the commons example. Thus the major subapparatus for that role (notably the treasury) remains a central function. Implementation of the policies – in the provision of infrastructure, for example – may be handed to local governments to administer (perhaps with some slight flexibility). Similarly, functions that cover several rôles, such as education, may be administered locally – to promote consensus and integration, but will be controlled centrally, especially with regard to their contribution to the production rôle (hence the many problems of attempts at local economic policies: see Boddy & Fudge 1984).

The local state is not autonomous, therefore. Like the central state it is a territorial unit, with a defined centre and reach. But it is not a necessity, and the functions that it performs are determined centrally. As with the state itself, it has some empirical appearance of autonomy, in that those running local administrations take decisions, but they are constrained by the rules governing the operation of local elements of the state apparatus, and by the fact that the local elements can be dissolved by the central state – only the latter has autonomous power.

Running the state

In many presentations relating to the state, especially those more theoretical than empirical in their orientation, it has the semblance of a disembodied institution, not a locus of human decision making. How those decisions are made by the two groups involved – politicians and bureaucrats – is interpreted differently in the separate theoretical positions. According to the pluralist position, those making the decisions do so to reflect popular opinion; they do what the people want them to do – and no more. According to the managerialist position, they act autonomously according to their own interpretations, and they are openly accountable in a very general sense. According to the instrumentalist position, they act according to the dictates of the powerful class within the mode of production.

The pluralist position is not only a theory of how the capitalist state operates, it is also part of its legitimation; the ideology of many states is that they are run not only for the people but also by them, through their elected representatives. The nature of that representation is somewhat confused, however, as is the process of representative selection (which contains a major geographical element: Taylor & Johnston 1979). Bogdanor (1985), for example, has identified four potential rôles for representatives: (a) as representatives of the residents of territorially defined constituencies; (b) as representatives of certain partisan or ideological aims; (c) as protectors of particular interests; and (d) as legislators of policies. In many countries, the relative importance of these rôles leads to confusion in the creation of an electoral system (Johnston 1985).

Attempts have been made to relate these representative rôles to actual practice. With regard to the central state, for example, it is argued that its primary concern with securing the conditions of production means that the last two of the rôles are paramount, and that elected representatives can only act for consitutent and partisan interests within the constraints which these dominant concerns set. Indeed, it is argued that as legislators, elected representatives are much more influenced by the arguments of interest groups not directed at the electorate than they are by those who voted for them. This produces what is known as a corporatist model of politics. The degree of flexiblity within it reflects the structure of the electoral and legislative system: US representatives are much better able to serve constituency interests, via the porkbarrel (Johnston 1980b) than are British MPs (Hoare 1983), for example. In the local state, on the other hand, the first two rôles are much more important. The major functions of local government relate to legitimation (securing social consensus) and securing social integration (through the provision of public services). Elected representatives are able to serve both partisan and, if they wish, territorial interests (Pattie 1986): parties may contest elections on platforms relating to the level of local service provision, and individual elected members can both represent constituency interests (Newton 1976) and seek to win benefits for the territories that returned them to power. Thus whereas the central state is characterized by the operation of the corporatism model, the local state gives a greater opportunity for operation of the pluralist model. This produces what Saunders (1986) terms the dual state model: the politics of production, centred on class interests and private property rights, are focused on the central state, where the corporatist model best represents the nature of interest group interaction and instrumentalism provides the most apposite theory; the politics of consumption, centring on consumption sector interest and citizenship rights, on the other hand is contested in the local state, where the competitive model prevails and the pluralist theory is relevant.

Of course, the local state, as emphasised above, is constrained by both central state and the disciplines of the marketplace so that, for example, recent British central governments have reduced local spending and activities in order to curb total public spending, in line with anti-inflation policies, and US city governments have come near to bankruptcy as they experience local fiscal crises (David & Kantor 1979). (Note that Harvey (1985a) has pointed out that the bourgeoisie plays relatively little part in local government; its main interest is in the operation of the corporatist model, and it disciplines local spending via central legislation.)

The geography of liberal democracy

Both the corporatist and the pluralist model of government operation assume a democratic system, whereby the state (central or local) is run by an elected government which responds to pressures and is accountable to the electorate. But only a minority of the states of the world have elected governments, the majority of them (as Johnston (1986b) shows) in the core of the world economy. How, then, do we understand the running of government in the rest of the world?

The simplest (and a misleading) answer to this question is provided by modernization theory (see Taylor 1986a), which argues that as economic development takes place so people are mobilized into new forms of social and political behaviour, including democracy; with modernization, accompanied by education, people take greater control over their own destinies (see Deutsch 1961, Coulter 1975). Thus, if we accept capitalist manifestos (such as Rostow's (1971)), we assume that eventually the whole world will adopt the democratic form of government (Taylor 1985, 1989).

An alternative answer not only sees individual countries as part of the world economy rather than controllers of their own destinies, but incorporates the many experiments with democracy that have occurred in the so-called developing world (experiments which, according to Coulter's model, are premature). Democracy is a characteristic of core countries, and its use is part of the legitimation of the capitalist system; the ideology of democracy – and its equation with freedom – is very powerful in the core. People are given the empirical appearance of control (an appearance whose validity is clearly queried by the corporatism model). It carries potential disadvantages, because that empirical appearance must have some substance: democracy must be seen to give people control, which means that (with a universal franchise) the proletariat should be able to win benefits from capital via their control of the state. Capital can allow this to happen, because it can afford to do so in the core – and because, ultimately, it retains control since the corporatist model, not the pluralist, dominates politics. (See Johnston (1984c, 1986b) on the use of electoral reform arguments in Britain during an economic crisis to promote, implicitly, corporatism over pluralism.)

In the periphery of the world economy, on the other hand, such benefits are not readily afforded. Whereas in the core, democracy engenders stability in social relations, in the periphery it is frequently argued that democracy stimulates instability, the creation of proletarian demands which, if met, would frighten investment away. Stability, it is argued, requires strong, consistent government that can provide the sort of environment within which investment is attracted and development occurs; democracy can be provided later. But where the development is not forthcoming, and those denied what they identify as their political rights (the freedom to organize, to vote, and to stand for public office) see no substantial benefits from the constraints, popular resentment against the state may emerge. The state may seek to repress this, at a cost (i.e. in the repressive apparatus) that has an impact on its economic policies and may eventually lead it to yield and allow a democratic form of government. In turn, this too may fail, because investment flees (as from Jamaica (Mandle 1982)), and the forces of capital (linked almost invariably to the armed forces) will remove democracy in order to create a more stable situation.

This outline suggests a continuous cycle of democracy–dictatorship–democracy in peripheral countries, with irregular transfers of executive power and periods of military rule (Johnston 1984c, 1986b; see also Giner 1985). Many countries apparently fit this model, providing a clear core/periphery distinction to what Taylor (1985) terms the geography of liberal democracy. Some do not, however. Some have never experienced democracy, presumably because mobilization of the potential electorate has been insufficient to counter a repressive state. And a few have experienced substantial periods of democracy, as in India; this comes about, according to Osei–Kwame & Taylor (1984), and Taylor (1986) because parties are able to mobilize different sectors of the electorate at successive elections, so that their failure to deliver promises to one sector is countered by the offer of a new set to another sector – they term this as the politics of failure (see also Taylor 1986; a somewhat similar trend in core countries experiencing economic crisis is termed dealignment: Johnston 1987).

The concept of democracy promoted in capitalist ideology is that of *liberal democracy*; it has people in control of the superstructure but not of the economic base. An alternative conception is of *popular democracy*, which involves the dictatorship of the proletariat in classless societies; under it, people have control over all aspects of their lives. According to Marxian theory, popular democracy should see the withering away of the state, under communism. Socialism is an intermediate state towards that situation, with the state promoting the interests of all, and being responsive to all. There is only one interest group – everybody – and therefore only one party; competing interest groups and thus competing parties are not needed.

The pattern of states

At present, the world's surface is divided into about 150 states. They exhaust all of the land surface, with the exception of Antarctica where international agreement (challenged by some states) has led to the establishment of spheres of influence but no formal sovereignty. For most, their sovereignty (i.e. right to exist) is recognized; exceptions include Israel, whose position is challenged by most of its neighbours but sustained by other outside powers, and the so-called independent states (Transkei, Venda, etc.) established by South Africa but recognized by no other. For many, although their existence is recognized, their exact spatial parameters are not. There are many contentious inter-state boundaries, for example, including maritime boundaries on those parts of the Earth's surface that until recently have not been formally incorporated into state territories (Prescott 1986). And there are both intra-state (as with the Corsicans in France) and inter-state (the Kurds, for example) movements whose goal is the creation of new states.

In Western Europe some of the states predate the development of capitalism, so that their boundaries reflect the outcome of processes of state formation under feudalism (see Mann 1986; for a brief review see Johnston 1982). They provided the containers within which the first regional alliances of capitalism were established. As capitalism extended its spatial influence, so a new pattern of states was created both within Europe – the colonization of Ireland from England, for example – and, much more importantly, outside that continent. Through colonialism, new states were established throughout both American continents, Africa, Asia, and Australasia, reflecting the interests of external powers and in many cases overriding (sometimes obliterating) pre-existing, non-capitalist state territories. A century or more later, those new states were organizing containers for anti-colonial movements. But once independence from the colonial powers was achieved, in many cases this was succeeded by nationalist movements promoting secessionist claims for certain parts of the new states, as in Nigeria and Zimbabwe. In order to counter colonialism, it was possible to sustain a national identity, but the colonizing states had not entirely removed the pre-existing cultural variety of nations, however, providing the foundations for nationalist movements seeking to rewrite the post-colonial map.

Colonialism was one form of orgainizing the spatial structure of international capitalism and imposing the core nations upon the periphery (on the difference between intra- and inter-state core/periphery relations see Blaut 1986). Its relative absence at present does not mean that the core states – especially the United States – do not seek some form of hegemony over other parts of the world. Indeed, the establishment of such hegemony is central to Taylor's (1985) portrayal of global geopolitics (for an alternative view, see Modelski 1978) which links the Kondratieff cycles of economic activity to the creation and decline not only of spheres of influence but also of super-power status within the core of the world economy. This provides a theoretical structure for understanding geopolitics (much superior to the earlier, environmental determinist approach (Parker 1985)), and thereby much of the geography of violence, warfare, and premature death (O'Loughlin 1986, Johnston *et al.* 1987). The creation of superstates, both economic (e.g. the EEC) and military (e.g. NATO), can also be accounted for in terms of this model.

The state, political geography, and geography: in summary

In the conventional division of labour within the social sciences, economics deals with the operation of markets, sociology with social relations, and political science with the state apparatus. For geographers, this implies subdisciplines of economic, social, and political geography, dealing with the separate spatial (including environmental) elements of each. And during the last three decades this has indeed been the situation, with three relatively separate subdisciplines, albeit of unequal importance (economic geography was strongest in the 1950s and 1960s; social geography gained in strength in the 1970s; political geography has experienced a revival in the 1980s). There have also been sub-subdisciplines (such as transport geography, a part of economic geography), and spatial subdisciplines (notably urban geography, which itself has spawned separate urban social and urban political geographies).

Such subdisciplines are necessary to some extent, because of the need to focus detailed research (especially empirical research) on particular topics. But their separation, and the lack of any synthesis of their work, provides a major impediment to the advancement of understanding (Johnston 1986c, 1986d). This chapter has shown that the state and the state apparatus are proper subjects for study, but that they cannot be divorced from the study of markets and social relations. One of the problems of many analyses is that they look at one component of a place only, when the focus should be on all three (identified in Johnston (1986c) as: position in the spatial division of labour; social relations; institutional apparatus). Together the three comprise the culture of the place (unfortunately, cultural geography focuses almost exclusively on human artifacts), and without an appreciation of all three (as Johnston argues) understanding is partial and, probably, of little value. Hence, although the development of a theory of the state as a territorial, autonomous unit may be the task of political geographers, their work must fully incorporate analyses of economic and social phenomena (as the work of Harvey (1985b, 1985c), Massey (1984) and others so clearly shows) and must be used to inform the development of holistic geographic theory. Political geography is a means to an end, not and end in itself.

To date, the development of political geography and a viable theory of the state as a territorial unit has not proceeded very far, depite a few substantial efforts (notably Clark & Dear (1984) and Taylor (1985)). As noted at the outset, the routes taken in that development tend to fall into three main groupings: the pluralist, managerialist, and instrumentalist theories of the state. The present review has suggested that a fully fledged theory will incorporate elements of all three. The instrumentalist contribution will inform that part of the theory which sees the rôle of the state as necessary to the functioning of the capitalist (or some other) mode of production, whereas the pluralist and managerialist contributions will inform studies of how those functions are performed. The state is not only necessary to capitalism, it is a geographical necessity; how it fulfils its necessary rôles will depend on the operation of human agency.

Note

I am grateful to Gordon Clark, Phil Cooke, Ruth Fincher, John Lovering, Dick Peet, and Nigel Thrift for their comments on a draft of this chapter. I fear that I have been unable to deal with them as fully as they deserve – to do so would have resulted in a chapter much longer than the editors would countenance.

References

Alford R. R. & R. Friedland (1985). *Powers of theory: capitalism, the state and democracy.* Cambridge: Cambridge University Press.

Ardrey, R. 1969. *The territorial imperative.* London: Fontana.

Blaut, J. M. 1986. A theory of nationalism. *Antipode* **18** 5–10.

Blomley, N. K. 1986. Regulating legislation and the legitimation crisis of the state. *Enviroment and Planning D, Society and Space* **4**, 183–200.

Boddy, M. & C. Fudge (eds) 1984. *Local socialism.* London: Macmillan.

Bogdanor, V. (ed.) 1985. *Representatives of the people?* Aldershot: Gower.

Brams, S. J. 1975. *Game theory and politics.* New York: Free Press.

Chase-Dunn, C. 1982. Introduction. In *Socialist states in the world-system*, C. Chase-Dunn (ed.) 9–20. Beverly Hills: Sage Publications.

Clark, G. L. 1981, Law, the state, and the spatial integration of the United States. *Environment and Planning A* **13**, 1197–227.

Clark, G. L. 1985. *Judges and the city*. Chicago: University of Chicago Press.

Clark, G. L. 1986. Making moral landscapes. *Political Geography Quarterly* **5**, S147–S162.

Clark, G. L & M. J. Dear 1984. *State apparatus: structures and language of legitimacy*. Boston: Allen & Unwin.

Coulter, P. 1975. *Social mobilization and liberal democracy*. Lexington, Mass.: Lexington Books.

David, S. M. & P. Kantor 1979. Political theory and transformations in urban budgetary arenas: the case of New York City. In *Urban Policy Making*, D. R. Marshall (ed.) 183–220. Beverly Hills: Sage Publications.

Davis, H. & R. Scase. *Western capitalism and state socialism*. Oxford: Basil Blackwell.

Deutsch, K. W. 1961. Social mobilization and political development. *American Political Science Review* **55**, 494–505.

Downs, A. 1967. *Inside Bureaucracy*. Boston: Little, Brown.

Duncan, J. S. & D. Ley, 1982. Structural marxism and human geography: a critical assessment. *Annals of the Association of American Geographers* **72**, 30–59.

Giddens, A. 1985. *The nation–state and violence*. Oxford: Polity Press.

Giner, S. A. 1985. Political economy, legitimation and the state in southern Europe. In *Uneven Development in Southern Europe*, R. Hudson & J. Lewis (eds), 309–50. London: Methuen.

Habermas, J. 1976. *Legitimation crisis*. London: Heinemann.

Hardin, G. 1968. The tragedy of the commons. *Science*, 162, 1243–48.

Harvey, D. 1982. *The limits to capital*. Oxford: Basil Blackwell.

Harvey, D. 1985a. The geopolitics of capitalism. In *Social relations and spatial structures*, D. Gregory & J. Urry (eds), 128–63. London: Macmillan.

Harvey, D. 1985b. *The urbanization of capital*. Oxford: Basil Blackwell.

Harvey, D. 1985c. *Consciousness and the urban experience*. Oxford: Basil Blackwell.

Hoare, A. G. 1983. Pork-barrelling in Britain: a review. *Environment and Planning C, Government and Policy* **1**, 413–38.

Johnston, R. J. 1979. *Political, electoral and spatial systems*. Oxford: Oxford University Press.

Johnston, R. J. 1980a. Political geography without politics. *Progress in Human Geography* **4**, 439–46.

Johnston, R. J. 1980b. *The geography of federal spending in the United States*. Chichester: Wiley.

Johnston, R. J. 1981a. Political geography without dogma. *Progress in Human Geography* **5**, 595–8.

Johnston, R. J. 1981b. British political geography since Mackinder: a critical review. In *Political studies from spatial perspectives*, A. D. Burnett & P. J. Taylor (eds), 11–32. Chichester: Wiley.

Johnston, R. J. 1982. *Geography and the state*. London: Macmillan.

Johnston, R. J. 1984a. Marxist political economy, the state and political geography. *Progress in Human Geography* **8**, 473–92.

Johnston, R. J. 1984b. *Residential segregation, the state and constitutional conflict in America metropolitan areas*. London: Academic Press.

Johnston, R. J. 1984c. The political geography of electoral geography. In *Political geography: recent advances and future directions*, P. J. Taylor & J. W. House (eds), 133–48. London: Croom Helm.

Johnston, R. J. 1984d. The region in twentieth century British geography. *History of Geography Newsletter* **4**, 26–35.

Johnston, R. J. 1985. People, places, votes and parliaments. *Geographical Journal* **151**, 327–38.

Johnston, R. J. 1986a. The general good of the community: some perspectives on town planning and residential segregation. *Planning Perspectives* 1, 131–45.

Johnston, R. J. 1986b. Individual freedom and the world-economy. In *A world in crisis? Geographical perspectives* R. J. Johnston & P.J. Taylor (eds), 173–95 Oxford: Basil Blackwell.

Johnston, R. J. 1986c. Four fixations and the quest for unity in geography. *Transactions of the, Institute of British Geographers* NS11, 449–53.

Johnston, R. J. 1986d. *On human geography*. Oxford: Basil Blackwell.

Johnston, R. J. 1986e. Placing politics. *Political Geography Quarterly* 5, 563–78.

Johnston, R. J. 1987. Dealignment, volatility and electoral geography. *Comparative Studies of International Development*.

Johnston, R. J., J. O'Loughlin & P.J. Taylor 1987. The geopolitics of violence and premature death. In *The quest for peace*, R. Vayrynen (ed.). Beverly Hills: Sage Publications.

Laver, M. 1981. *The politics of private desires*. London: Penguin.

Laver, M. 1986. *Social choice and public policy*. Oxford: Basil Blackwell.

Leeming, F. A. 1986. Chinese industry – management systems and regional structures. *Transactions of the Institute of British Geographers* NS10, 413–26.

Lovering, J. 1986. Localities in a militarised world economy. Unpublished paper. University of Bristol, School for Advanced Urban Studies,

Lovering, J. 1987. Militarism, capitalism and the nation state: towards a realist synthesis. *Environment and Planning D, Society and Space* 5, 283–302.

Mandle, J. R. 1982. Jamaican democratic socialism and the strike of capital. In *Socialist states in the world-system*, C. Chase-Dunn (ed.), 219–38. Beverly Hills: Sage Publications.

Mann, M. 1984. The autonomous power of the state: its origins, mechanisms and results. *European Journal of Sociology*. 25, 185–213.

Mann, M. 1986. *The sources of social power*. Vol. 1. Cambridge: Cambridge University Press.

Massey, D. 1984. *Spatial divisions of labour*. London: Macmillan.

Modelski, C. 1978. The long cycle of global politics and the nation state. *Comparative Studies of Society and History*. 20, 214–35.

Newton, K. 1976. *Second city politics*. Oxford: Oxford University Press.

O'Connor, J. 1973. *The fiscal crisis of the state*. New York: St Martin's Press.

O'Loughlin, J. 1986. Spatial models of international conflicts: extending current theories of war behavior. *Annals of the Association of American Geographers* 76, 63–80.

Osei-Kwame, P. & P. J. Taylor 1984. A politics of failure: the political geography of Ghanaian elections, 1954–1979. *Annals of the Association of American Geographers* 74, 574–89.

Parker, W. H. 1985. *The development of western geopolitical thought*. London: Croom Helm.

Pattie, C. 1986. Positive discrimination in the provision of primary education, Sheffield. *Environment and Planning A* 18, 1249–58.

Peet, R. 1983. Relations of production and the relocation of United States manufacturing industry since 1960. *Economic Geography* 59, 112–44.

Peet, R. 1986 Industrial devolution and the crisis of international capitalism. *Antipode* 18, 78–95.

Pickles, J. 1985. *Phenomenology, science and geography*. Cambridge: Cambridge University Press.

Prescott, J. R. V. 1986. The maritime political boundaries of the world. London: Methuen.

Rawls, J. 1972. *A theory of justice*. New York: Oxford University Press.

Rostow, W. W. 1971. *The states of economic growth*. Cambridge: Cambridge University Press.

Sack, R. D. 1983. Human territoriality: a theory. *Annals of the Association of American Geographers* **73**, 55–74.

Sack, R. D. 1986. *Human territoriality*. Cambridge: Cambridge University Press.

Saunders, P. 1986. Reflections on the dual politics thesis: the argument, its origins and its critics. In *Urban political theory and the managment of fiscal stress*. M. Goldsmith (ed.). Aldershot: Gower.

Shaw, D. J. B. 1986. Spatial dimensions in Soviet central planning. *Transactions of the Institute of British Geographers* **NS10**, 401–12.

Szymanski, A. 1982. The socialist world-system. In *Socialist states in the world-system*, C. Chase-Dunn (ed.). 57–84. Beverly Hills: Sage Publications.

Taylor, M. 1976. *Anarchy and cooperation*. Chichester: Wiley.

Taylor, P. J. 1982. A materialist framework for political geography. *Transactions, Institution of British Geographers* **NS7**, 15–34.

Taylor, P. J. 1985. *Political geography: world-economy, nation–state and locality*. London: Longman.

Taylor, P. J. 1986. A world-systems interpretation of political parties. *Political Geography Quarterly* **5**, S5–S20.

Taylor, P. J. 1989. The error of developmentalism. In *New horizons in human geography*, D. Gregory & R. Walford (eds). London: Macmillan.

Taylor, P. J. & R. J. Johnston, 1979. *Geography of elections*. London: Penguin.

Taylor, P. J. & R. J. Johnston 1984. The geography of the British state. In *The human geography of contemporary Britain*, J. R. Short & A. Kirby (eds). 23–39. London: Macmillan.

Tiebout, C. M. 1956. A pure theory of local expenditures. *Journal of Political Economy*. **64**, 416–24.

Van der Laan, L. & A. Piersma 1982. The image of man: paradigmatic cornerstone in human geography. *Annals of the Association of American Geographers* **72**, 411–26.

Whiteman, J. 1983. Deconstructing the Tiebout hypothesis. *Environment and Planning D, Society and Space*. **1**, 339–54.

12 *The geography of law*

Gordon Clark

The geography of law[1]

Analyzing the spatial impacts and consequences of law is an increasingly important field of research in geography. A paired set of review articles was recently published surveying the field (see Blacksell *et al*. 1986, Economides *et al*. 1986), and special sessions on the topic have been held at major conventions (special sessions have been arranged at the Annual Meeting of the Association of American Geographers).[2] There is a steady, albeit small, stream of articles on the topic in the journals (see, for example, Blomley 1986, Clark 1986a, Johnston 1986). And, most importantly, two books explicitly devoted to analyzing the rôle of American courts in structuring geographical outcomes were published in the last few years signalling the intellectual intersection between these fields of academic enquiry (Clark 1985a, Johnston 1984). Even lawyers have attempted to introduce geographical context into their analyses (see, for example, Finch & Nagel 1983).

There are, of course, other studies which have considered legal issues when analyzing the spatial impacts of government regulation. For example, Platt *et al*. (1983) and Walker & Heiman (1983) have studied environmental and land use management issues from a law and geography perspective. In fact, there are many studies covering a wide variety of topics which begin with a geographical problem, like the spatial patterns of political elections, and then introduce the regulatory and legal environment as a way of explaining observed patterns.[3]

Even so, the geography of law is not a mainstream topic of research. The dominant theoretical perspective on cities supposes that cities are much the same the world over. Based on an ahistorical mode of reasoning, neoclassical economic theorists argue that urban structure can be described by universal principles such as land rent, distance costs, and individual preferences. Alonso's (1964) model was one of the first treatments of urban structure utilizing this approach, and has been recently extended by Thrall (1987) through what that author called the consumption theory of land rent. These models allocate competing land uses on the basis of relative prices, given individuals' preferences. Local context is eschewed in favour of a standard image of the city, applicable to all cities.[4] If it is, nevertheless, maintained that there are significant differences between different cities, it is sometimes suggested that these so-called uncomfortable facts are evidence of the need for further research (Mills & Hamilton 1984). A more plausible strategy is to suggest that the underlying institutional structure of many cities is so different that interurban differences are sustained outside the logic of the neoclassical model.

Much of the literature on the geography of law is premised upon this more plausible strategy. Explicitly or implicitly it is assumed that different institutional structures foster different geographical outcomes. It is also assumed that the conventional neoclassical approach is fundamentally inadequate as an explanation of urban form and processes. Johnston's (1984) study of the impact of the judiciary on the US urban scene is a good example of this approach. Writing from a British perspective, Johnston argued that the particular political culture of the US, coupled with distinctive institutions like the US Supreme Court, combine to create an American spatial organization of capitalism.[5] As neo-Marxist conceptions of the state have come to dominate our understanding of government policy, urban structure, and the rôles of local institutions (like law and the local state; see the chapters by Johnston and Fincher in this volume) have been similarly interpreted (see Clark & Dear (1984) on state theories and the legal apparatus).[6]

Unfortunately, despite significant advances in theorizing the nature of state functions and state institutions like law, geographers have neglected law as a mode of discourse. Studies of the spatial impacts of law ignore some of the deeper and potentially rich interrelationships between geographical and legal reasoning.[7] An important goal of this chapter is to prompt recognition of the links between these reasonings.[8] In the first section, the reader is introduced to theories of law, and the possible rôles of the social sciences in studying law and legal reasoning. A second goal of the chapter is to describe the essential elements of current approaches to the geography of law. This is accomplished next. There are few studies which go beyond the simplest conception of the geography of law. Through a critique of this literature, an agenda for future research is developed. In the third section, emphasis is placed on the methods of law, and the ways in which 'Law's Empire' (a phrase used by Dworkin (1986)) is legitimized. It would be a mistake, though, to imagine that there is a hidden agenda in this chapter for replacing geographical reasoning with legal reasoning. In point of fact, there are many reasons to be sceptical of the power of legal reasoning. The chapter therefore concludes with a discussion of the limits of law.

One qualification ought to be noted. It will become apparent that much of the empirical work on law and geography has drawn inspiration from studies of American law.[9] Perhaps one reason for this focus is the wealth of detail typical of American judicial decisions. Another reason for this focus is the central rôle of the judiciary in American political discourse. But, perhaps the crucial reason is the practice of law: at issue is the English common law tradition as opposed to American legal constitutionalism (cf. Dicey (1959) with Tribe (1985)).[10] It could be argued that political discourse in America has been overtaken by the language of law; political debates have been framed in terms of constitutional rather than class or sectional imperatives.[11] It is inevitable that this chapter reflects the focus of the literature, and the central preoccupations of the American legal system.

What is law?

In this section two general topics are considered. One topic is the theory of law, an issue which is more problematic than might be first imagined. The second topic has to do with the interrelationships between social science and legal theory. Some attention is also paid to the rôle and status of the judiciary in modern society. An understanding of these issues is necessary if the reader is to appreciate the significance of current debates in the law and society literature, as well as the relative utility of alternative approaches to studying the geography of law.

Theories of law

Conventionally, there are three different types of theory of law.[12] The positivist tradition, as exemplified by Hart (1961), defines law through an empirical test.[13] A law is a law to the extent that society at large freely obeys certain statutes and regulations. It is not enough for institutions of society to assert that statutes and regulations have the status of law; laws can only be recognized as such by society. In this kind of theory law has a special moral status derived, implied, or deliberately assigned, according to the behaviour of citizens. Notice that the moral status of law in this positivist vision does not rest on any deeply embedded normative conception of society – what society is or what it might be. In point of fact, Hart made no special claim about how or why society ought to be arranged.

The positivist conception of law aims to be neutral with respect to outcomes; in Hart's theory laws are procedures of social action, their legitimacy is not derived from anything that law would achieve (Raz 1985). In this sense, Hart attempted to separate the procedures of law from otherwise idealistic moral imperatives which he believed clouded recognition of the underlying attributes of law as a system of rules.[14]

Hart sought to distinguish the positive theory of law from more traditional conceptions based on the natural rights thesis.[15] Natural rights theorists define law by reference to some divine, non-empirical conception of the good society. This theory of law rests on a simple supposition: there are inalienable rights which we as human beings deserve and have the right to claim regardless of social contingencies (see Finnis (1980) for a seminal contribution). While conventionally framed with reference to human rights like freedom and justice, it has served historically as a means of legitimating the authority of élites, especially the crown.[16] Hart's theory continues an intellectual tradition developed by Jeremy Bentham;[17] while the ultimate warrant of natural rights theories derives from a non-contextual theory (or theories) of the inherent rights of citizens, institutions, and authority.

Neither theory of law is adequate although each theory has certain advantages (see Brink (1985) for a comparison). The positivist theory of law provides for a testable empirical definition of law. Natural rights theorists can only assert what is law by reference to a moral order which itself may be quite problematic and subject to deep disagreement. On the other hand, Hart's theory is strangely idealist in the sense that it presupposes the very conditions which law is designed to obtain. That is, for the positivist theory to hold, we must assume society is composed of individuals capable of freely expressing and articulating their preferences. Yet, without some social order and some laws (rules and regulations) regarding proper conduct, these conditions may be impossible to achieve. Clearly, social action may still be a useful test of the plausibility of law. But, it does raise the question as to the design of laws in the first place, a question that liberal theory in general has found difficult to rationalize given that the state is simply assumed to be a derivative symbol of democratic politics. Natural rights theorists have one advantage over Hart's positive theory. The origin of law is outside the immediate social context. Consequently, there is no need to find law in the day-to-day behaviour of individuals; it is enough to assert that law exists as an epiphenomenon.

While both the positive and natural theories of law have their supporters, a third type of theory of law is preferable to both options. This type of theory assumes that law is an institution, an organized social practice maintained and reproduced by the state, and bound by customary rules of discourse which are themselves fostered in élite institutions. As such, there are close and intimate links between those charged with the responsibility of law making and the purposes of law as defined by those élites. What law is, and what it means in certain circumstances, is a product of the judicial apparatus. This is a realist theory of law.[18]

The realist theory of law supposes that law is much less than claimed by natural rights theorists, and much more than claimed by positivists. Compared to natural rights theorists, the realist theory of law does not claim that law has a moral virtue or an origin separate from society. Being a social practice, it reflects the (hierarchical) structure of society and reproduces the structure of society. In this respect, realist theories of law assume that natural rights claims about good societies are rhetorical devices aimed at legitimating the legal apparatus and ultimately the state. Compared to positive theories, realist theories suppose that law is recognized as such by virtue of the coercive force of the legal apparatus. That is, law is defined as such by requirements that citizens obey prescribed rules and regulations governing proper behaviour.[19]

This is obviously a state-centred theory of law. In this sense it is consistent with recent neo-Marxian state theories which begin with institutions of state authority and then work through to the form of structure of contemporary society (Clark & Dear 1984). There is one crucial difference, though, between realist legal theories and related neo-Marxian state theories. In realist theories of law, the judiciary is an active agent of structure by virtue of its rôle as interpreter of statutes and regulations, and by virtue of its coercive enforcement powers. Many neo-Marxian state theories fail to identify agents of state power, preferring instead to invoke a structural metatheory of the primacy of the capitalist mode of production over the actions of state agents (cf. Poulantzas (1978) with Harden & Lewis (1986)), or simply to dismiss the legal apparatus as just another tool of capitalist domination (Edelman 1979). The realist theory of law supposes that social structure is indeterminate, that the rules of the relations between capital and labour are interpretable, and that social structure can be made and remade by the legal apparatus (Clark 1985a).[20]

Law and social science

All kinds of theorists, from all kinds of political perspectives, readily admit the significance of the state in everyday life. There are those on the right and left who fear the state (for a related discussion see Clark & Dear 1984, Ch. 9), believing that élites typically control the state's policies, while others are more optimistic in principle, if not in practice, about the state's ability to regulate society in the best interests of all its citizens. Regulation of the conduct of business, enterprises, and competition is pervasive (MacAvoy 1979). Similarly, regulation of individual behaviour, social relationships, and morality is very extensive (Dworkin 1985). In short, the legal apparatus of the state is an ever present part of life.

The study of law might reasonably be restricted to lawyers. But, in the language of a popular aphorism, law is too important to be left to the lawyers. To some extent, social science has taken heed of this imperative. There are academic societies devoted to the study of law and society, as there are journals (especially in economics) with interests and titles like *Journal of Law and Economics*, and *Journal of Legal Studies*. There can be no doubt as to the basic thrust of the economics journals in this field; they publish studies where standard economic principles are applied to legal problems, thereby demonstrating the available alternative decision frameworks. In recent years, there has been an explosion of interest in the study of law by the humanities, especially by those engaged in studying English literature and rhetoric. One of the principal suppositions of this latter group is that law is like literature: it is essentially an hermeneutic or interpretive endeavour. Legal reasoning, the literary theorists argue, is but a special case of literary criticism – no more and no less.[21]

Laws, as in statutes, regulations, constitutions, and the like, are both the intermediary variables between individuals and corporations, and the determinants of behaviour. Law as the judicial institution adjudicates disputes between citizens, and law as a system of rules structures their behaviour. As noted in Clark (1981, p. 1197), laws on contracts both define the process and context of commercial exchange and the degree of interdependence and obligations between entrepreneurs. At one level, laws are formal imperatives defining correct behaviour. But, once enacted, laws become the underlying (even unrecognized) rules and standards of social behaviour. Thus, law can be thought of as directly coercive and, at the same time, part of the fabric of customary behaviour. Note that laws are also rules *and* standards. That is, laws are formal directives *and* moral imperatives. By defining correct behaviour, one way or another, implicitly or explicitly, laws set standards for good behaviour. Essentially, then, laws are standards by which judges evaluate the behaviour of those who come before the court.

These assumptions are important in a methodological sense. But, though there has been a good deal of debate over the relationship between law and society,[22] the focus of research has shifted over the course of the 20th century from substantive matters of law (like the obligations and duties of individuals) to procedural issues of the administration and enforcement of law. This is particularly apparent in the law and economics literature. Much of the academic enquiry in this field is dominated by concern for the economic impacts of anti-trust regulations, legislation regarding competition, and similar bodies of regulations (Panzor 1980). Likewise, there has been considerable attention given to the impacts of environment legislation, health and safety legislation, and many other related laws.[23]

The impact of regulation (and deregulation, see Persons (1984)), has been an important theme in law and economics. It has been taken up in geography and regional science, but only to a limited extent. Teitz's (1978) paper is notable in two respects. It was one of the first attempts explicitly to relate the legal structure to the spatial context. In doing so, Teitz sought to demonstrate the importance of the field, and the necessity for further work in this area. Even so, there have been few attempts to develop his ideas. Indeed, one looks in vain in the geographical literature for evidence that this paper had any real impact. A second notable aspect of the paper was its scope: like the law and economics literature, Teitz's paper dealt with regulatory aspects of law, not underlying substantive issues. Put more plainly, the regulatory tradition either presumes an underlying normative intent of law (but ignores it) or refuses to recognize that law has an underlying substantive structure. Either way, the regulatory tradition deals with the impacts of law from another perspective – economic efficiency. While Teitz was not so concerned with the efficiency issues, he nevertheless evaluated the impacts of law on regional systems from the perspective of the discipline(s), not the underlying normative structure of law.

In recent years, a new approach to the study of law and society has evolved. Although there are different versions of this approach, the basic thrust has been to consider the substantive structure of law, as opposed to its procedural image. Posner (1977) claimed that the underlying principles of American jurisprudence are economic; that is, the arrangement of law, its normative intent, is premised upon a fundamental interest in sustaining national economic efficiency. While he did not claim that judges have always explicitly recognized that this is the substantive reference point for adjudication, he nevertheless claimed that such a reference point has always existed behind the practice of adjudication. Many theorists disagree with Posner (see Tribe 1985). There are disagreements concerning the evidence that Posner chooses to introduce to justify his assertions, and there have been disagreements over Posner's positivist conception of legal reasoning (see Michelman 1979). But Posner's assertion, that the underlying substantive structure of law is fundamental if the practice of law is to be understood, finds favour with others (like this author) who, though disagreeing that economic efficiency is the ultimate reference point, nevertheless assert that such a substantive reference point is fundamental to the practice of law.[24]

Geography and law

In this section, literature on the geography of law is reviewed, paying particular attention to the extent to which the literature recognizes the underlying substantive bases of law. Three different approaches are reviewed. The first is the simplest: land use regulation, where the focus is upon the geographical impacts of regulations like zoning and rent control. The second approach is more complex; it deals with the interaction between the judiciary and geography. The third approach is the most complex: the substantive bases of law, and alternative interpretations of those foundations. This review is not meant to be exhaustive so much as illustrative.[25]

Spatial impacts of law

Of the few studies of the geography of law, the overwhelming majority are about the spatial impacts of land use laws on urban structure. For example, in a recent case study of land use changes in British Columbia, Everitt (1984) documented the effects of land use regulations designed to control urban sprawl on the rural fringe of the metropolitan area. His study, like many others, combined a brief legislative history of the regulations with aspects of the geographical circumstances in which these regulations were introduced. Although of a quite different type, a study of retail development in Belgium by Dawson (1982) was similarly interested in the spatial ramifications of government regulations. A related study was Stutz & Kartman's (1982) analysis of spatial variations in housing prices. Part of its focus was on how local land use regulations created a geographically differentiated pattern of housing prices.

These studies are typical of geographers' attempts to describe the geography of law. Methodologically, they are similar to House's (1982) conception of the geography of public policy. House argued that geographers ought to study the spatial impacts of policy, as one might study the impacts of firms on the landscape; emphasis is to be on spatial form, not process. In this setting the geography of law simply describes the landscape in terms of the impacts of laws. Geography is a passive stage on which laws, like other public policies, are distributed. Not only is geography passive, laws are assumed little different from policies; law is just a means of policy implementation.

Some urban economists and planners approach the issue in a similar manner. For example, in the literature on housing codes, zoning, and rent control, the issue is often simply one of documenting the impacts of such laws on the urban housing market. Hirsch & Law (1979) noted that the regulations governing acceptable standards and terms of accommodation have tended to shrink the availability of rental housing. Similarly, Kiefer (1980) argued that rent control and housing codes have quite specific (negative) effects on housing and neighbourhood stability. In a related context, Simmie & Hale (1978) considered the distributional consequences of growth control policies, suggesting that in their case study area such policies adversely affect lower-income people. As with geographers' studies of the spatial impacts of law, these kinds of studies are limited to documenting the ramifications of policies in a given setting.

But most urban economists and planners are more ambitious than this. Most seek to establish the welfare costs of housing regulations relative to an imaginary market solution, absent from government regulation. So, for example, Segal & Srinivasan (1985) studied the impacts of suburban growth restrictions on US housing price inflation and concluded that such regulations added significantly to the rate of inflation. The presumption is that without such regulations, housing price inflation would have been much less than that actually measured. Dowall (1984), in a book devoted to studying the patterns of house price inflation in the San Francisco Bay area, argued more forcefully that suburban growth restrictions (laws) were to blame for declining prosperity in the region. He suggested that these regulations increase the cost of housing by restricting the supply of land, limiting the availability of services, and narrowing the range of house types. As a consequence, Dowall argued, such policies make it difficult for firms to attract new employees to the area. In the long run this means that firms with significant labour needs will relocate to less restricted areas, thereby pushing the region into a downward economic spiral.

The reference point for Dowall was, again, an ideal market solution, unfettered by regulation. Not surprisingly, the policy implications drawn from these studies are that land use regulations and growth controls should be abolished, thereby removing the aggregate welfare costs of these regulations.

This kind of analytical paradigm is, of course, consistent with those who would argue that government regulation is always inefficient; that the proper rôle of law is just to protect individuals' freedom (Nozick 1974). While these conclusions are quite drastic, not all urban economists would necessarily agree that the market solution is the only possible solution. White & Wittman (1979) demonstrated that long-run economic efficiency can be attained in the urban land market provided that local governments develop an appropriate mix of liability rules (laws), taxes, and zoning. Their work is representative of a significant body of work in law and economics which deals with the problems of externalities and transaction costs in relation to an otherwise efficient market. For White and Wittman, spatial externalities make short-run efficient market solutions problematic. In their view, laws are needed to facilitate the efficient operation of the market.

Generally, geographers and economists have analyzed the spatial impacts of law from their own disciplinary perspectives. That is, geographers are concerned with describing the spatial patterns of law, in a variety of settings and with respect to various aspects of the local economy. This focus represents geographers' traditional concerns for the patterns of the landscape. So, for example, Morrill's (1981) review of the process and nature of political redistricting begins with a geographical perspective on regionalization and applies it to the judicial problem of adjudicating malapportionment of political districts. There have been, of course, some celebrated cases relating to electoral gerrymandering, and it is clear that the courts have found it difficult to resolve disputes in ways that have satisfied all litigants. But in reading Morrill's analysis, the legal issues are lost amongst goegraphical technique. And, he ends his review by suggesting that 'legal constraints make it impossible to devise ideal regions' (Morrill 1981, p. 63). The implication is that the courts (and legal niceties) are an impediment to geographical efficiency.

Urban economists are similarly preoccupied with applying their techniques to the geographical scene. They take the basic rules of the market as their reference point, and then consider how laws and regulations affect the market. Few urban economists have considered how laws might facilitate the urban land market; most deal with the impacts issue alone. As well, economists like some geographers, claim that from their superior reference point they can evaluate the efficacy of judicial decision making. In both instances, geographers and economists rarely deal with the internal logic of law, its place in society, the kind of society it seeks to represent, and its special rules of adjudication and decision making.

Implied in some studies of the geographical impact of law is a very conservative conception of society. As noted with respect to Dowall's study, the image of society seemed consistent with the narrowest interpretation of liberalism. But, of course, in most instances the underlying model of society embodied in the laws studied is never articulated. If the law is ever evaluated, it is not evaluated in its own terms. Rather, it is evaluated in terms of the respective disciplines and debates within those disciplines regarding the utility of (for example) various theories of urban structure and market efficiency. Such narrow reading of the courts' functions and the legal bases of decision does little justice to the complexity of law. Like law and economics theorists, the position implied by Morrill suggests that social scientists are the ultimate arbiters of the correctness or otherwise of judicial decision making. This is a wholly unwarranted position, especially since law embodies more than simple social decision making (albeit in a certain institutional context). As we noted above, law has a substantive (moral and ethical) base which involves much more than efficiency.[26]

Geography and the judiciary

R. Johnston's (1984) book on the US Supreme Court takes the study of geography and law further than the previous literature. At one level, his interest is in understanding the power of the courts in affecting the urban mosaic. He demonstrates the impacts of the court on local policy, and argues that we can identify the agents of spatial structure in their decisions. Thus at a deeper level, Johnston provides an introduction to the institutional context of American urban life. By focusing upon the courts and conflicts over local policies within the judicial context, he provides an important perspective on points of tension in contemporary America. In this respect he breaks with conventional theories of urban structure – economic, geographical, and sociological. He mounts a critique of preference theory by indicating the structuring rôle that institutions play in defining the choices of local residents.

Johnston's theory of law and society begins with the basic rules of capitalism – private property, class, and power – and adds two American conditions – race and jurisdictional fragmentation. What is so interesting about this book is Johnston's explicit treatment of law. This issue becomes evident when he asks why progressives were unable to use the courts fundamentally to restructure relations between cities and suburbs. His answer has to do with his theory of law and his conception of state power. According to Johnston, laws are *texts*, to be interpreted and read for meaning, as opposed to having the meaning of law found in the words of statutes – by this theory of law, meaning is made not found in the text. In the course of reviewing various judgements by the US Supreme Court, Johnston suggests that the courts have kept to a narrow interpretation of federal statutes, and have based their interpretive practice upon an ideology that both promotes the interests of private property owners and glorifies local democracy. He argues that the urban mosaic is deliberately structured and maintained by an élite in the interests of a narrow constituency located in the suburbs. Thus, his interpretation of the geography of law is directly linked to the nature of law and the political constituency of the judiciary.

As noted in a review of his book, there are problems with his analysis (Clark 1985b). On the basis of the discussion of the nature of law in the previous part of this chapter, it is apparent that the proposition that law is like literature can be readily agreed to. However, in presenting his case material Johnston implies that the meanings of judicial decisions are obvious and without their own internal indeterminacies. For instance, he makes a strong claim that local governments are more than rhetorical shells. He argues like a social scientist, appealing to data, facts, and the like. In this respect he uses the external logic of geography and political science to evaluate the geography of law. The issue of interpretation, while central to his *conception* of the law, is apparently less important than the social sciences when it comes to analyzing the *impacts* of law. Yet, it is also plausible that social science should be conceived as an interpretive device (Taylor 1979) – as a strategem designed to create order out of disorder. If this is the case, then the interpretation of law from a disciplinary perspective is just as problematic as the judiciary's interpretation of law.

These issues have been developed in a couple of recent studies of judicial decision making by Clark (1985a, 1986a). In a full length treatment of American courts, Clark (1985a) sought to explain judicial adjudication in terms of an interpretive mode of analysis. Here the subject matter was the courts' interpretations of local government autonomy, historically and in the present. In a series of case studies, Clark advanced two arguments. First, lawyers and the courts have extraordinary discretionary power in choosing the terrain on which local disputes are to be adjudicated. For example, in a dispute over local jobs in Boston, judicial adjudication so transformed the issue that the very meaning of the original dispute was restructured in favour of legal convention and the language of legal discourse. In this respect, the courts not only adjudicate social disputes, they actually determine the meaning (linguistically and figuratively) of political discourse. In a world of heterogeneity, this determination of meaning is a fundamentally important function of the judiciary – without a final determination of meaning there can be no concerted social action.

A second theme sought to integrate geographical context with judicial reasoning. Judicial outcomes are inevitably based on an interaction between legal doctrine and the problem at hand. By itself, legal doctrine was argued to be indeterminate; the words of doctrine cannot provide an internal reference point for determination. This argument was premised upon an assumption regarding language and meaning; an assumption of heterogeneity of the meaning of given words and conceptions of the proper interpretation of common ideological symbols (like local autonomy). Thus, there is an innate geographical and historical relationship with the process of adjudicating the meaning of doctrine. In this sense, judges not only influence society and geography with their decisions, their very mode of decision making is contextual. Ultimately, this argument favours a relativistic as opposed to a universal conception of meaning.[27] For judges, this implies that decision frameworks are best interpreted as rhetorical devices situated in time and space.[28]

In Clark (1986a) an attempt was made to link the judicial adjudication of land use disputes to the internal structure of principles, and the context at hand. Here, the argument was that to the extent that adjudication was premised upon American liberalism, context provided the point at which determination of meaning was made possible. It was also noted that any resolution of the inherent ambiguity of liberal principles was contextual, relative to the situation at hand. In this paper, as in the previously noted book, an explicit attempt was made to link the substantive bases of interpretation with the context of decision making.[29]

There have been a couple of other attempts to integrate the substantive bases of adjudication with context. Johnston's (1986) paper dealt with two decisions of the US National Labor Relations Board concerning the rights of management to move production from one site to another during the course of a labour contract. Under the Carter Presidency, the Board found that relocation could not proceed without the permission of labour, given that labour had been required to choose between two equally unpleasant choices (either accepting a wage cut or allowing relocation). However, a later Reagan Board found that such relocations were legal, and relied upon a strict literal interpretation of the contract (see also Clark 1986c, 1988).

Johnston linked differences in interpretation of contract to different conceptions of the law, and different conceptions of the proper arrangement of social structure held by Board members. She was able to show how the Board decisions were structured; in terms of law, the Carter Board interpreted contract in a way that emphasized the common law heritage of contract theory. That is, the meaning of contract was interpreted from the standpoint of customary behaviour, the established (albeit unstated) rules of convention. In contrast, the Reagan Board interpreted contract in an extraordinarily formal and exclusive manner. If there were no express provisions regarding relocation, the Board held that management had the right to relocate at any time. They justified their ruling on the property rights of owners.

Underlying these alternative conceptions of contract were two arguments. One had to do with the nature of legal reasoning. The Reagan Board members held that the former Board had erred in not paying strict attention to the specific language of the contract. They contended that good legal practice depends upon an interpretation of the words, not the context, for an understanding of the obligations of the parties to one another. But, of course, to make this argument they also had to assert that language has its own internal meaning – a presumption that finds little favour in the legal literature (see Tushnet 1983), or in the literature of rhetoric and language (see Fish 1983). Moreover, it became apparent in their interpretation of the dispute that they sought to justify their position with reference to a different context – the competitive position of the firm, and ultimately of the nation. As was mentioned above, context need not be considered as a passive aspect of adjudication: the courts' power derives in part from their capacity to choose the relevant context unchallenged by those who would disagree with their conception of the issues.

The second argument had to do with the perceived powers of the parties to influence the contract. The Reagan Board, as in other instances (Clark 1986c), asserted the dominant property powers of owners. The Carter Board asserted that these powers are, and can be in other instances, constrained by bilateral relationships. Thus, at one level the argument was technical – the nature of law. But at a deeper level the argument was about the substantive bases of adjudication – the nature of capitalist society. It is little wonder that the Reagan Board had so many enemies amongst labour; it reintroduced a particularly narrow image of capitalism, justified by notions of efficiency and sheer necessity.

Substantive bases of federalism

In the previous sections it was suggested that understanding the substantive bases of adjudication is a necessary condition for an intellectually mature geography of law. It is apparent, however, that there may be significant disputes over the proper substantive bases of adjudication, and the outcomes which flow from such choices. In this section, we reconsider the significance of a basic conception of American society, political liberalism. This theory of society can be found in many court decisions as the substantive reference point justifying particular interpretations of the law. Nevertheless, it is not the only substantive reference point, nor the only legitimate reference point, despite appearances to the contrary. Actually, for all of its ideological significance, liberalism as *the* substantive reference point is under attack from the right and the left. As a point of reference, these issues will be considered in relation to the geography of federalism.[30]

In an early paper, Clark (1981) argued that an interpretation of the history of US Supreme Court decisions could reasonably explain such decisions as a deliberate policy of fostering spatial integration. Granted, this interpretation might be one among many, and may not find ready acceptance with some constitutional scholars.[31] Even so, as a substantive interpretation of the decisions of the Supreme Court, it has the advantage of going beyond a simple analysis of geographical impacts of law, as if these impacts were unanticipated, by integrating the decision frameworks of the courts with the geographical structure of constitutional power. The argument was that despite an elaborate rhetoric of decentralized liberalism, the Supreme Court has systematically denied the relevance of state and local economic growth policies in favour of the growth of the whole economy. The basis of this judicial policy was argued to lie in two doctrines: one related to the commerce clause of the Constitution, the other related to fragments of a liberal conception of social structure.

With respect to legal doctrine, it was argued that the commerce clause, the privileges and immunities clause, and other lesser clauses and statutes (like the Depression era reconstruction law enacted by the Roosevelt Administration), were the terrain chosen by the Court to sustain its interpretation of the geography of federalism. This involved bolstering the powers of the national government over those of the states, and developing the geography of economic intercourse most consistent with a powerful national government. In this interpretation, the geographical structure of the economy was deliberately fostered by the Court to further its own political aspirations, and those aspirations of its constituents (the Executive Branch). To sustain the argument, a doctrinal review was undertaken, going as far back as to Chief Justice Marshall's court, and the early debates over banking powers. This process accelerated through the 1930s and the most recent era largely because of the activist rôle taken by the government after the 1930s Depression. Spatial integration was conceived as a fundamental underlying substantive conception of the nation–state, though one conception amongst many.

It was also argued that this substantive conception was often justified or legitimated by reference to a form of liberalism. Here, the imperative of decentralization was eschewed in favour of the imperative of centralization. That is, local and state governments were often argued to place unwarranted restrictions on the free mobility of their citizens. That these restrictions interfered with the current of commerce was one reason to find them unconstitutional. But in legitimizing these decisions, the economic imperative was not enough; the courts invoked the centralization imperative as a means of referencing their decisions to a desirable substantive conception of society. Thus, the courts did not deny in absolute terms the relevance of decentralization, rather they emphasized the logic of economic centralization in relation to the dangers of decentralization.

Notice the importance attached to both the substantive interpretation of adjudication, and the rationale used by the courts to justify their decision making. The geography of the economy, the geography of state and local powers, and the geography of government regulation are conceived in relation to an underlying goal and a conception of the proper form of society. We might interpret the goal as derived from the imperatives of accumulation, and the class interests of the judicial élite. But, it would be misleading to suggest that this was the sole logic that sustained the judicial policy. To be effective, it had to be legitimated by reference to an ideal conception of society. This conception was anything but a class conception; rather it was a variation on utilitarianism. Given the nature of American society (its ideal conception), it is difficult to imagine such a judicial policy going unopposed without such an ideological shell. But notice the order of importance attached to the ideological image, and the judicial policy. In these terms, liberalism as *the* substantive reference point was used as a means of legitimization, not as the ultimate rationale for judicial policy.

This is not necessarily the best interpretation of the geography of American law. Other interpretations may be reasonable. In fact, an attack from the right on the significance of liberalism as the substantive reference point comes in unlikely guise: the law and economics movement. Nominally, this movement supports many of the basic tenets of liberalism: maximum personal freedom and decentralized decision making. It also supports a legal order that would be premised upon rules of economic efficiency, as opposed to social justice. At one level the arguments with respect to the geography of federalism appear similar to those just noted above. For example, Easterbrook (1983) suggested that the Supreme Court had consistently used anti-trust regulations to centralize control of the economy, in the interests of maximizing national economic growth. He argued through a series of case reviews that the Court had deliberately denied the states a rôle in economic regulation, and had fostered a nationally integrated system of economic regulation. Thus, at one level his conception of the underlying substantive goal of the Court was much the same as Clark's (1981).

However, Easterbrook went on to argue a case for decentralization of economic regulation. There were two parts to his argument. First, he suggested that state regulation of economic activity would foster competition amongst the states for relative economic position. In contrast, he suggested that centralization has created a strong nation–state unfettered by internal competition between states. The implication he draws from this is that the nation–state is relatively inefficient when regulating business. By returning responsibility for regulation to the states, inter-state competition would encourage more efficient regulation, thus a more prosperous economy overall. Second, to justify this idea, he uses the liberal conception of decentralization as a means of placing his judicial policy in a plausible ideological framework. But notice, like Clark (1981), the substantive reference point for the geography of federalism is economic, not ideological. Liberalism is used to justify the policy as a rhetorical device.

Of course, there are fundamental differences between Easterbrook and Clark with respect to their conceptions of the economic rationale at issue. For Easterbrook, the issue is one of economic efficiency; this is the ultimate reference point for law and economics.[32] Easterbrook would have all judges reassess their decision logic as a matter of intellectual integrity. For Clark the practice of law, with regard to the geography of federalism, was also conceived as an economic issue. But, the motivation for such a conception comes from an interpretation of the judiciary's position in society.

Legitimating 'Law's Empire'

In this discussion of the geography of law, we have avoided explicit discussion of legal methods. However, it should be apparent that the argument and critique of the related geographical and economics literature in previous sections reflect issues which concern legal theorists. Especially important are questions relating to the logic of legal reasoning, and the substantive bases of law, especially liberalism, which justify or legitimize its power. Both issues are now directly addressed.

Legal formalism

Given the importance of substantive law, how might we understand the nature and common practices of legal reasoning in contemporary Anglo-American culture? Analytically, the dominant tradition is called legal formalism. Unger (1983) identified legal formalism by reference to two basic characteristics of modern jurisprudence, objectivism and formalism. He defined objectivism in the following terms: the belief that the authoritative legal materials – the system of statutes, cases, and accepted legal ideas – embody and sustain a defensible scheme of human association. He defined formalism as a mode of decision making which invokes impersonal purposes, policies, and principles as an indispensable component of legal reasoning. And, he suggested that formalism in the conventional sense – the search for a method of deduction from a gapless system of rules – is merely the anomalous, limiting case. Legal formalism is then the combination of these two dimensions of legal practice.[33]

Objectivism and formalism presuppose the existence of the other. That is, judges who use objectivism need formalism if they are to sustain their claims of neutral non-contextual decision making. Similarly, judges who use formalism need objectivism if they are to justify the outcomes of their decisions. Put another way, if judges are to maintain their legitimacy they must claim that the method through which they reach their decisions is above reproach. Indeed, they must be able to distinguish judicial decision making from ordinary politics because politics is the origin of disputes that come before them. The status of the judiciary depends on a presumption that the judiciary is *above* politics. Likewise, the judiciary must also claim that the logic of its decisions arises from a principled adherence to justice, in contrast to ordinary politics which is presumed to operate on the basis of subjective self-interest.

Why should the judiciary be so concerned to legitimate its decisions using a form of legal discourse like legal formalism? There are two obvious and one less obvious reason. In contemporary societies like the US and Canada, the judiciary has a great deal of power and status. Whether deserved or not, this power provides the judiciary with the right to intercede in the very fabric of society. The fact that so few citizens are qualified (by reason of training in a few élite universities) for the position guarantees that judges will be extraordinarily careful to ensure their élite social position. Yet this social status is vulnerable precisely because of the élite connotations embodied in the position. Most judges are appointed, not elected. Most judges do not answer to any legislature, and have tremendous discretion in how they choose to consider issues. Thus, it is readily apparent that the judiciary must be conscious of its power and vulnerability. These two issues have been noted by many scholars of various political persuasions.[34]

More critically, the judiciary has a rôle at the very heart of society. It adjudicates disputes in instances where social cohesion is most fragile; where conventional modes of dispute resolution have been unable to deliver a determinate solution. Here its involvement in social conflict makes the judiciary liable for the resulting conclusion. It is little wonder then that the judiciary clings to devices like legal formalism. It is a means of transforming disputes from the immediate texture of a dispute to a structured discourse controlled by the judiciary. In this respect legal formalism is a means of protecting the judiciary from the tensions of any one dispute. Thus, the less obvious reason for the significance of legal formalism to the judiciary has to do with the tensions within society itself.

These remarks should not be taken as implying that rules are unambiguous, or fully determined independently of the dispute at hand. Like ordinary language, rules are open-textured; meaning is ascribed to rules not found within the fabric of rules. Like Austin (1975), it is contended that there must inevitably be dispute over the very meaning, indeed relevance, of interpretive rules, whether in literature or in the adjudication of disputes. Some legal theorists while debating this point would contend, in any event, that as long as these rules were interpreted in a way consistent with principles there need not be any problem of unbounded judicial discretion (see Tushnet (1983) for an extended treatment of these issues). But if rules depend upon principles for meaning, and if these principles require their own interpretation – which is surely the implication of Austin's argument that language is open-textured – then the application of legal formalism to any dispute could be quite a problematic process. Nevertheless, because the judiciary controls the form of legal discourse, how rules are applied and how they are interpreted can quite radically transform the terms of any dispute.

Principles of liberalism

Legal formalism is more than the application of rules to circumstances. Rules by themselves do not necessarily provide determinate solutions to disputes (Dworkin 1972). The issue is as much the interpretation of rules as their application in particular instances. Likewise, a structured set of rules regarding local authority, for example, need not provide an unambiguous blueprint for the allocation of powers between contending agencies and groups. Rules require interpretation; for interpretation they depend upon underlying principles.

As there are many rules, there are many principles. Since the previous discussion of the geography of federalism considered aspects of liberalism, the interrelation between legal formalism and substantive principles will be illustrated with reference to liberalism. Unger noted that the claim of objectivism depends on a defensible scheme of human association. Thus, although there are many possible principles that could be invoked as the substantive reference points for adjudication, practically speaking only those principles which are at the very centre of social life would qualify as grounds for judicial defence. Notice, however, that there is a great deal of presumption involved in deciding what principles are at the centre of society. One could imagine more radical positions than liberalism being claimed to represent the centre of society. For instance, Posner (1981) has suggested that the appropriate central principles should be economic efficiency and wealth maximization. This is, of course, a fairly conservative form of the radical position which has been subject to a great deal of debate (see Ackerman (1984) and Dworkin (1980) for just two critiques). Others might claim social justice as the central principle (see Rawls 1971, Clark & Dear 1984).

The point is that the judiciary has extraordinary powers in choosing the terrain on which to interpret rules. It is this terrain which is likely to give rules different meanings if different principles are used to interpret similar rules. So, for example, if a judge were to interpret necessary powers of local government from a socialist perspective where private property was not protected, a local government might legitimately appropriate property according to its defined functions. On the other hand, a more conservative judge, using a liberal perspective to interpret local powers, might hold that any appropriation of private property is illegal; that is, outside a standard interpretation of necessary powers. It is quite obvious that urban outcomes would be radically different in both instances even if localities were nominally under control of the same rule.

Liberalism, as described by theorists such as Lowi (1979), Ely (1980), and Sandel (1982), begins with individuals as the very basis of society. In contrast to structuralist notions of society, liberals assume individuals exist prior (in logical time) to society. Individuals are complete as rational and emotional beings; geographical context provides a stage in which to act and find fulfilment (Pred 1984). More extreme versions of this theory suppose that individual utilities are unstructured by social factors, and that individual self-interest is a natural phenomenon (see the more detailed discussion and critique by Sen & Williams (1982)). Of course, this is an ideal image. Once material circumstances are introduced even the most optimistic liberal is likely to acknowledge that society can radically affect peoples' desires (Rawls 1971).

In fact it is precisely this possibility that has lead some liberal theorists to argue that individuals are essentially untrustworthy. Choper (1980) argued that when individuals' selfish interests are combined in a group, especially a majority, others will inevitably be adversely affected. It is for this reason that liberal theorists often use original positions as analytical devices to separate individuals from immediate material interests. Rawls (1971) begins his analysis by locating individuals in a non-material context, behind a veil of ignorance. From that vantage point, he then asks individuals to choose a set of rules that would protect them in the event that they end up in an inferior social position. This strategy is utilized by Rawls to ensure a just solution to basic entitlements, without recourse to material circumstances. The liberal world is one where individuals have fundamental status, despite their often undesirable behaviour.

Not only does liberalism have a highly articulated vision of individual motivations (even though these individuals do not appear to have any social personality; see Sandel 1982, Clark 1985a), it also claims a particular conception of the proper rôle of government. Most obviously, a liberal state should protect the rights of individuals. After all, if individuals are so fundamental, their potential for action must be fully realized. Otherwise, if individuals were compromised in the exercise of their rights, their whole integrity would be at risk. Just as obviously, individuals must be protected from those who would not respect the rights of others. But, this is not the last word on the rôle of the liberal state for a number of theorists, past (including de Tocqueville and Locke) and present (including Nozick (1974) and Taylor (1982)) have argued for a particular spatial configuration of state powers.

Taking Nozick (1974) as the paradigmatic case, some liberals have suggested that decentralized government is the most appropriate form of government. At this level, it is argued, human association is most convivial. De Tocqueville suggested that this is because the small town is closest to nature, mans' original position. Nozick also suggested that having a set of small towns can allow like-minded people to find and consume their true preferences. As a consequence, decentralized homogeneous communities might also limit the tendencies for individual, selfish exploitative behaviour. De Tocqueville invoked God to justify his vision of decentralized life (a natural rights theory of law). Nozick justified his spatial geometry by the fundamental assumptions of liberal philosophy: individuals' self-interest and their fundamental integrity (also a natural rights theory, though with different roots than de Tocqueville's). This liberal vision of community life is what has been termed elsewhere the imperative of decentralization (Clark 1985a).

This imperative does not, however, stand alone. It is counter-posed by another imperative, that of centralization.[35] Because liberals do not trust individuals' actions in social groups, isolated individuals must be protected. That is, there must be a mechanism by which those individuals who feel victimized in a community can appeal to some other authority. Also, communities themselves must have protection from other communities which may seek to dominate them. Consequently, a centralized review agency, like the courts or some other higher tier agency, would be necessary to retain the integrity of the whole system of communities. Inevitably, there is a tension between the two imperatives – decentralization and centralization – and a large rôle for the courts in adjudicating the relative significance of these two imperatives in different situations. Thus, when Easterbrook (1983) argued for a particular conception of federalism, referencing liberalism as his substantive claim for legitimacy, he was referencing a very particular theory of society. This theory, though, may or may not be attractive to others. In this sense, the underlying substantive rationale for any judicial decision must be part of any analysis of the geography of law. If not, the geography of law is doomed to superficiality.

Whither the geography of law?

The last part of this chapter is concerned with placing the geography of law into a critical context. In the first instance, this involves a discussion of the significance of law in modern society. In the second instance, it is argued that the rule of law is less perfect than is often appreciated. Thus, as students of the geography of law we need not accept the letter of the law as inherently plausible.

The rule of law

Recent interest in law and society is not a new event. What is different is the application of the techniques and methods of the social sciences to legal problems. For most theorists, law is just a framework for social decision making. As Atiyah (1983, p. 147) noted, liberal theory held (and still holds) that the purpose of law was simply to provide a framework within which human beings could pursue their own salvation. The law and economics tradition exemplifies this assumption. By the application of economic principles to legal disputes, they hope to provide a better framework for human association. For geographers interested in the law this is a tempting model to mimic. Even so, law is more and less than this simple conception would allow.

Law is more than this conception would allow because it represents an ideal of human association. To be consistently regulated, to be under the shelter of non-arbitrary decision making is one of the most appealing aspects of the rule of law. Its appeal is lodged in a dim memory of the absolute power of the monarchy, and the not so dim memory of dictatorship. In this sense, law is valued in its own right, as a substantive end, not simply as a means to an end. Further, it is often supposed, as Unger (1983) noted, that the rule of law embodies a defensible conception of human association. In this sense, law is perceived by lawyers and non-lawyers alike as the *moral* locus of society. As a consequence, the (conservative) law and economics movement fundamentally mistakes the malleability of the law in assuming that law is an empty vessel into which the tools of economic analysis can be placed. In point of fact, the rule of law is extraordinarily powerful as a substantive value. As a result it has tremendous power as a structuring force on society. This has been lately recognized (albeit somewhat implicitly) by Brennan & Buchanan (1985).

But, law is also less than this simple conception. Morally, as opposed to technically, the power of law is confined within a tense dialectic. On the one hand it draws its power from its representation and adjudication of the rules of social association. On the other hand, its power is limited by the heterogeneity of social values, social positions, and social differentiation. To suppose that one value, or one mode of adjudication is capable of integrating society is at best naïve (representing society as cohesive and collaborative) and at worst Machiavellian. Law, as an institution, is powerful because of the heterogeneity of society, but this does not mean that it can, or should, dominate society with one rule of conduct. Again, the (conservative) law and economics movement mistakes the power of law in supposing that the moral ambiguity of law can be replaced with standard economic principles. The geography of law as an intellectual movement will surely fail if it models itself on the law and economics movement.

Limits of judicial reasoning

More technically, though, there are other reasons for being cautious of the power of law in completely structuring society. It is one thing to imagine a set of unambiguous decision rules that would structure judicial adjudication; legislation and constitutions attempt to do just that, presuming a wholly ordered world (Goodman 1984). Practically, though, such planned determinacy is never achieved. Three reasons for indeterminacy can be identified.[36] The first reason can be termed judicial incapacity. Despite legal theorists' desires (see Richards 1979), judges are not supermen or superwomen. Applying rules to situations inevitably requires judgement; each new situation is a challenge to the judiciary because rules are general not specific. Rules are designed to cover many circumstances. It is for this reason that there is often debate over the appropriateness of different rules.

The second kind of reason is termed the incoherence of principles. Rules require principles for their design and interpretation as events change and new circumstances are confronted. There must be some intent behind the design of a rule, otherwise the rule would be meaningless. Of course, it is plausible that rules are poorly conceived in terms of their justificatory principles. But more problematic are instances where the principles are themselves incoherent. Any system of rules premised upon a confused set of principles is inevitably compromised. Elsewhere, I have argued that this is the case for liberal principles; there is an in-built contradiction in liberalism which makes any set of liberal rules appear arbitrary and capricious. Thus, any appeal to such principles as justification for a particular decision will likely generate its own internal inconsistencies. While these internal problems may not be readily apparent, it has been shown that consistent use of such principles can lead to a significant instability in adjudication, even though the problem may be relatively stable over time (Clark 1985a, Ch. 8).

The third, and most powerful, reason for judicial indeterminancy has to do with the methodological separation between theory (principles) and practice (rules). This kind of reason might be termed analytical abstraction. Because principles are conceived as abstract analytical statements, empirical rules will always be distant from their original locations. Rules attempt to provide guidelines for action; principles eschew action for simple clarity. Essentially, rules are the boundaries of principles. Quine (1953) suggested that there is an inevitable dissonance between abstract principles and empirical reality. Because principles depend upon abstraction for their integrity it is likely that no system of principles will be immediately applicable to specific circumstances. It is their very abstraction which makes principles desirable, but it is precisely this kind of abstraction which makes principles difficult to relate to circumstances. There will always be some form of interpretive dilemma as judges move from abstraction to practice and back to abstraction.

Finale

For those interested in the geography of law, the lessons of this chapter are threefold. First, to study the spatial impacts of law requires an appreciation of the substantive structure of law. While the regulation of economic activities, and their spatial impacts, for example, is an area of obvious importance, to understand these impacts adequately we must link regulations with their substantive foundations. Second, it is also apparent that there may be widespread disagreement over the significance of different substantive principles. This might be thought of as just an empirical problem. That is, can we discriminate between Posner's economic efficiency foundations and others' liberal foundations? But this is hardly simple. Embedded in diverse conceptions of substantive structure are very different conceptions of society. It is extraordinarily difficult, if not impossible, to imagine a simple empirical test which would discriminate between these normative images (Putnam 1981). Third, it should be apparent that substantive foundations of liberalism, to take one example, are not unambiguous. Judicial reasoning, although based on these kinds of premises, may also be compromised by tensions embedded in the chosen substantive foundations of adjudication.

It might be argued that the geography of law should be less ambitious, less concerned with integrating notions of legal practice with social-cum-geographical theory, and just concerned with documenting the spatial impacts of judicial decisions. That is, the proper domain for geographers' analysis of the law should be drawing maps of judges' actions. However appealing this stance might be in the sense that it holds in abeyance larger social issues, it is too simple. The links between law and society, between law and geography, are indissoluble since as law is drawn from society it also reproduces society. And, as law is structured by context, it structures context.

In this respect, the geography of law is part of a larger intellectual project which has as its basic problem the understanding of the structuration of the landscape of experience. The only intellectually viable geography of law is one which takes this statement to heart.

Notes

1 This chapter is based on my work on the American judiciary, and has benefited over the years from advice from David Kennedy, Gerry Suttles, and John Whiteman. Nick Blomley, Dick Peet, Nigel Thrift, and Ron Johnston made useful comments on a previous draft of this chapter. I benefited also from discussions at Exeter University on the interactions between law and geography. Thanks to Kim Economides and Felix Driver for comments on my presentation. They should not be held responsible for this latest incarnation.

2 In a presentation at the Association of American Geographers' Annual Meeting, Thompson & Wijeyawickrema (1986), suggested that geographers' concerns with the spatial impacts of law could be traced back to Ellen Semple (around 1911). Even so, by their own reckoning there were few publications that would count as geography of law studies until the mid-1970s.

3 See Taylor & Johnston (1979) on the significance of boundary selection on parties' electoral performance, and Palm (1979) on the impact of government policies on housing and urban land prices.

4 While economic theories of urban structure have been influential in geography, so too have sociological theories. The standard ecological model owes its roots to the Chicago School, championed by Park and Burgess (see Kurtz 1984). As used in contemporary settings, this model is generally applied (like the neoclassical economic model), whatever the socio-political and cultural context. Hence the many attempts to apply Park and Burgess type models around the world (for a general introduction see Berry & Kasarda 1977).

5 Of course, not all researchers are so concerned with the spatial organization of capitalism *as a mode of production*. For instance, a less radical perspective was utilized by Levy *et al.* (1974) when they considered the impact of the bureaucratic structure of policy implementation on local public service provision.

6 In a related context, Dear (1980) summarized this mode of research in the phrase 'the public city' (see Kirby (1983) for an extended commentary on the concept). Here, this notion is assumed to mean that since institutions have a fundamental rôle

in structuring society institutions have a fundamental rôle in structuring geography.

7 As exceptions to this observation see the studies by Mercer (1985, 1987) who has considered Australian constitutional issues in relation to, respectively, environmental management and Aboriginal land rights.

8 As will become apparent, the perspective of this chapter is itself interpretive and thus hermeneutic. Taylor (1985, p. 15) suggested that interpretation

> is an attempt to make clear, to make sense of, an object of study. This object must, therefore, be a text, or a text-analogue, which in some way is confused, incomplete, cloudy, seemingly contradictory. . . . The interpretation aims to bring to light an underlying coherence or sense.

There are many different versions of this perspective, deriving from Marxism, critical theory, and Habermas's new rationalism (Held 1980). It is a crucial theme in contemporary American legal theory (see Unger (1983) for a general introduction and the study by Frug (1980) of cities' legal powers as an example of the perspective in action).

9 Most obvious exceptions have been Blomley's (1985, 1986) studies of the Shops Act in Britain, and Johnston's (1983) more theoretical analysis of judges' and bureaucrats' political organization of space.

10 There is evidence that the opposition between these two systems is declining. Atiyah (1987) has suggested that the two systems are becoming more alike.

11 See Epstein (1982) for an argument to the effect that disputes over common law rules are more likely settled in the political arena than in the judicial arena. In contrast, Epstein suggests that by the nature of constitutional theory, disputes are more likely settled in the judicial arena than the political arena.

12 There is a massive and growing literature on the philosophy of law. In such a brief sketch of the theories of law, we cannot do justice to the subtleties and nuances of each theory, and their relationships one to the other. Dworkin (1986) provides a useful introduction to the problems of the literature, Tuck (1979) provides an historical perspective on natural rights theories, and Raz (1979) provides an updated defence of legal positivism. For a comparison of Dworkin, Hart, and the new legal realists, see Yanal (1985).

13 For an appreciation of the contribution of Hart to jurisprudence see the volume edited by Gavison (1987).

14 The idea that procedures and outcomes can be so separated is a crucial claim of liberal jurisprudence. A recent example of this theory in action is to be found in Ely's (1980) attempt to identify a judicial interpretive procedure which would be neutral with respect to judges' values and competing social blueprints.

15 For a comparison of positivist and natural rights theories of law see the review and assessment by Beyleveld & Brownsword (1985).

16 Blackstone used a natural rights thesis to great effect in legitimating the prerogatives of the English crown some two centuries ago (Jones 1973). It would be no exaggeration to claim that henceforth the natural rights thesis has been interpreted as a conservative theory, legitimating the status quo (see Kennedy's (1979) deconstruction of Blackstone's *Commentaries* along these lines). It is also an important theme in contemporary arguments relating to the integrity of private property against claims that property ought to be a public good (Clark 1982).

17 Hart (1982) has published a series of essays on Bentham's influence on contemporary legal theory. See Clark (1984) for an example of Bentham's legal theory as applied to the issue of local autonomy in American cities.

18 The term realist is used relatively generally. In the 1930s, American jurisprudence was radicalized by a realist movement which sought to analyze law as *practised* as opposed to *idealized* (see Twining (1973) for a critical assessment). According to Ackerman (1984), the realist movement also sought recognition for new terms of

legal discourse related to the New Deal state. In a sense, realism was empirical and social, eschewing moralism for the practice of law. However, the realist movement was not a positive theory of law as we might associate with Hart (1961). By Dworkin's (1986) interpretation, the realist movement also argued for a context-relevant theory of adjudication. In this respect, the realist movement aimed to reflect the interaction between cases and judges. Thus, the realist movement was not a movement with a hermeneutic method. The realist theory of law referred to in this chapter combines a concern for the institutions of law with an interpretive perspective based upon a more recent development in American jurisprudence – the critical legal studies movement (see Unger (1983) and the *Stanford Law Review* (1983) for overviews of the essential elements of this movement).

19 By this definition law is a principled force. Coercion is at the very heart of law, but is distinguished from barbarism by virtue of the general principles that provide law with its political legitimacy (see Dworkin (1978) on the nature of principles and policies).

20 A somewhat different perspective on the rôle of law in capitalist (English) society is provided by O'Hagan (1984). He sought to combine Marx with J. S. Mill in a theory combining a concern for radical equality and individual rights. Implicitly, and in contrast to recent critical legal studies theorists, O'Hagan assumes law to be unproblematic as a social discourse. In this sense, O'Hagan's theory of law is more about society than about law.

21 See, for example, the recent issue of the *Texas Law Review* (1984) on the politics of interpretation.

22 Including theorists such as Marx, Weber, and Durkheim; see Edelman (1979) and Trubek (1972) for general overviews to the literature.

23 See Walker & Storper's (1978) study of the Clean Air Act as a rare example of geographers analyzing the impact of recent legislation.

24 So for example, writers such as Horwitz (1977) and Kennedy (1976) associate the practice of legal reasoning and adjudication with the imperatives for capitalist reproduction. They do not agree with Posner that law is efficient, or that the law ought to mimic the market (see Kennedy & Michelman 1980), but they do agree that understanding judicial adjudication can only be done with reference to the substantive structure of law.

25 In order to maintain the themes introduced previously, and spare the reader comments on a set of diverse and unrelated fields of research, the review is largely limited to issues relating to urban structure. There are other kinds of research on law and geography which deserve explicit treatment. Especially important is the literature on the geography of crime. Much of this literature is concerned with the spatial patterns of crime (Harries 1984, 1980), the rôle of the environment in the incidence and control of crime (Herbert 1979), and the deterrent effects of law with respect to the spatial incidence of crime (Pyle *et al*. 1974). Only Lowman (1982) has attempted to deal with the legal aspects directly, though even in this case, the author approaches the problem from a disciplinary perspective as opposed to a mixed legal and geographical perspective.

26 Tribe (1985, p. 187) argued that law must serve an emancipatory function. It should protect those who have insufficient clout in the marketplace, those who are trapped in large institutions, and those who have no property but work for a living. Tribe supports an administrative–interventionist state; a state which uses the legal apparatus to foster redistribution. By this mandate, law is more complex and political than implied by disciplinary interests in the relative efficiency of legal decisions. For an interesting review of Tribe in relation to the law and economics movement and critical legal studies, see Tushnet (1986).

27 Relativism is a difficult notion to define, and defend. Williams (1972) criticized a vulgar form of relativism that supposes that there can be no morally correct

decision; because everything is relative no position can claim dominance over others. This kind of relativism is akin to nihilism which was defined by Rosen (1969) in the following terms. If there are many values, all with their own virtues and with no internal order of significance, then if all values are equally desirable, no single value is likely to find support. This is not what is meant by relativism in this chapter. Here, relativism describes a world in which there are many different interests, all competing for power. Since there is no ready made world, an empirical reality which is found as opposed to made through interpretation, those who control the interpretive apparatuses (like the courts) control social definition of right and wrong (cf. Goodman 1984).

28 In this respect, the argument advanced in Clark (1985a) parallels recent attempts by Thrift (1985) and others to develop a contextual theory of meaning.

29 Goodrich (1986) and White (1985) have each attempted similar extended treatments of the interpretive bases of law. Both deal with the rhetorical devices of law, and the hermeneutic tradition in legal practice. White is especially good on law as discourse.

30 Bennett (1980) provides a useful discussion of the geography of federalism in a variety of countries. Notice that his interest is in the public finance dimensions of inter-governmental grants, between different spatial tiers of government.

31 See the volume edited by Friedman & Scheiber (1978) for a set of related essays on the historical evolution of American jurisprudence. They emphasize the complexity and spatial variety of state legal systems and the tortuous path to an integrated federal judicial system. In this respect, the essays in that volume provide a deeper and more varied account of the patterns identified in Clark (1981).

32 Recently, Easterbrook (1984) argued that the Supreme court has become more sophisticated on matters of law and economics. He suggested that a well developed and articulated vision of the economy and the rôle of law in relation to the economy can be interpreted in the Court's recent decisions. According to Easterbrook, this increasing sophistication is not a political matter, rather it reflects a growing appreciation of economics as a mode of thinking. At the time, Easterbrook was a professor of law at the University of Chicago. He is now a federal district judge, sitting with a previous law professor, Richard Posner. If the courts were reticent about the use of economics in deciding legal questions before these two law professors were appointed, this is sure to change.

33 This discussion is based on the methods of critical legal studies. For critical assessments of this literature see Finnis (1987) and Hunt (1986).

34 See Wolff (1971) for a radical view, Ely (1980) for a middle-of-the-road view, and Choper (1980) for a conservative view.

35 This is a structural feature of liberalism as a mode of social thought (Kennedy 1979), and as a logic for a particular spatial configuration of human association.

36 See Clark (1986a) for the text upon which this section depends.

References

Ackerman, B. A. 1984. *Reconstructing American law*. Cambridge, Mass.: Harvard University Press.

Alonso, W. 1964. *Location and land use*. Cambridge, Mass.: Harvard University Press.

Atiyah, P. 1983. *Law and modern society*. Oxford: Oxford University Press.

Atiyah, P. 1987. *Pragmatism and theory in English law*. London: Stevens.

Austin, J. L. 1975. *How to do things with words*, 3rd edn. Cambridge, Mass.: Harvard University Press.

Bennett, R. J. 1980. *The geography of public finance*. London: Methuen.

Berry, B. J. L. & J. Kasarda 1977. *Contemporary urban ecology*. New York: Macmillan.

Beyleveld, D. & R. Brownsword 1985. The practical difference between natural-law theory and legal positivism. *Oxford Journal of Legal Studies* **5**, 1–32.

Blacksell, M., C. Watkins & K. Economides 1986. Human geography and law: a case of separate development in social science. *Progress in Human Geography* **10**, 371–96.

Blomley, N. 1985. The Shops Act (1950): the politics and the policing. *Area* **17**, 25–33.

Blomley, N. 1986. Regulatory legislation and the legitimation crisis of the state: the enforcement of the Shops Act (1950). *Environment and Planning D, Society and Space* **4**, 183–200.

Brennan, G. & J. Buchanan 1985. *The reason of rules: constitutional political economy*. Cambridge: Cambridge University Press.

Brink, D. O. 1985 Legal positivism and natural law reconsidered. *Monist* **68**, 364–87.

Choper, J. H. 1980. *Judicial review and the national political process*. Chicago: University of Chicago Press.

Clark, G. L. 1981. Law, the state, and the spatial integration of the United States. *Environment and Planning A* **13**, 1197–32.

Clark, G. L. 1982 Rights, property, and community. *Economic Geography* **59**, 120–38.

Clark, G. L. 1984. A theory of local autonomy. *Annals of the Association of American Geographers* **74**, 195–208.

Clark, G. L. 1985a. *Judges and the cities: interpreting local autonomy*. Chicago: University of Chicago Press.

Clark, G. L. 1985b. Review of *Residential segregation, the state, and constitutional conflict in American urban areas* by R. J. Johnston. *International Journal of Urban and Regional Research* **9**, 593–5.

Clark, G. L. 1986a. Adjudicating land use disputes in Chicago and Toronto: legal formalism and urban structure. *Urban Geography* **7**, 63–80.

Clark, G. L. 1986b. Making moral landscapes: John Rawls' theory of justice. *Political Geography Quarterly* **5**, (supplement), S147–S162.

Clark, G. L. 1986c. Restructuring the US economy: the NLRB, the Saturn Project and economic justice. *Economic Geography* **62**, 289–306.

Clark, G. L. 1989. A question of integrity: the NLRB and the relocation of work. *Political Geography Quarterly* (forthcoming).

Clark, G. L. & M. Dear 1984. *State apparatus: structures and language of legitimacy*. Boston and Hemel Hempstead: Allen & Unwin.

Currie, D. P. 1985. The constitution in the Supreme Court: the protection of economic interests, 1889–1910. *University of Chicago Law Review* **52**, 324–88.

Dawson, J. A. 1982. A note on the law of 29 June 1975 to control large scale retail development in Belgium. *Environment and Planning A* **14**, 291–6.

Dear, M. 1980. The public city. In *Residential mobility and public policy*, W. A. V. Clark & E. Moore (eds). Beverly Hills, CA: Sage Publications.

Dicey, A. V. 1959. *An introduction to the study of the law of the constitution*, 10th edn. London: Macmillan.

Dillon, J. 1911. *Commentaries on the law of municipal corporations*, 5th edn. Boston: Little, Brown.

Dowall, D. 1984. *The suburban land squeeze: land conversion and regulation in the San Francisco Bay area*. Berkeley: University of California Press.

Dworkin, R. 1972. Social rules and legal theory. *Yale Law Journal* **81**, 855–90.

Dworkin, R. 1979. Taking rights seriously. Cambridge, Mass.: Harvard University Press.

Dworkin, R. 1980. Is wealth a value? *Journal of Legal Studies* **9**, 191–242.

Dworkin, R. 1985. *A matter of principle*. Cambridge, Mass.: Harvard University Press.

Dworkin, R. 1986. *Law's empire*. Cambridge, Mass.: Harvard University Press.

Easterbrook, F. H. 1983. Antitrust and the economics of federalism. *Journal of Law Economy* **26**, 23–50.

Easterbrook, F. H. 1984. Forward. The court and the economic system. *Harvard Law Review* **98**, 4–60.
Economides, K., M. Blacksell & C. Watkins 1986. The spatial analysis of legal systems: towards a geography of law. *Journal of the Law Society* **13**, 161–81.
Edelman, B. 1979. *Ownership of the image: elements for a marxist theory of law* (translated by E. Kingdom). London: Routledge & Kegan Paul.
Ely, J. 1980. *Democracy and distrust: a theory of judicial review*. Cambridge, Mass.: Harvard University Press.
Epstein, R. A. 1982. The social consequences of common law rules. *Harvard Law Review* **95**, 1717–51.
Everitt, J. 1984. Influences on land use change in the urban fringes: the case of Surry, British Columbia. *Geographical Perspectives* **53**, 1–14.
Finch, M. & T. Nagel. Spatial distribution of bargaining power: binding arbitration in Connecticut school districts. *Environment and Planning D, Society and Space* **1**, 429–46.
Fincher, R. 1989. The political economy of the local state. In *New models in geography*, R. Peet & N. Thrift (eds). London and Winchester, Mass.: Unwin Hyman.
Finnis, J. 1980. *Natural law and natural rights*. Oxford: Clarendon Press.
Finnis, J. 1987. On the critical legal studies movement. In *Oxford essays in jurisprudence*, J. Eekelaar & J. Bell (eds). Oxford: Oxford University Press.
Fish, S. 1983. *Is there a text in this class? The authority of interpretive communities*. Cambridge, Mass.: Harvard University Press.
Friedman, L. & H. Scheiber (eds) 1978. *American law and the constitutional order*. Cambridge, Mass.: Harvard University Press.
Frug, G. 1980. The city as a legal concept. *Harvard Law Review* **93**, 1057–154.
Gavison, R. (ed.) 1987. *Issues in contemporary legal philosophy: the influence of H. L. A. Hart*. Oxford: Clarendon Press.
Goodman, N. 1984. *Of mind and other matters*. Cambridge, Mass.: Harvard University Press.
Goodrich, P. 1986. *Reading the law: a critical introduction to legal method and techniques*. Oxford: Basil Blackwell
Harden, I. & N. Lewis 1986. *The noble lie: the British constitution and the rule of law*. London: Hutchinson.
Harries, K. D. 1974. *The geography of crime and violence*. New York: MacGraw-Hill.
Harries, K. D. 1980. *Crime and the environment*. Springfield, Illinois: Charles Thomas.
Hart, H. L. A. 1961. *The concept of law*. Oxford: Oxford University Press.
Hart, H. L. A. 1982. *Essays on Bentham: jurisprudence and political theory*. Oxford: Oxford University Press.
Held, D. 1980. *Introduction to critical theory: Horkheimer to Habermas*. Berkeley: University of California Press.
Herbert, D. T. 1979. Urban crime: a geographical perspective. In *Social problems and the city: geographical perspectives*, D. M. Herbert & R. Johnston (eds). Oxford: Oxford University Press.
Hirsch, W. Z. & C.-K. Law 1979. Habitability laws and the shrinkage of substandard rental housing stock. *Urban Studies* **16**, 19–28.
Horwitz, M. 1977. *The transformation of American law 1789–1860*. Cambridge, Mass.: Harvard University Press.
House, J. W. 1982. The geography of public policy. *Transactions of the Institute of British Geographers* **NS7**, 23–36.
Hunt, A. 1986. The theory of critical legal studies. *Oxford Journal of Legal Studies* **6**, 1–45.
Johnston, K. 1986. Judicial adjudication and the spatial structure of production: two decisions by the National Labor Relations Board. *Environment and Planning A* **18**, 27–39.
Johnston, R. J. 1983. Texts and higher managers: judges, bureaucrats and the political organization of space. *Political Geography Quarterly* **2**, 2–30.

Johnston, R. J. 1984. *Residential segregation, the state and constitutional conflict in American urban areas*. New York: Academic Press.

Johnston, R. J. 1986. The general good of the community. Some perspectives on town planning and residential segregation: a Mount Laurel case study. *Planning Perspectives* **1**, 131–45.

Johnston, R. J. 1989. The state, political geography, and geography. In *New models in geography*, R. Peet & N. Thrift (eds). London and Winchester, Mass.: Unwin Hyman.

Jones, G. (ed.) 1973. *The sovereignty of law: selections from Blackstone's commentaries on the laws of England*. Toronto: University of Toronto Press.

Kennedy, D. 1976. Form and substance in private law adjudication. *Harvard Law Review* **89**, 1685–778.

Kennedy, D. 1979. The structure of Blackstone's commentaries. *Buffalo Law Review* **29**, 205–381.

Kennedy, D. & F. Michelman 1980. The efficiency of contract. *Hofstra Law Review* **8**, 711–70.

Kiefer, D. 1980. Housing deterioration, housing codes, and rent control. *Urban Studies* **17**, 53–62.

Kirby, A. 1983. A public city: concepts of space and the local state. *Urban Geography* **4**, 191–202.

Kurtz, L. 1984. *Evaluating Chicago sociology: a guide to the literature, with an annotated bibliography*. Chicago: University of Chicago Press.

Levy, F. S., A. J. Meltsner & A. Wildavsky 1974. *Urban outcomes*. Berkeley: University of California Press.

Lowi, T. 1979. *The end of liberalism*. New York: Norton.

Lowman, J. 1982. Crime, criminal justice policy and the urban environment. In *Geography and the urban environment: progress in research and applications*, vol. 5, D. T. Herbert & R. Johnston (eds). New York: Wiley.

MacAvoy, P. W. 1979. *The regulated industries and the economy*. New York: Norton.

Mercer, D. 1985. Australia's constitution, federalism and the Tasmanian dam case. *Political Geography Quarterly* **4**, 91–110.

Mercer, D. 1987. Patterns of protest: native land rights and claims in Australia. *Political Geography Quarterly* **6**, 171–94.

Michelman, F. 1979. A comment on 'Some uses and abuses of economics'. *University of Chicago Law Review* **46**, 307–15.

Mills, E. S. & B. W. Hamilton 1984. *Urban economics*, 3rd edn. Glenview, IL: Scott, Foresman.

Morrill, R. L. 1981. *Political redistricting and geographic theory*. Washington, DC: Association of American Geographers.

Nozick, R. 1974. Anarchy, state, and utopia. New York: Basic Books.

O'Hagan, T. 1984. *The end of law?* Oxford: Basil Blackwell.

Palm, R. 1979. Financial and real estate institutions in the housing market: a study of recent house price changes in the San Francisco Bay area. In *Geography and the urban environment: progress in research and applications*, vol. 2, D. Herbert & R. Johnston (eds). New York: Wiley.

Panzor, J. C. 1980. Regulation, deregulation, and economic efficiency. *American Economic Review* **70**, 311–15.

Persons, G. A. 1984. Deregulation policies and the issue of income distribution. *Urban League Review* **8**, 67–77.

Platt, R., G. Macinko & K. Hammond 1983. Federal environmental management: some land use legacies of the 1970s. In *United States public policy: a geographical view*, J. House (ed.). Oxford: Oxford University Press.

Posner, R. 1977. *Economic analysis of law*. Boston: Little, Brown.

Posner, R. 1981. *The economics of justice*. Cambridge, Mass.: Harvard University Press.

Poulantzas, N. 1978. *State, power, socialism*. London: New Left Books.
Pred, A. 1984. Place as historically contingent process: structuration and the time-geography of becoming places. *Annals of the Association of American Geographers* **74**, 279–97.
Putnam, H. 1981. *Reason, truth, and history*. Cambridge: Cambridge University Press.
Pyle, G. 1974. *The spatial dynamics of crime*. Research paper 159. Chicago: University of Chicago, Department of Geography.
Quine, W. 1953. *From a logical point of view*. Cambridge, Mass.: Harvard University Press.
Rawls, J. 1971. *A theory of justice*. Cambridge, Mass.: Harvard University Press.
Raz, J. 1979. *The authority of law: essays on law and morality*. Oxford: Clarendon Press.
Raz, J. 1985. Authority, law and morality. *Monist* **68**, 295–324.
Richards, D. A. J. 1979. The theory of adjudication and the task of a great judge. *Cardozo Law Review* **1**, 171–218.
Rosen, S. 1969. *Nihilism: a philosophical essay*. New Haven, Conn.: Yale University Press.
Sandel, M. 1982. *Liberalism and the limits of justice*. Cambridge: Cambridge University Press.
Segal, D. & P. Srinivasan 1985. Impact of suburban growth restrictions on US housing price inflation 1975–1978. *Urban Geography* **6**, 14–26.
Sen, A. & B. Williams (eds) 1982. *Utilitarianism and beyond*. Cambridge: Cambridge University Press.
Simmie, J. & D. J. Hale 1978. The distributional effects of ownership and control of land use in Oxford. *Urban Studies* **15**, 9–21.
Stanford Law Review 1983. **36**, 1–674.
Stutz, F. P. & A. E. Kartman 1982. Housing affordability and spatial price variations in the US. *Economic Geography* **58**, 221–35.
Taylor, C. 1979. Interpretation and the sciences of man. In *Interpretive social science: a reader*, P. Rabinow & W. Sullivan (eds). Berkeley: University of California Press.
Taylor, C. 1985. *Philosophy and the human sciences*. Cambridge: Cambridge University Press.
Taylor, M. 1982. *Community, anarchy, and liberty*. Cambridge: Cambridge University Press.
Taylor, P. & R. Johnston 1979. *Geography of elections*. London: Penguin.
Teitz, M. 1978. Law as a variable in urban and regional analysis. *Papers of the Regional Science Association* **41**, 29–41.
Texas Law Review, 1984. **60**, 373–586.
Thompson, G. & C. Wijeyawickrema 1986. The growing affinity between geography and law. Paper presented at the annual meeting, Association of American Geographers, Minneapolis, Minnesota.
Thrall, G. I. 1987. *The consumption theory of land rent*. London: Methuen.
Thrift, N. 1985. Flies and germs: a geography of knowledge. In *Social relations and spatial structures*, D. Gregory & J. Urry (eds). London: Macmillan.
Tribe, L. 1985. *Constitutional choices*. Cambridge, Mass.: Harvard University Press.
Trubek, D. M. 1972. Max Weber on law and the rise of capitalism. *Wisconsin Law Review* **3**, 720–53.
Tuck, R. 1979. *Natural rights theories: their origin and development*. Cambridge: Cambridge University Press.
Tushnet, M. 1983. Following the rules laid down: a critique of interpretivism and neutral principles. *Harvard Law Review* **96**, 871–927.
Tushnet, M. 1986. The unities of the constitution. *Harvard Civil Liberties and Civil Rights Review* **55**, 285–306.
Twining, W. 1973. *Karl Llewellyn and the realist movement*. London: Weidenfeld & Nicolson.

Unger, R. M. 1983. The critical legal studies movement. *Harvard Law Review* **96**, 561–675.

Walker, R. & M. Heiman 1983. Quiet revolution for whom? *Annals of the Association of American Geographers* **71**, 67–83.

Walker, R. & M. Storper 1978. Erosion of the Clean Air Act of 1970: a study in the failure of government regulation and planning. *Boston College Environmental Affairs Law Review* **7**, 189–257.

White, J. B. 1985. *Hercules' bow: essays on the rhetoric and poetics of the law*. Madison: University of Wisconsin Press.

White, M. & D. Wittman 1979. Long-run versus short-run remedies for spatial externalities: liability rules, pollution taxes, and zoning. In *Essays on the law and economics of local government*, D. L. Rubinfled (ed.). Washington, DC: The Urban Institute.

Williams, B. 1972. *Morality: an introduction to ethics*. New York: Harper & Row.

Wolff, R. P. (ed.) 1971. *The rule of law*. New York: Simon & Schuster.

Yanal, R. J. 1985. Hart, Dworkin, judges, and new law. *Monist* **68**, 388–402.

13 *The political economy of the local state*

Ruth Fincher

Dramatic changes in urban government activities and alliances have occurred over the last decade in the advanced capitalist countries. This chapter on the political economy of the local state analyzes why these changes occur and their implications. Of special interest are the interactions of the state apparatus with different class groups. The local state is the set of governmental institutions acting in a locality: the combination of government departments and agencies (federal, state, urban or regional in their spatial jurisdictions) that take action with respect to a locality. This definition of the local state is a straightforward, descriptive one: other definitions are also used in political economy (Fincher 1987).

Three crises in local government informing the concerns of the literature on the local state are described in this chapter. The chapter then discusses the local state literature in the light of these research themes, and concludes with a survey of contemporary trends in urban politics suggesting how the theory of the local state might proceed.

Three crises in local government

The government of cities since World War II has varied between the United States, Britain, and Australia. For example, the responsibility for funding urban programmes and built environment projects is allocated to different levels of government in the three countries. Hence, urban governments (and often their constituents) suffer from the out-migration of commerce and residents if they rely heavily on the local tax base for revenue, as in the United States, but less so if senior levels of government take major responsibility for urban expenditures, as in Australia. Class groups interacting with local governments also vary: industrialists interact with tiers of government responsible for production, property developers with those in charge of land use planning and regulation. Yet there are also similarities in the advanced capitalist parliamentary democracies, in terms of relations between local states, the broader state apparatus, and dominant class groups. Three case studies are examined below, from the United States, England, and Australia. New York City almost went bankrupt in the mid-1970s, whereupon an alliance of state government, financiers, and businesspeople assumed control of local spending. In the early 1980s, local governments in London and Sheffield developed socialist policies. The conservative central government increased its control over, and restrictions on, local spending. And in Melbourne in 1981, a decade of conflict between the Melbourne City Council and the state government over central city land use planning culminated in the state government's sacking the council, and replacing it with unelected commissioners.

New York City in the 1970s

For older United States central cities, surrounded by suburbs that are separate political and fiscal jurisdictions, the period since World War II has seen severe fiscal strain. Productive capital moved from northeastern to western and southern regions of the country (Perry & Watkins 1977), and the movement of people, jobs, and investment to the suburbs accelerated (Harvey 1977). As a result, property tax revenue became scarcer at the same time as it was increasingly needed to pay for expensive central city public services. New York City was affected by all these trends, and its particular crisis must be placed in this more general context.

Between 1964 and 1974, there was a 7.5 per cent decline in the number of private-sector job holders in New York City, with the dollar value of the city's economic activities declining correspondingly. Yet the city's budget tripled in the same decade (Zevin 1977, pp. 22–3). The creativity of City officials, at a time of changing federal–local fiscal relations, enabled this expansion. Newly available grants and subsidies were applied for, while money was also sought from the New York State government. Operating expenditures were sometimes treated as capital expenditures (Zevin 1977, p. 24). New York City more than tripled its outstanding debt by the end of the period: debt service payments totalled one seventh of the city budget (Zevin 1977, p. 24). In 1975, the city hovered on the brink of default, unable to meet its payroll and debt-financing commitments. Banks and bond market investors, who had bought New York City securities in the previous decade, now refused to lend (Alcaly & Bodian 1977, p. 30).

The State government advanced the City $400 million in April and again in May 1975 to meet its obligations. In June 1975 the State changed its assistance strategy and created the Municipal Assistance Corporation (MAC – soon known as Big Mac) to act as interim borrowing agency for New York City. MAC was to raise $3 billion in loans. Its members were leading financiers (the City's creditors) and businesspeople, this justified by the need to foster investor confidence in the City (Alcaly & Bodian 1977, pp. 31–2). The City's creditors, as well, were legally permitted to reorganize the City's spending procedures:

> In New York, the creditors have had all the advantages of receivership without the problems of a technical default. And the city is being reorganized: cutbacks have been made, entire city agencies threatened, home rule has been relinquished, and new agencies and management 'panels' are springing up. Predictably, the changes have hurt labor and the city's poor and middle-income residents the most, the creditors the least.
>
> This initial phase of re-organization served several functions. While maintaining the illusions that default was on the horizon and that the suspension of democratic rule was a fair price to pay for the good offices of the investment community, the creditors bought time to maneuvre into

positions of greater strength. The city gained only month-to-month survival while its fiscal plight provided a vehicle for a continued attack on social services and labor unions. And meanwhile, the public became accustomed to their city being managed by creditors (Alcaly & Bodian 1977, pp. 31–3).

How can this situation be explained? The view presented in the business media was that of a crisis occurring because of 'growing welfare costs, public-sector worker militance and bureaucratic inefficiency' (Mollenkopf 1977, p. 113). By contrast, rather than being the fault of the city's workers, political economists argued that the causes of the city's problems lay in the city's loss of jobs, without which large sections of its workforce were unemployed and municipal services could not be sustained. In turn, job loss was linked to 'the cyclical nature of our economy and . . . secular trends brought about by public and private decision structures, which minimize private costs and ignore externalities, specifically the social costs of development patterns' (Tabb 1978, p. 246). New York's fiscal problems were seen as typical of those of most large cities in the older industrial regions of the United States. The extent of its overborrowing was the product of particularly high costs incurred by a city pushed by finance capitalists to serve as 'the headquarters city of the giant U.S.-based international corporations' (Tabb 1978, p. 245).

Political economists also stressed the differences in class interests embodied in the restructuring resolutions to the urban crisis. As Tabb (1978) asks, why should a city's response to economic crisis be made according to a banker's logic? Budget cuts in public service provision, the loss of democratic local government, and the firing of city workers are short-term solutions to budgetary difficulties for they have their own negative multiplier effect – they mean 'less adequate services, a deteriorated environment, more flight from the city, a smaller tax base, the need to fire more workers' (Tabb 1977, p. 320). Solutions responding more to the needs and interests of most city residents, rather than to the economic logic of the finance establishment, argues Tabb (1978, p. 259), would embrace the social control of investment, planned full employment, and price controls, all these measures being planned by consumers and ordinary working people rather than by corporate financiers.

London and Sheffield in the 1980s

Britain relinquished its position as the world's leading industrial producer in the late 19th century, but its relative decline has hastened since 1950. (Anderson *et al.* 1983, p. 5; Massey & Meegan 1982). Britain's general economic decline has a distinct spatial face. Old industrial regions and the inner cities have experienced the most profound decline. Reduction in the size of the manual working class, the traditional supporter of socialist Labour Party policies, was one reason for the defeat of the British Labour Party in the elections of 1979 and 1983. A map of the 1983 national election results shows 'Labour with only two clear geographical bases, the declining regional, industrial heartlands and the inner cities' (Boddy & Fudge 1984, p. 4). The prominence of certain 'socialist' municipalities in Britain was a part of the efforts of the Labour Party left to extend its appeal to a broader constituency and to take positive and innovative initiatives rather than rely upon defensive policies. This section emphasizes the policy initiatives taken by Labour Party-controlled local authorities in London and Sheffield in the early and mid-1980s.

The Labour Party took control of the Greater London Council (GLC) after the May 1981 election. The GLC was established in 1965, as an upper tier local authority comprising 32 borough councils and the City of London Corporation – it was therefore equivalent to the county level of local administration elsewhere in Britain (Boddy & Fudge 1984, p. 261). Its responsibilities included strategic planning, policy and financial control of urban public and private transport, waste disposal, and parks. Housing, social services, and local planning were functions of member boroughs.

The GLC was prominent in the development and implementation of socialist policies. Its London Industrial Strategy is the most complete statement of GLC intentions and philosophy:

> It focuses on industry and employment and also draws in many of the other policy priorities . . ., bringing together women, ethnic minorities and other 'new social forces' into an alliance capable of bringing about more fundamental change. The attempt by the new municipal socialist authorities to tackle problems of employment in a way which challenges the market marks them off from other more 'mainstream' councils (Cochrane 1986, p. 188).

The strategy expressed the GLC's concerns for matters like job creation, the survival of neighbourhood shops, transport and services, the availability of quality childcare. It proposed that the necessary economic restructuring be conducted in the interests of labour rather than capital (Rustin 1986, p. 77; Goodwin & Duncan 1986), and that this should be achieved through local government agencies regenerating their localities' productive sectors, using central government funds (Rustin 1986, p. 78). Guidelines on which restructuring recommendations were based included 'job-creation, socially-useful production, and distribution of products or services in favour of the disadvantaged' (Rustin 1986, p. 79). Expansion of social control over investment and socially useful production was organized by popular planning (workplace and community control over restructuring plans and their implementation) and technology networks (expertise in tertiary education institutions combined with that of trade unionists to plan new products and forms of enterprise) (Boddy 1984a, pp. 172–3). Other initiatives of the GLC related to race, equal opportunity, and the police. It attempted through its contracts to encourage non-racist and non-sexist social practices; it advocated popular control of the police, who would work in the community of their recruitment.

Sheffield City also developed radical economic strategies, following the election of a group of young leftist councillors to the traditionally Labour-controlled council in 1980. The council represents a metropolitan district within South Yorkshire County with responsibilities for housing, planning, education, and social services. Faced with higher than national average unemployment rates in the early 1980s, it became the first such council to establish an Employment Department (Boddy & Fudge 1984, p. 243), whose purpose was

> to co-ordinate everything the City Council can do (alongside trades unions, employers' and community organizations), (i) to prevent further loss of jobs in the City, (ii) to alleviate the worst effects of unemployment and to

> encourage the development of new skills, (iii) to stimulate new investment, to create new kinds of employment and to diversify job opportunities, (iv) to explore new forms of democracy and co-operative control over work (Boddy 1984a, p. 166).

Sheffield established industrial sector working parties in the City's steel and engineering industries; it offered assistance to workers left unemployed by disinvesting multinational firms while also working with them to save the plant; it used the purchasing power of the local authority as a market for locally produced products (Boddy & Fudge 1984, p. 252).

The conservative central government did not ignore these developments in local government. (The central government's actions are also consistent, however, with the stated objectives of all British central governments in the 1970s: to ensure that 'local government's spending plans are consistent with the Government's economic objectives' (Boddy 1984b, p. 228).) The Thatcher Government attempted to control and reduce local government spending. In 1983 a plan was presented for central determination of the rates (local property taxes) set by local authorities. A central government manifesto pledged to abolish the GLC and the metropolitan county councils and, early in 1986, the GLC was indeed abolished, its functions largely being turned over to member boroughs.

Boddy (1984b, p. 234) interprets the central government's actions in the light of overall state policy on public expenditure. *Local* government expenditure and taxation were reduced, while public expenditure as a whole was not so constrained.

> This makes it hard to see central government's attack on townhall spending as simply part of an overall plan to cut public spending and taxation. Particularly since what has evolved is a system of control directed at individual high-spending authorities, above all Labour-controlled metropolitan authorities, inner London boroughs and the GLC (Boddy 1984b, p. 234).

Such central government policies affected many people reliant on local government services – cheap public transport, public housing, social services, and education. As it forced reductions in local collective services, the central government in Britain embarked on local economic policies of its own, including: Enterprise Zones where firms get ten years' exemption from taxes and duties; Urban Development Corporations where firms escape local planning requirements, being accountable only to the central government; and Freeports near airports or docks where firms comply less than normally to tax, tariff, and customs rules (Goodwin & Duncan 1986, pp. 28–9). These policies are presented as operating in the national interest. Their capacity to reduce local economic autonomy is clear.

In the wake of the June 1987 British election the Conservative Government continued its efforts to control local government, using a major Cabinet co-ordinating committee to address inner-city problems. One media report claimed that 'the target will be left-wing councils which the Government blames for discouraging business and creating conditions for inner-city depression (Stevens 1987). For example, more urban development corporations have been created with the powers to override local government in redevelopment planning.

Melbourne in the 1980s.

In 1980 a crisis in intergovernmental relations occurred in Melbourne. Relations between the elected Melbourne City Council (MCC) and élite economic interests in the central city had seriously deteriorated. In early 1981 the Victoria State government (a conservative Liberal Party government) dismissed the MCC, replacing it for an unspecified period with three appointed commissioners (Saunders 1984a, p. 93): a former town clerk and two senior executives of private companies. Their charge was simultaneously to carry out the functions of the MCC and make proposals for its reform (Saunders 1984a, p. 91).

A decade of conflict had preceded the 1980 crisis. Melbourne City Council has authority over 31 square kilometres at the original centre of the metropolitan area – this includes the central business district and a number of inner-city residential areas, many now gentrifying. Social consumption expenditures are largely made at other levels of government and infrastructural expenditures on utilities, metropolitan-wide planning, roads, and public transportation are made by statutory authorities.

Conflicts over land use planning, the main function of the MCC, precipitated the 1980 crisis. In the early 1970s, CBD landowners and retailers, concerned at the growth of Melbourne's suburbs and their commercial centres, urged the State Premier to stem the expected decline of the central business district. The MCC was asked by the state government to plan for this: its 1974 Strategy Plan was the result. The plan was supported by inner-area residents, urban fringe environmentalists, and CBD retailers and landowners, because it recommended the restriction of intensive office development to the existing CBD area (Saunders 1984a. p. 95; Logan 1982). But it was opposed by developers and landowners interested in developing commercial centres outside the central city: they enlisted the support of the statutory authority responsible for metropolitan-wide planning, and an alternative plan was proposed by this authority to represent these interests. In 1980, neither plan had been agreed to, causing uncertainty for developers, landowners, and planners. When a committee of resolution set up by the State Planning Ministry failed, the minister drew up his own plan. Planning for the central business district was removed from the MCC's control, and the minister's plan implemented.

Why, asks Saunders, was the MCC dismissed when the planning issue had been decided in this way? Business domination of the Council had eroded; and it was likely that residential, environmentalist, and even labour interests might soon control the MCC:

> When the premier bemoaned in Parliament the 'lack of vigour' of the Council, when the state government complained of 'poor administration' in Melbourne and when city centre business interests decried the 'lack of certainty' in local policy making, what they were all coyly referring to was the decline of the CBD control over the local political process. So, like a spoilt child who finds himself losing a game of Monopoly which he had

expected to win, the big retailers, insurance companies and property owners of the CBD simply kicked the board over and decided to begin the game again under different rules (Saunders 1984a, p. 101).

The election of a Labour state government in 1982 removed the mandate of the appointed commissioners to redraw the boundaries of Melbourne City, and therefore the immediate likelihood that inner residential neighbourhoods, with their anti-business lobbies, would be in different local planning jurisdictions from the CBD. Nevertheless, business interests were strongly encouraged in the city while the commissioners ruled it; in the four months following the MCC's dismissal a record number of planning permits was issued to developers (Saunders 1984a, p. 102). Since the restoration of an elected MCC, in 1982, however, the Labour state government has re-evaluated the planning function in Melbourne. The statutory authority previously responsible for metropolitan-wide planning was made part of the State Planning Ministry. The state government also asserted its authority over the city's parklands by overruling on several occasions the MCC's wish to preserve parks for inner-city residents. Melbourne's central business district remains the finance centre of the metropolitan area and also of the State of Victoria. The Victorian State government has come to 'see the city, and particularly the central city as the focus for any growth oriented state economic strategy' (Collins 1986, p. 10). This explains its growing interest in planning the CBD itself, reducing the capacity of the MCC to do so. In 1987 the New South Wales State government came to the same conclusion, sacking the Sydney City Council and replacing it with appointed commissioners. This also is attributed to differences between the State government and the City Council over planning the central business district (Mowbray 1987, p. 3).

Research themes raised by local government crises

Such times of crisis pose stark choices for the alliances and interest groups concerned with local government and policy change. Clearly, local government is far more than the politically neutral provision of public services, and associated administrative tasks. Rather, an ever changing, conflicting set of relations exists between different levels of government, and between government and local community members, alliances, and groups. This section examines the three crises, linking their conflicting causal relationships to policy outcomes. Two general research themes emerge. These are then used to classify the local state literature.

New York City's fiscal crisis, it can be argued, was the product of conflicts between finance capitalists and municipal government, where municipal government was defending the economic interests of public sector employees and the wellbeing of residents reliant on public services. Conflict between State and City governments was also significant: the State government sided with representatives of finance capital to control municipal spending and priorities. The concerns of financiers prevailed to end the crisis, preserving an international investment community in New York with a local government more attuned to its requirements. Even more clearly demonstrated in the New York

City crisis is the class basis of urban financing. Class lines were clearly drawn between on one side groups from the capitalist class (financiers, other business people) and their ally on this occasion (the State government), and on the other groups from the City's working class (public service union members, recipients of public services) and their defender (the city government).

The conflict between local socialist and central conservative governments in Britain in the 1980s was conducted as a debate within the state apparatus, yet different class interests are present in the conflicting stands of the two levels of government. In the London docklands, for example, central government policy encouraged large-scale property capital investment and gentrification, while the GLC's Popular Planning Unit 'provided support for communities living in London's increasingly derelict docklands, which have fallen prey to big business and luxury property developers oblivious to the needs of local people' (Palmer 1986, p. 117). That socialist alternatives are possible in local government was demonstrated by the GLC, though the problems of implementing them were clear as well (Palmer 1986, Rustin 1986). It is an achievement in the interests of non-capitalist groups that, because of the socialist GLC, 'a significant number of people . . . have proved to themselves that there are practical and reliable solutions to their problems and aspirations' (Palmer 1986, p. 122).

The emergence of local socialism in certain British cities and the abolition of the GLC by the Conservative central government focuses attention on the tensions between local and central governments. It shows that direct opposition by local government is unlikely to be sustained in the political arena when the central government has power to suspend the conditions allowing local government disagreement to emerge. It also indicates that different class allegiances are the source of profound intergovernmental tensions.

The Melbourne crisis had its roots in conflict between commercial and retail investors in the central business district and property developers and suburban entrepreneurs in the rest of the metropolitan area. The central city group wanted to preserve the rôle of the CBD as Melbourne's chief investment and retail site; the other group wished to extend profit making in these activities to the metropolitan limits. Groups outside the capitalist class (environmentalists, inner-city residents, small businesspeople) sided with one or other group. The conflict also drew different parts of the state apparatus to either side of the dispute. The Melbourne case poses the same issues of intergovernmental conflict and the class basis of policy positions and implementation strategies. Perhaps more than the other cases, however, it underlines the complexity of the class relations present in government. It shows that different groups within the capitalist class can be allied to different parts of the state apparatus and can be in conflict with each other, and that local–central clashes may not be depicted simply as local government representing working-class groups and central government capitalist class interests.

The three cases together make two important points. First, choice and implementation of policy by government in localities is not a neutral matter, but arises from conflict between, and within, class groups, and between parts of the state apparatus. Second, despite differences in the responsibilities of local government in different countries, there are similarities in the control exercised over local government by the upper tiers of the state apparatus, and by finance capitalists and real estate entrepreneurs. It appears from these three crises that the most severe disciplining of local government by state or central governments will occur when a major class clash is obvious in the conflicting stands of different levels of government, when the legitimacy of one government's support for capitalist enterprise is clearly called into question by the actions of another government body.

The local state has therefore been an important site of conflict in the last decade and, in response, a literature has developed, largely in North America and Britain. Writers have questioned the actions of local governments and their distributional and political implications. The past decade also saw the political economy of the capitalist state take root as a topic of interest in radical geography (Harvey 1978, Clark & Dear 1981, Dear 1986). Political turmoil in cities forced academic geographers and others to confront urban inequality, with Marxist theory adopted as explanation (Peet 1977). Urban geographers re-examined their environments, revising their interpretations of matters previously ignored or taken for granted (e.g. local government operations), and focused on the political economy of the local state.

Themes in local state research

Before assembling the local state literature to show its major themes, it is useful to consider two forceful contributions to early urban political economy which greatly influenced subsequent research: O'Connor's *The fiscal crisis of the state* (1973) and Cockburn's *The local state* (1977). The rest of this section then deals with local–central government relationships and the class context of the local state. That such issues concern political economists is clear from the three cases of local government crisis discussed above.

Early urban political economy

While O'Connor's book proposes a theory of the capitalist state as a whole rather than solely the local state, it was so significant in the development of urban political economy that its argument must be noted here (see also Schwartz 1983). Cockburn's study was an important application of a (particular) political-economy approach to local government change. Both books have been criticized and new viewpoints developed partly in response. The central arguments made by O'Connor and Cockburn, are assessed here and reasons offered as to why their frameworks are no longer entirely accepted.

O'Connor argued that state expenditure increasingly forms the basis for growth in production in advanced capitalist economies. He proposed that the capitalist state tries to perform two functions: accumulation and legitimation. Through its expenditures, the state must promote the conditions for profitable capital accumulation; at the same time it must maintain social harmony, often by offsetting the socially divisive effects of continued capital accumulation. Various types of social capital expenditure lower production costs and the reproduction costs of labour and thus facilitate capital accumulation by reducing capitalists' costs. These constitute the state's accumulation function. Social expenses are incurred by the state in performing its legitimation activities. State expenditure in both these categories is increasing, but the state is less and less able to finance them. The relationship of public and private sectors in capitalist economies is increasingly one in which the profits of joint public and private investment are appropriated privately and the costs are socialized through the state.

O'Connor (1973, Ch. 5) uses this framework to situate government spending in cities as 'social consumption' expenditures lowering the costs of reproducing labour power. He classifies these expenditures into two groups: 'goods and services consumed collectively by the working class' – roads, schools, hospitals, home mortgage subsidies, urban renewal projects; and 'social insurance against economic insecurity' – unemployment insurance, social security benefits, health insurance (O'Connor 1973, p. 124). Wealthy urban communities, usually suburbs rather than central cities in the United States, can afford higher quality social services, paying for them largely with local property taxes. Many wealthy suburbs enforce 'zoning ordinances to preserve their exclusive character and to bar housing that fails to pay its own way in property tax revenue' (O'Connor 1973, p. 130).

O'Connor's framework has been challenged on several grounds (Schwartz 1983), but two points of criticism are relevant here. First, O'Connor's notion that the state always acts in order to preserve capital accumulation, and that its other actions are mainly legitimizing, denies the variety of rôles carried out by different parts of the state apparatus – many of these actions cannot be interpreted (and are certainly not intended by the state) as enhancing accumulation or legitimizing capitalism. The continued and complex creation of the state and its policies by groups and alliances within and outside it is not captured by O'Connor's formulation. Second O'Connor's point that government expenditures in cities primarily reproduce labour power ignores the sizeable efforts of local states to underwrite the costs of production in their communities, to compete for, and develop, local expansion in commodity production.

Cockburn's (1977) study, more directly aimed at local government activities, coined the term 'local state' to reflect the fact that all branches of government are part of an over-arching capitalist state (1977, p. 2). Cockburn recognized the politics of studying localities:

> A good deal of contemporary research, with warm-hearted intentions, studies working-class groups and situations on behalf of those who make policy. This book begins with a different point of view about political change: that it stems from the working class. So, this is a study of urban managers and urban management situations and techniques from the viewpoint of those who are managed (1977, p. 2).

Like O'Connor, she aimed to identify and understand the contradictions of (local) government. How, for example, does it happen that 'housing departments evict families from council flats for falling into arrears and the next day have to fulfil their statutory obligations to rehouse them' (Cockburn 1977, p. 1)? From a study of Lambeth Council in inner London in the 1970s, Cockburn proposed that such circumstances arise because of local government's conflicting involvement in modern corporate management and community participation. Modern management methods were introduced by city authorities in the 1970s to keep down costs at a time when public concern about government spending was growing but rising urban poverty (and so the need for such spending) was also rising (Cockburn 1977, p. 65). Central governments sponsored the new management methods, because 'the growing difference (between local council expenditures and revenues) was made up by the centre' (Cockburn 1977, p. 65). But unemployment, housing, and family income problems continued to increase in Lambeth. Working-class militancy grew. At a time when the local council had no more resources to provide services for a needy population

> it faced the militancy of a working class whose interests in the borough's life were in a direct conflict with those of capital but who addressed their anger mainly at the council (which) was as dilatory over repairs and improvements as their old local state and local people were continually threatening to evade the proper relation to authority (Cockburn 1977, p. 72).

Participatory democracy and community development were then introduced, again at central government suggestion, to help implement policies which 'reproduce the relations of authority' (Cockburn 1977, p. 131) and provide information about the community for local managers. Greater community participation, it was anticipated, would legitimate local state priorities and actions. However, community action in the real interests of working-class Lambeth residents could not take place on terms set by the state. State-organized community action made the struggle over reproduction seem classless, territorially based, and related to problems within the state rather than to economic allocations (Cockburn 1977, Ch. 6). Local authorities like Lambeth found themselves in a contradictory position because, on the one hand, they tried to minimize costs through efficient management methods, while on the other hand spending to build up inefficient methods of decision making using community participation.

Objections to Cockburn have generally been raised about her concept of local government as a mere aspect of national government, which in turn is simply a tool of capitalist domination (see Duncan & Goodwin 1982a, pp. 169–70). First, no account is taken of the differences in priorities and class backing between central and local levels of the state apparatus, nor of the complexity of class and group alliances around the policies of different parts of the state. Second, explaining the nature of the local state as a logical outcome of capitalist domination is inadequate. State characteristics are historical products, which were never inevitable, and which were produced by groups struggling to effect political change. The fact that such conflict occurred in a capitalist context is relevant, but is not sufficient as explanation. Cockburn's empirical work, however, has been praised as more attentive to complexities and historical circumstances than her conceptual starting point promised (Duncan & Goodwin 1982b).

O'Connor and Cockburn taken together demonstrate the contradictory position of the contemporary capitalist state – O'Connor at the national level, analyzing expenditures, Cockburn at the local level assessing the implementation of the policy in urban service provision. Cockburn's book points to the important of local–central government relations in Britain, a theme taken up in the literature on the local state. O'Connor's contribution was to indicate how much state expenditure directly furthers capital accumulation and so the interests of dominant groups within the capitalist class. (The class basis of much local state activity, including expenditure, is the second theme in the local state literature described below.) Though criticized, O'Connor's and Cockburn's contributions to the local state debate were significant in structuring later work.

Local–central government relations

A number of political economists took up the relations between local government and other tiers of the state apparatus. These relations contribute significantly to the nature of the local state in particular places and times, as was clear in the three cases discussed earlier. Local-central governmental relations have been classified by some authors, using typologies. Others have focused instead on the independence permitted local governments within particular local–central governmental structures.

Saunders (1979, 1981, 1985) develops a typology of the functions of different levels of the state apparatus. He begins with four traditional activities of the British state as a whole over the last century: welfare support or collective consumption; foreign policy; internal control; and guaranteeing property rights (1979, pp. 142–3). All have expanded in the period. Local government has increased in significance as well, in the functions it retains, especially those of land use planning and the provision of social welfare services, housing, education, and roads. Combining O'Connor's classification of state expenditures with Cockburn's insights into the functions performed by British local government, Saunders conceives a taxonomy of the major functions of local government (what he terms the local state) in Britain:

Sustenance of private production and capital accumulation

(a) through the provision of necessary non-productive urban infrastructure (e.g. road developments);
(b) by aiding the reorganisation and restructuring of production in space (e.g. planning and urban renewal);
(c) through the provision of investment in 'human capital' (e.g. education in general and technical college education in particular);
(d) through 'demand orchestration' (e.g. local authority public works contracts).

Reproduction of labour power through collective consumption

(a) by means of material conditions of existence (e.g. low rent local authority housing);
(b) by means of the cultural conditions of existence (e.g. libraries, museums, recreation parks).

Maintenance of order and social cohesion

(a) through the means of coercion (e.g. police);
(b) through the support of the 'surplus population' (e.g. social services and other welfare support services such as temporary accommodation);
(c) through support of the agencies of legitimation (e.g. schools, social work, 'public participation') (Saunders 1979, pp. 147–8).

More recently Saunders (1985) presents a further typology distinguishing central, local, and (to some extent) regional government functions in Britain. His analysis shows a movement away from Marxist conceptualization of the local state, to a more eclectic 'dual politics' explanation and classification, in which different theories are presented to explain the activities of the two major levels of the British state apparatus. The point is repeated, from the earlier work, that collective consumption provision is made by local rather than central government. But the new formulation also separates the social base, mode of mediation, and dominant ideology of the two levels of government (central and local), all the while classifying that group of characteristics into which central government involvement falls as a 'politics of production' (better able to be explained with Marxist theory), and that in which local government is contained as a 'politics of consumption' (less able to be explained with Marxist theory) (1985, p. 153). Saunders's newer work also expresses empirical interest in the variation *between* localities in the political forms that develop within the state as local or regional government, or as both.

In a Canadian analysis of the functions of local branches of the state apparatus Dear (1981) defined the local state as 'any government . . . having a political and spatial jurisdiction at less than a national scale' (1981, p. 187). Following O'Connor, Dear noted that the local state produces three types of output: production related services; services that reproduce the labourforce; and conflict prevention services. Their distribution between levels of government indicates that in Canada the local state is most heavily involved in the second function – reproducing the labourforce. Dear shows how, between 1950 and 1976, local municipal revenues and expenditures increased at a greater rate than their federal counterparts. Furthermore, three quarters of local expenditures were on the provision of goods and services, compared to one quarter of federal and one third of provincial expenditures (Dear 1981, p. 187). In later work, Clark & Dear (1984, p. 133) amended Dear's earlier definition of the local state to government with jurisdictions less than those of a state or province. These parts of the state apparatus have a crisis avoidance function, but also serve heterogeneous local needs 'in keeping with the principles of local self-determination so important in American democracy' (Clark & Dear 1984, p. 133). In this account, local state or municipal government functions centre on the maintenance of political stability to ensure continued social reproduction.

Researchers of urban political economy first classified local and central government functions by expenditures, following O'Connor in concept and empirical focus. There are two difficulties with using a description and classification of local government expenditures as the basis of a theory of the local state. First, functions and expenditures change with different local–central governmental relations, as was evident in the discussion of the three crises of local government. An expenditure typology, accompanied by generalizations about what is normally the case, may present the situation, inaccurately, as static, without class context, and natural (see Fincher 1981). Second, even if expenditures are made locally, decisions determining them may be made elsewhere in the state apparatus. An aspect of local–central governmental relations not revealed by expenditure outcomes alone is the degree of independence of local governments to make financial decisions.

Political economists have indeed investigated the independence of local government's political priorities from those of central government, despite the flow of funds from centre to municipality. Saunders (1979) contributed usefully to this subject, reviewing the ecological, political, and economic constraints on urban managers (local government councillors and bureaucrats) in Britain. Ecological constraints are those that spatially differentiate localities by the ability of the local labourmarket to provide accessible jobs, the distribution of negative externalities due to land use patterns, and the like. Local government attempts to reduce inequalities due to these features vary between places in necessity, cost, and success. Economic constraints link local governments to the market. They include the reliance of local governments on the private sector for revenue, either through the taxation of land-holders or loans from finance capital. Political constraints are the limits imposed on local government by central government, and also relations within the local government structure – for example, those between elected and bureaucratic personnel. With regard to the first of these political constraints, as we have already noted, Saunders explains that though British local government powers have been lost to the centre over the last century, the significance of remaining responsibilities has increased along with spending on them (1979, p. 193). Also, local government autonomy in particular policy areas remains unchallenged.

The complexity of the autonomy question is not unravelled by identifying the general political constraints on local government. Saunders (1979, p. 196) argues that the decisions of urban managers are limited but not determined: the precise scope of local government decision-making discretion is a matter for empirical investigation of particular cases.

Interest in the relative independence of local government has persisted through the 1980s in Britain, especially given the prominence of socialist municipalities and the restructuring of central–local relations in 1979. Boddy (1983, p. 129) describes the latter, noting that 'the impact of the Thatcher government has . . . produced a major shift in the balance of central–local relations, radically restructuring the geography of political power relations within the framework of state institutions.' The financial clauses of the Local Government, Planning and Land Act 1980 have expanded central control over local affairs by limiting the total capital expenditure of local government and enforcing limits on the current revenue expenditures of individual local authorities. Other changes have concerned housing. Local discretion over whether council (public) housing should be sold was removed by the Housing Act of 1980, giving tenants the right to buy their council houses. Local authorities have been required to list land that could be sold to the private sector (Boddy 1983, p. 131). Like Saunders, Boddy (1983, p. 134) finds that the degree of local government autonomy in Britain remains unclear. Further study is required not only to show where autonomy can exist generally within the new framework of central–local relations established in the 1980s, but also to examine how individual local governments might create more progressive policies within this context. Assessments of the recent introduction of socialist policies within several British municipalities are of interest here.

A collection of articles edited by Boddy & Fudge (1984) sets the scene, focusing on

> the attempts of more radical, Labour councils to maintain and defend the collective provision of services in the face of cuts, controls and pressures to privatise public provision; to develop new initiatives and alternatives; to mobilise popular support and build alliances behind progressive policies; and to explore and develop the role of local government in a viable socialist alternative (Boddy & Fudge 1984, pp. 2–3).

Interviews with the leaders of the Sheffield City Council and the Greater London Council (two prominent socialist authorities) showed their optimism that community controlled, local economic development programmes, developing better race and gender relations, would both transform the ways people regarded the possibility of locally led change and provide a broader electoral base for the Labour Party left (Boddy & Fudge 1984, Ch. 10). But other authors were less optimistic, focusing particularly on the political and organizational difficulties of radical local authorities resisting central government policies (Saunders 1984b, Bassett 1984). Although tensions between radical Labour councils and the Conservative central administration have been most publicized, difficulties have occurred within the Labour Party as well, over questions like 'at what stage is it justifiable for local interest to unite to defy a democratically elected central government in the name of local democracy?' (Bassett 1984, p. 98).

Attributing more importance to 'how state institutions do things' (the social relations of the state) than to the specific functions of local or central government, Goodwin & Duncan (1986, p. 16) distinguish traditional local economic policies that aid capital from local socialist economic policies which restructure in the interests of labour. They evaluate a case of the latter – the efforts of Sheffield City Council's Employment Department to make local employment initiatives. Requiring better rates of pay and improved conditions for employees doing the same work as previously, the Department failed to revive a declining cutlery firm, despite joining with unions, employers, and financial institutions in the effort. Sheffield Council then moved away from the private sector and from direct attempts to increase employment, to concentrate on areas of traditional public sector involvement and expertise, like local planning (Goodwin & Duncan 1986, p. 23). Instead of improving the working lot of private sector employees, it supported the interests of its own workforce. The council had had political as well as economic aims, wanting to encourage political mobilization for socialist economic alternatives by demonstrating their viability. But its reorientation from Sheffield's steel and engineering workforce, those it had originally aimed to mobilize, to its own personnel, undermined its political aspirations.

Both Rustin (1986) and Cochrane (1986) evaluate the GLC's London Industrial Strategy. Sympathetic to its intentions, they nevertheless point to problems caused by the complexity of its context and the difficulty of its task. They also identify the deficiencies of the sector-by-sector approach taken (Cochrane 1986), the assumption that socially-useful production is a clearly specified term, and the view that decisions taken in the public sector will necessarily be more democratic than those taken in the private sector (Rustin 1986).

British research interest in the local state over the last decade, particularly its context of local–central governmental relations, has included documentation of collective consumption functions and analysis of prospects for local government autonomy in production and financial decisions. The latter concern is obviously linked to the restructuring of the local state during the Thatcher Administration. North American researchers have not had such a dramatic political incentive to extend their analyses of local government autonomy in the 1980s. Nevertheless, Clark (1984; see also Clark & Dear 1984, Ch. 7) paid attention to local government autonomy in the United States from a legal point of view. After assessing municipal scope to initiate regulation and legislation, and the susceptibility of local government to the rulings of other parts of the state apparatus, he concludes that 'local autonomy has been systematically diminished over the past 200 years' (1984, p. 199). Based on interpretations of legal texts in British and American adjudications of local–central responsibilities, Johnston (1983) finds that the autonomy of the local state in practice may differ from what it appears to be in law. British records show how judges' rulings have permitted central restriction of local government autonomy, whereas the US evidence is of decisions in favour of continued local autonomy (Johnston 1983).

Class and the local state

The class relations of the local state are not generally addressed in studies concentrating on local–central governmental relations. However, another theme in the political economy of the local state has explicated the class relations of the local communities of which the local state is part and to which it contributes. Two major strands of research exist. First, early North American work investigates the implications of particular capitalist crises for local government form and functions. Here the class relations of the local and national economy are seen as a general context for local government. Local government form and functions are an *outcome* of this exogenous context. Second, contemporary British research on the social and class relations of localities hypothesizes the existence of social relations particular to a region, emerging from the unique history of the labourmarket, political institutions and cultural traditions found there. The local state *contributes* to these local social relations and expresses them, in the way it conducts its activities as much as in its expenditures or functions.

Economic crisis has been identified as a significant influence on local state form and activities. Ironically, this produces fiscal crisis within the state too. Much of the early North American work on the local state followed O'Connor in identifying the implications of capitalist economic crisis for (local) government expenditure. Friedland *et al.* (1977) hypothesize that at times when accumulation and legitimation demands on the state are heavy, municipal management structures experience considerable pressure:

> The electoral representative arrangements which underpin municipal governments make them vulnerable to popular discontent, and also limit their ability to employ extraordinary strategies of collective mobilisation or repression to cope with discontent. At the same time, municipal authorities are helpless to intervene in the economic developments which may have triggered discontent and, indeed, find it difficult to resist even new demands arising from the private sector on which they are fiscally dependent (Friedland *et al.* 1977, p. 449).

But institutional arrangements deflect such problems. Decentralization or centralization of government functions occurs while economic and political functions in municipalities are separated institutionally. In dissipating conflict, however, this increases government costs. Periods of fiscal crisis inside the state replace social conflicts outside it.

Politically fragmented metropolises experience particular fiscal stress. Markusen (1978) related the contemporary fiscal difficulties of north-eastern United States cities to their metropolitan fragmentation, itself the product of class-based battles in the 19th century over whether local or state government would control the granting of contracts to develop land. By the end of the 19th century, urban home rule 'had become a universal state constitutional feature, creating a political structure which was to impede metropolitan political integration in the future' (Markusen 1978, pp. 96–7).

Goodman (1979) noted the difficulties faced by American regional and local governments competing for investment by the footloose plants of multinational corporations by offering financial inducements. In this research, then, the nature of local government is an outcome of class struggles and economic crises outside it.

The recent (and increasingly debated) British focus on the nature of the locality is a second research strand that considers the class relations of the local state (Murgatroyd *et al.* 1985). Rather than viewing local government as an outcome of class conflict conducted elsewhere, this research treats local government as one institution that expresses the range of social relations particular to a locality, including class relations. The local state's form and activities also contribute to the continuing formation of these social relations.

The conceptual stimulus to this research is the understanding that people's experience of economic restructuring and the directives of central government is mediated through local class relations and work arrangements, but also through local modes of consumption, political and cultural traditions, and family structures. Accordingly, Rees (1985, p. 5) says that 'the spatial unevenness of productive relations coalesces with the local particularities of other dimensions of the social structure to generate characteristic forms of political expression in such communities.'

Cooke (1986) provides a summary of theoretical and empirical considerations in contemporary British localities research. Defining localities schematically 'in terms of specific intersections of labour market types . . . and socio-spatial types', he points out that 'there is a key nexus at the local level between Work–State–Family which it is necessary to explore in order to grasp the extent to which change brought about by economic restructuring can be accommodated' (Cooke 1986, p. 246).

The characteristics of the local state, its concerns and modes of operation, are part of the social relations of a locality and may be quite peculiar to it (Fincher 1987). However, this recognition that local political institutions and expressions are related to local employment circumstances, local social class patterns, and the characteristics of institutions like the local housing market, has made it no easier for researchers to show the means by which such factors are linked. Savage (1987) makes this clear in his efforts to determine whether localities are the basis of contemporary political alignments (expressed through voting patterns) in Britain. He finds 'little evidence that "local political cultures" have become more important in recent years, and indeed considerable evidence that they now hold much less importance in determining contemporary voting patterns' (Savage 1987, p. 72).

This difficulty in linking the characteristics of local labourmarkets with those of local political and social institutions is reflected in the concentration of most British localities research on the class and gender relations of local workplaces alone. Though local political institutions are included as contributors to the range of social relations in localities, the local state (and indeed the whole literature on its political economy) is rarely referred to explicitly.

The British localities work has been widely discussed. But it is in American research on the contemporary significance of local class alliances that the embeddedness of the local state in local class relations is given the best airing. Harvey (1985, p. 140) identifies urban regions as having 'structured coherence', each with a particular mix of consumption patterns, labour processes, and expectations about quality of life and style of living. Local class alliances form between the owners of capital (especially those committed to the fixed built environment), groups within the working class (especially those who have access to home ownership), and the local state (Harvey 1985, p. 149). Such alliances are unstable and often contradictory, changing as their members take up different positions and issues and react to the pressures of circumstances beyond the locality. Harvey's work emphasizes the fact that local governments liaise with a variety of class groups within their jurisdictions.

A contradictory pattern emerges from studies of the activities of local states in advanced capitalism and the possibilities for local democracy. Some argue that people in advanced capitalist countries increasingly identify with their locality as a site of political expression and an arena in which an acceptable quality of life can be preserved. Urry (1981, p. 464) attributes this to an increase in the external control of local economies, an increase in the degree to which the distribution of conflict affects the locational distribution of government expenditure, and an increase in the politicization of local inhabitants affected by economic change. Others indicate an erosion of local control over local affairs. The more localities rely on 'peripatetic multinationals' (to use Harvey's (1985, p. 149) felicitous phrase), the more uncertain their economic prospects appear, especially when they are in areas of disinvestment. Research on central–local governmental relations, including cases like those described above, indicates that central governments are reluctant to accept local government decisions that undermine a locality's economic potential and competitiveness or present a different example to counter the national commitment to a particular mode of economic restructuring.

Conclusion: research frontiers

Two avenues of investigation open out from these circumstances, both having important class implications. The first research task is to document and evaluate the different forms of economic influence sought by local governments. This demands more case studies of local economic policies that facilitate restructuring for capital or for labour, in Goodwin & Duncan's (1986) terms. The second research task is to analyze the contribution to the formation of local class and other social relations (like gender) made by the local state, especially in its collective consumption provisions. The changing functions, expenditures, and modes of operation of the state in localities do affect local social relations, but at present there is little indication of how.

Local governments seek to influence local economies in different ways; this has not been adequately reported in the local state literature. In large part the particular local economic policy chosen reflects the class alliance within the local state, and the perception held by that alliance of externally imposed economic and political constraints and opportunities. In the declining industrial regions of the United States, for example, local and regional governments compete to attract property investors or manufacturing plants. They seek to establish direct links with capitalist producers, and to maximize the number of jobs brought to the area through new investment. In offering inducements to capture wavering investors, often allowing firms concessions if they locate in the local jurisdiction, local states facilitate restructuring in the interests of capital. The socialist municipalities of Britain tried to take a different path in the search for local economic wellbeing. They tried to stem the flow into their localities of capitalist relations of production, to build 'socialist islands in a sea of capitalism' so that their residents would have greater control over their working lives. They were restructuring in the interests of labour, it has been claimed (Goodwin & Duncan 1986).

Political economists are interested in the implications of these styles of economic policy for class relations in communities and regions. Various angles need to be investigated. There needs to be more study of the precise effects of restructuring for capital on local divisions of labour and their class and gender relations. It is unclear at present, for example, whether different groups within the working class – perhaps more marginal ones – are created in particular places where the local state gives priority to new investment at all costs. Studies are needed to reveal cases in which restructuring for capital strategies in local state economic policy has resulted in benefits for the local working class, and the political and economic circumstances in which this was possible. In the same vein, more detailed investigation might be made of the processes and outcomes of restructuring in the interests of labour in different places. It is unlikely that all working-class groups will benefit equally from local socialist employment initiatives. Problems in the efforts of the City of Sheffield have already been noted, for example, and hints were given in that analysis that Sheffield City's support for women in paid work left something to be desired, women not being well represented in the unions of traditional manufacturing workers allied with the local state (Goodman & Duncan 1986). Which groups within the working class benefit and suffer in the implementation of socialist local economic policy, and under what circumstances, needs clarification.

The classification of local economic initiatives as 'for labour' or 'for capital' is an important step. But this does not inform us of the varying class implications of the processes and outcomes of such local state economic initiatives, which arise in different economic, political, and spatial circumstances. Neither does an analysis of the relative independence of local government from central legal and financial control indicate adequately the likelihood that economic policies of net benefit to a wide range of working-class groups will be implemented. There is need, then, for research that documents and assesses the class implications of local economic policies in different places, under a range of economic and political conditions. This is not simply a call for more case studies along the lines of those already carried out; for description of the class outcomes of local economic initiatives and why they are occurring, in any detail, has hardly begun.

A second major research emphasis needs to be placed on the contribution made by local states to the formation of class and gender relations in local communities. This issue arises directly from contemporary localities research, and is something such research has failed to address adequately (Fincher 1987). The conceptual position that local political and social institutions combine with the characteristics of local workplaces and the economy to create *together* the significant social relations of the locality is attractive. It is also more accurate than the view that class relations are formed in the workplace, and non-class social relations outside it. But localities research to date has failed to show empirically how the local state *combines* with specific workplace and cultural practices to have this result; the emphasis so far has been on economic restructuring alone as the major determinant of local social relations.

There are various ways in which the activities of the local state could contribute to the marginalization or improvement in position of different working-class groups, exacerbating or offsetting the influences of other institutions or factors. Its economic interventions have been mentioned above. But its particular policies for public service provision, the traditional local government collective consumption function, can influence the nature of local class relations. Local governments continue to have considerable responsibility for social consumption expenditures, even when relying on funds from higher levels of government to pay for them. Local governments usually determine the types of services to be provided and their precise spatial allocation. At a time of cutbacks in the welfare spending of advanced capitalist countries (e.g. Piven & Cloward 1982, pp. 16–19), local responsibility for social consumption seems to be growing and the division of functions between local and central levels of the state apparatus (expenditure on reproduction versus production), minor local economic initiatives notwithstanding, becomes more entrenched. It is important to monitor the distributional consequences of local government's assuming greater responsibility for social reproduction. When accompanied by reduced central government funding, political and financial difficulties will ensue for even the most progressive local administration.

Within this context, the precise manner of provision of collective consumption goods in different places, and its implications, requires study. Though the bulk of local public services remain government funded and delivered in most capitalist countries, the rôle of non-government organizations is growing. Increasingly, non-profit self-help and charity-based groups, as well as for-profit businesses, provide local public services. Worthy of investigation are changes in the availability and quality of public services associated with private sector provision. The development of new forms of social consumption in localities as relationships between local governments and non-government service providers change, can create, marginalize, or advance different gender and class groups. An emphasis on the distributional consequences for women of changes in the form of the local state is particularly important, especially at times (like the present) when women are participating in the paid labourforce in increasing numbers. Changes in local public service provision may increase women's household responsibilities, or even reduce their flexibility to take paid work outside the household. That lack of available childcare can change the class position and experience of women, for example, is clear. Related to this, further documentation of the rôle of women in forming, through their participation in local politics, a local state that makes adequate public service provision, is an important research task (cf. Mark-Lawson *et al.* 1985).

If Melbourne, New York, London, and Sheffield were to experience their local state crises again, after completion of research such as that suggested above, interpretation of the crises would be advanced in two ways. First, we would understand a range of circumstances in which progressive and regressive outcomes had resulted from local state economic intervention. Second, we would be able to estimate the degree to which changes in class and other social relations could be offset or encouraged by local state interventions in collective consumption. Both sorts of knowledge might inform appropriate political responses to the crises.

References

Alcaly, R. & H. Bodian 1977. New York's fiscal crisis and the economy. In *The fiscal crisis of American cities*, R. Alcaly & D. Mermelstein (eds). New York: Vintage Books.

Anderson, J., S. Duncan & R. Hudson (eds) 1983. *Redundant spaces in cities and regions*. London: Academic Press.

Bassett, K. 1984. Labour, socialism and local democracy. In *Local socialism*, M. Boddy & C. Fudge (eds). London: Macmillan.

Boddy, M. 1983. Central–local government relations: theory and practice. *Political Geography Quarterly* **2**, 119–58.

Boddy, M. 1984a. Local economic and employment strategies. In *Local socialism*, M. Boddy & C. Fudge (eds). London: Macmillan.

Boddy, M. 1984b. Local councils and the financial squeeze. In *Local socialism*, M. Boddy & C. Fudge (eds). London: Macmillan.

Boddy, M. & C. Fudge (eds) 1984. *Local socialism*. London: Macmillan.

Clark, G. 1984. A theory of local autonomy. *Annals of the Association of American Geographers* **17**, 195–208.

Clark, G. & M. Dear 1981. The state in capitalism and the capitalist state. In *Urbanization and urban planning in capitalist society*, M. Dear & A. Scott (eds). London: Methuen.

Clark, G. & M. Dear 1984. *State apparatus*. London: Allen & Unwin.

Cochrane, A. 1986. What's in a strategy? The London Industrial Strategy and municipal socialism. *Capital and Class* **28**, 187–93.

Cockburn, C. 1977. *The local state*. London: Pluto Press.

Collins, T. 1986. Planning in the fast lane. *Australian Society* **5**, 10–12.

Cooke, P. 1986. The changing urban and regional system in the United Kingdom. *Regional Studies* **20**, 243–51.

Dear, M. 1981. A theory of the local state. In *Political studies from spatial perspectives*, A. Burnett & P. Taylor (eds). London: Wiley.

Dear, M. 1986. Editorial: Thinking about the state. *Environment and Planning D, Society and Space* **4**, 1–5.

Duncan, S. & M. Goodwin 1982a. The local state and restructuring social relations: theory and practice. *International Journal of Urban and Regional Research* **6**, 157–86.

Duncan, S. & M. Goodwin 1982b. The local state: functionalism, autonomy and class relations in Cockburn and Saunders. *Political Geography Quarterly* **1**, 77–96.
Fincher, R. 1981. Analysis of the local level capitalist state. *Antipode* **13**, 25–31.
Fincher, R. 1987. Space, class and political processes: the social relations of the local state. *Progress in Human Geography* **11**, 496–516.
Friedland, R., F. Piven & R. Alford 1977. Political conflict, urban structure and the fiscal crisis. *International Journal of Urban and Regional Research* **1**, 447–73.
Goodman, R. 1979. *The last entrepreneurs*. New York: Simon & Schuster.
Goodwin, M. & S. Duncan 1986. The local state and local economic policy: political mobilization or economic regeneration. *Capital and Class* **27**, 14–36.
Harvey, D. 1977. Government policies, financial institutions and neighbourhood change in the United States. In *Captive cities*, M. Harloe (ed.). London: Wiley.
Harvey, D. 1978. The Marxian theory of the state. *Antipode* **8**, 80–9.
Harvey, D. 1985. *The urbanization of capital*. Baltimore, MD: Johns Hopkins University Press.
Johnston, R. 1983. Texts, actors and higher managers: judges, bureaucrats and the political organization of space, *Political Geography Quarterly* **2**, 3–19.
Logan, W. 1982. The future shape of Melbourne. In *Conflict, politics and the urban scene*. K. Cox & R. Johnston (eds). London: Longman.
Mark-Lawson, J., M. Savage & A. Warde 1985. Gender and local politics: struggle over welfare policies, 1918–1939. In *Localities, class and gender*, L. Murgatroyd, M. Savage, D. Shapiro *et al*. London: Pion.
Markusen, A. 1978. Class and urban social expenditure: a Marxist theory of metropolitan government. In *Marxism and the metropolis*, W. Tabb & L. Sawers (eds). New York: Oxford University Press.
Massey, D. & R. Meegan 1982. *The anatomy of job loss*. London: Methuen.
Mollenkopf, J. 1977. The crisis of the public sector in America's cities In *The fiscal crisis of American cities*, R. Alcaly & D. Mermelstein (eds). New York: Vintage Books.
Mowbray, M. 1987. Sydney's untimely sacking. *Australian Society* May, 3–4.
Murgatroyd, L., M. Savage, D. Shapiro, J. Urry, S. Walby, A. Warde & J. Mark–Lawson 1985. *Localities, class and gender*. London: Pion.
O'Connor, J. 1973. *The fiscal crisis of the state*. New York: St Martin's Press.
Palmer, J. 1986. Municipal enterprise and popular planning. *New Left Review* **159**, 117–24.
Peet, R. 1977. The development of radical geography in the United States. In *Radical geography*, R. Peet (ed.). Chicago: Maaroufa Press.
Perry, D. & A. Watkins 1977. *The rise of the Sunbelt cities*. Beverly Hills: Sage Publications.
Piven, F. & R. Cloward 1982. *The new class war*. New York: Pantheon.
Rees, G. 1985. Introduction: class, locality and ideology. In *Political action and social identity*. G. Rees, J. Bujra, P. Littlewood, H. Newby & T. Rees (eds). London: Macmillan.
Rustin, M. 1986. Lessons of the London industrial strategy. *New Left Review* **155**, 75–84.
Saunders, P. 1979. *Urban politics*. Harmondsworth: Penguin.
Saunders, P. 1981. *Social theory and the urban question*. London: Hutchinson.
Saunders, P. 1984a. The crisis of local government in Melbourne: the sacking of the city council. In *Australian urban politics*, J. Halligan & C. Paris (eds). Melbourne: Longman Cheshire.
Saunders, P. 1984b. Rethinking local politics. In *Local socialism*, M. Boddy & C. Fudge (eds). London: Macmillan.
Saunders, P. 1985. The forgotten dimension of central–local relations: theorizing the 'regional' state. *Environment and Planning C, Government and Policy* **3**, 149–62.
Savage, M. 1987. Understanding political alignments in contemporary Britain: do localities matter? *Political Geography Quarterly* **6**, 53–76.

Schwartz, A. 1983. Crisis of the fiscal crisis? *Antipode* **15**, 45–9.
Stevens, J. 1987. Dole queue cut gives Thatcher flying start. *Age*, Melbourne, 20 June.
Tabb, W. 1977. Blaming the victim. In *The fiscal crisis of American cities*, R. Alcaly & D. Mermelstein (eds). New York: Vintage Books.
Tabb, W. 1978. The New York City fiscal crisis. In *Marxism and the metropolis*, W. Tabb & L. Sawyer (eds). New York: Oxford University Press.
Urry, J. 1981. Localities, regions and social class. *International Journal of Urban and Regional Research* **5**, 455–74.
Zevin, R. 1977. New York City crisis: first act in a new age of reaction. In *The fiscal crisis of American cities*, R. Alcaly & D. Mermelstein (eds). New York: Vintage Books.

Index